Contributions to the UN Decade of Education for Sustainable Development

Environmental Education, Communication and Sustainability

edited by Walter Leal Filho

Vol. 33

PETER LANG
Frankfurt am Main · Berlin · Bern · Bruxelles · New York · Oxford · Wien

Fernando Gonçalves
Ruth Pereira
Walter Leal Filho
Ulisses Miranda Azeiteiro
(eds.)

Contributions to the UN Decade of Education for Sustainable Development

PETER LANG
Internationaler Verlag der Wissenschaften

Bibliographic Information published by the Deutsche Nationalbibliothek
The Deutsche Nationalbibliothek lists this publication in the Deutsche Nationalbibliografie; detailed bibliographic data is available in the internet at http://dnb.d-nb.de.

Layout:
www.kumpernatz-bromann.de

ISSN 1434-3819
ISBN 978-3-631-61347-4

www.peterlang.de

Table of Contents

Foreword

Our world is currently facing many kinds of problems: economic, social, political and environmental. These problems are different among the various world regions and may have a different impact on people's lives. The problems, and our understanding of them, are influenced by the way we see and perceive the complexities of life. Each day, many unsustaintable facts call for our attention. Sometimes we are aware of them and act. But to the same level, sometimes we are not and, as a result, fail to act.

In order to foster a global awareness about sustainable development, the United Nations General Assembly established a Decade of Education for Sustainable Development (2005-2014) which constitutes an urgent call for all people around the world to contribute to citizens' awareness and action guided by the idea of sustainability, to the benefit of the world community. But, without learning, *sustainable development* is impossible. Thus, it is important to foster sustainable development learning and enable people to participate in well-founded decision-making process, particularly (but not only) those concerning the environment, and take the appropriate course of action.

Bearing this in mind, this book presents essential learning approaches, introduces educational and training activities, as well as a various innovative methods aimed at the development of practical skills, in order to strengthen the continuous process of environmental education, and in particular, education for sustainable development (ESD). It does so by focusing on three dimensions: social, economic and environmental, as a means of achieving an effective change of behaviour. Similar to a previous book (volume 27, in this collection), this one "tries to bridge the gap between science and environmental education by describing a set of projects, initiatives and field activities".

The present book, after a brief review of the strategic action for ESD, educational policies and curricular orientations, analyses the obstacles that could explain the poor response to previous calls by the community of educators and researchers in science education. It also presents a global analysis on how higher education institutions are including sustainability issues in different areas (e.g. teaching, research, outreach and institutional management) and describes the progress made in this field during the last 10 years.

A special emphasis is given to teacher training programmes, conception, and implementation, highlighted the problems and barriers, which prevent development as far as integration of sustainability issues in higher education is concerned. Furthermore, this book presents instruments, related to ESD indicators, to help monitor and evaluate changes. The development of tools is crucial to clarify which educational approaches are necessary, and to evaluate the quality of these approaches in terms of their relevance and effectiveness to promote sustainable development.

This book makes a reflection on several environmental strategies and policies that reinforce a citizen's need for more collaborative and participant in actions towards biodiversity conservation, and the maintenance of the ecosystems' services. In particular, the use of fieldwork in environmental science programmes delivered through online methodology can be outlined, along with the concept of sustainability in fisheries, which pays a particular attention to the human dimension of fisheries.

In the framework of practical examples, other matters such as the implementation of Local Agenda 21, applying educational and environmental management concepts and ideas in the daily life of schools, problem-solving activities, environmental projects, participatory management, etc. are tackled. A special attention is given to natural sciences, because they are a far-reaching field for the development of innovative and diverse teaching strategies envisaging improvements in learning processes. Research-based recommendations on science and technology education are also highlighted, emphasizing good practices, at different levels, especially regarding science teachers' education, but also agronomists, civil engineers, architects and workers related to these sectors, to help to re-set education towards ESD.

In conclusion, this book intends to reshape science education in order to achieve sustainable development and to provide some concrete measures to address some of the current needs in the field, towards the of the decade and beyond.

Fernando José Mendes Gonçalves, Ruth Pereira,
Walter Leal Filho, Ulisses Miranda Azeiteiro
Autumn 2011

Chapter 1

The Need for Contributions to the Decade of Education for a Sustainable Future: an Ethical Commitment[1]

Amparo Vilches, Luís Marques, Daniel Gil-Pérez, João Praia

Abstract

Given the serious and urgent problems humanity is now facing, the United Nations General Assembly has adopted a resolution establishing a Decade of Education for Sustainable Development (2005-2014). This constitutes a new urgent call to educators of all levels and areas to contribute to citizens' awareness and understanding of the state of the world in order to enable them to participate in well-founded decision-making. Our aim in this paper is to analyze the obstacles that could explain the poor response to previous calls, in order to overcome these obstacles and make it possible to assume an ethical commitment on behalf of the community of educators and researchers in science education.

Introduction

We will begin by recalling the surprise some of us experienced during the first Earth Summit, held in Rio in 1992, when the United Nations asked educators from all levels and subjects to contribute to citizens' awareness and understanding of the current situation of planetary emergency in order to enable them to participate in well-founded decision-making. The use of the expression "*planetary emergency*" (Bybee 1991) gave the state of the world a dramatic overtone which we were not previously aware of.

Our surprise grew when a rapid inquiry revealed that such calls to educators had been made repeatedly for quite some time: for instance, as early as 1972 at the United Nations Conference on Human Environment in Stockholm. So why have most of these calls, along with the contributions from many environmental

1 This chapter has been conceived as a contribution to the Decade of Education for Sustainable Development, established by the UN General Assembly for the period 2005-2014.

educators and some researchers in the STS (science, technology and society) domain of science education, been ignored?

Despite receiving wide media coverage, the Rio Conference failed to involve educators in incorporating the state of the world topic into their teaching and research priorities (Vilches et al. 2003).

Consequently, the need for a long and intense campaign became clear and, ten years later, at the second Earth Summit (Johannesburg 2002), a Decade of Education for Sustainable Development (2005-2014) was proposed in order to create a social climate oriented towards involving all educators in making citizens aware of the current situation of planetary emergency and capable of participating in the decision-making processes necessary to stop degradation (Resolution 57/254, approved by the United Nations' General Assembly on 20 December 2002).

But how can this climate be created? The minimal success of previous calls compels us to anticipate serious obstacles that, if not taken into account, may hinder the new and ambitious initiative of the Decade. In other words, it is necessary to study what obstacles may exist and understand why the situation of planetary emergency has not received the attention of most educators up to now; because this lack of attention to the Earth's situation and its future affects all societies and includes most scientists, political leaders and educators.

In particular, research in science education has shown that the problem of the state of the world and sustainability education have been absent in most curricula until very recently, and even in curricula focusing on environmental education (Fien 1995). Orr's statement (1995), denouncing that we still educate the young as if there were no planetary emergency, continues to be valid, as well as the lamentation of the limited attention our educational systems pay to planning for the future (Hicks and Holden 1995; Anderson 1999). Most materials on environmental education focus exclusively on local problems without addressing the global situation (Hicks and Holden 1995; Vilches et al. 2008). Moreover, they show a reductionist approach which almost exclusively concentrates on natural resources, ignoring the strong connections between the natural environment and social, cultural, political and economic factors (Fien 1995; Tilbury 1995; Vilches et al. 2008). Summing up, the attention science education teachers *and researchers* pay to the state of the world is still minimal and constitutes a serious *missing dimension* in science education research and innovation (Vilches et al. 2003).

On the other hand, research has also shown that when a relatively in-depth, collective discussion is encouraged, most teachers correctly perceive the seriousness of the situation, brainstorm possible solutions and understand the necessity of contributing to citizens' education for a sustainable future (Gil Pérez et al. 2003). We will now describe how this discussion is organized and summarize the general results obtained.

Eliciting Science Teachers' Perceptions of the State of the World

We have organized numerous workshops for science teachers who are currently employed and studying the profession — grouped in teams of five – to discuss "the problems and challenges humankind will have to face in the near future, in order to construct a view of the current situation and measures to be taken which is as complete and correct as possible". In these workshops, the contributions made usually cover most of the aspects studied by experts:

Almost all of the teams indicated that the main problems facing humanity include:

- Environmental pollution and its consequences (acid rain, ozone layer depletion, development of the greenhouse effect, global climatic change, etc.)
- Depletion of natural resources (fossil energy resources, fertile soil, drinking water, etc.)
- Ecosystem degradation, destruction of biological diversity and desertification, etc.

We substantiate these contributions with texts from a variety of expert analyzes (World Commission on Environment and Development 1987; Worldwatch Institute 1984-2010; Mayor Zaragoza 2000; McNeill 2003; Lynas 2004; Gore 2006; Pearce 2006; Duarte 2006; IPCC 2007; Duarte Santos 2007).

A smaller number of teams made references to other *related* problems, such as:

- Increasingly disorganized and speculative urbanization (Girardet 2001; Vilches and Gil 2003, chapter 2)
- The destruction of *cultural* diversity (Folch 1998; Maaluf 1999; Vilches and Gil 2003, chapter 5; United Nations Development Programme 2004; Worldwatch Institute 2007; Burdett and Sudjic 2008)

Nevertheless, throughout the discussion there is general agreement on the importance of such problems *and the strong links among these problems*.

The same is true for the possible causes of the planetary emergency (which can be viewed as new problems). Teams make reference to:

- Economic growth guided by private, short-term interests (Meadows et al. 1972; Meadows, Meadows and Randers 1992; Brown 1998; Giddens 1999; Meadows, Randers and Meadows 2004; Sachs 2008)
- Over-consumption in "developed" societies and dominant groups as if the Earth's resources were infinite (Brown and Mitchell 1998; Folch 1998;

United Nations Development Programme 1998, 2003 and 2005; Diamond 2005; Sachs 2008)
- Demographic explosion on a limited planet. This aspect initially generates some controversy, but consensus is reached when some well-established data is presented: since the second half of the 20th century, more human beings have been born than in the whole of human history (World Commission on Environment and Development 1987; Orr 1994; Hubbert 1993; Ehrlich and Ehrlich 1990, 1993; Brown and Mitchell 1998; Folch 1998; Vilches and Gil 2003, chapter 9; Sartori and Mazzoleni 2003; Sachs 2008) and the *present* population would need the resources of three Earths to provide the standard of living of developed countries (United Nations 1997)
- Social inequalities, with billions of fellow humans scarcely able to survive in undeveloped countries and the exclusion of broad segments of the "first world" ... while one-fifth of the Earth's population follow the high-consumption model (World Commission on Environment and Development 1987; Vilches and Gil 2003, chapter 10; Sachs 2005)
- Conflicts and violence associated with these inequalities and the imposition of private interests and values (economic, ethnic, gender, cultural, etc.) via military conflicts, mafia activities, speculation of transnational enterprises beyond democratic control, terrorism, mass migration, etc. (Mayor Zaragoza 2000; Vilches and Gil 2003, chapter 11; Diamond 2005)

Most of these problems are mentioned by science teachers at all levels and in every workshop when they have the opportunity to freely discuss issues which demand global reflection, including:

- Enumerate the problems and challenges which, in your opinion, humankind will have to face in the near future, in order to construct a view of the current situation and measures to be taken which are as complete and correct as possible.
- Explore the problem of pollution in further detail, enumerating its various forms and consequences.
- Point out the resources whose depletion may be most concerning.
- Consider possible reasons why urban growth may prove to be a great concern.
- Discuss which aspects of environmental degradation may be most troubling.
- Try to identify the causes of the growing degradation of our planet.
- ...

Therefore, we must conclude that it is not difficult for science teachers to understand that we are in a serious state of planetary emergency, characterized by an ensemble of intimately related problems which demand global treatment (Morin 1999; Vilches and Gil 2003). Why then, is this problem absent in most curricula, teaching and research? What could the "hidden" obstacles be (Vilches et al. 2008)?

What Obstacles could Prevent Teachers from Addressing the Situation of Planetary Emergency?

An initial, and serious, obstacle for incorporating the state of the world into the curricula, teaching practice and research lies in the *lack of tradition in education when it comes to approaching these types of global problems*, which demand systemic treatment (Morin 1999). This is a serious obstacle, because, although each problem has a particular importance and deserves individual attention, none of them can be understood or treated without taking the entire group into account (World Commission on Environment and Development 1987; United Nations 1992; Fien 1995; Tilbury 1995; Mayor Zaragoza 2000; Vilches and Gil 2003).

However, as we have already mentioned, an explicit demand to analyze the Earth's situation on a global scale allows us to understand the close connection between the different problems, as well as their local and global repercussions. Consequently, we need to keep in mind the necessity of holistic approaches if we want to avoid the "natural" tendency towards local and isolated treatment as well as causal reductionism.

This tendency is reinforced by another tradition: *the consideration of our planet as immense and limitless*, which implies that human activities would have only local effects (Fien 1995). In fact, until the second half of the 20th century, while the Earth's population was much smaller than nowadays and technological developments had not globalized the planet, the effects of human activities remained locally compartmentalized. However, these compartments have begun to dissolve over the last few decades, and many problems (the development of the greenhouse effect, ozone layer depletion, migrations ...) have acquired a global dimension and the state of the planet has thus become subject to growing concern (Bybee 1991; Fien 1995; Colborn, Myers and Dumanoski 1997; Leakey and Lewin 1996; Lubchenko 1998; Broswimmer 2002; Diamond 2005; Sachs 2008).

Another deep-rooted tradition that hinders global approaches to the planetary emergency is the *defence of "ourselves"* (our family, our clan, our ethnic group, our country, our species ...) *against "the others"*, seen as enemies to defeat, following a "them or us" strategy. This results in limiting the attention paid to "our" problems, without considering the consequences for others or future generations (surely including our children's generation and probably our own generation). We have to understand that a sustainable future is incompatible with simplistic and Manichean "explanations" which attribute any difficulty to "foreign enemies" and the promotion of competitiveness, which is understood as a contest to achieve something *at the expense* of others who are pursuing the same objective (Mayor Zaragoza 2000; Vilches and Gil 2003).

We must also take into account certain *ideological and religious barriers* that make it difficult to comprehend the seriousness of problems such as the demographic explosion and the need to promote responsible family planning.

One pronounced obstacle stems from the *tendency to consider that individual actions only have negligible effects* on large problems such as resource depletion or environmental degradation. However, it is easy to demonstrate (using very simple calculations) that even though an individual can only save a very small quantity of energy or materials, the amount saved becomes quite large when these quantities are multiplied by millions of people. Consequently, there is a commensurate reduction in environmental pollution and degradation. In fact, in many cases the variety of individual actions have a broader effect than industry as a whole. This is what happens, for instance, with the development of the greenhouse effect: passenger cars produce more CO_2 than industries, even though only one-fifth of the human population has access to them (Vilches and Gil 2003).

We are not denying the responsibility of those who impose a model of development geared towards satisfying private interests, regardless of what may happen to others or to the planet. Nevertheless, it is necessary to avoid simplistic explanations which focus more on searching for culprits than in understanding causes and possible solutions. What we need is a sound comprehension of the situation to be able to participate in well-founded decision-making.

We need to analyze these and other barriers and pseudo-explanations which hinder the treatment of the state of planetary emergency. However, one of the most serious difficulties may result from *giving more attention to problems than possible solutions*: merely studying the problems provokes, at best, indignation and, at worst, despair (Hicks and Holden 1995). It is also necessary to study the possible solutions to the planetary crisis described, explore alternative approaches, and participate in activities which favour a sustainable future. Within this context, the *difficulty of understanding the meaning of sustainability* becomes another important obstacle.

As a result of the analysis of the state of the world that presents an unsustainable situation of planetary emergency (Bybee 1991) as a serious threat to the future of humankind, sustainability emerges as a negative concept.

"A threatened future" is, precisely the title of the first chapter of *Our Common Future*, the report compiled by the World Commission on Environment and Development (WCED 1987), where we find one of the first attempts to introduce the concept of sustainability through the definition of *sustainable development* as a development which meets present needs without compromising the ability of future generations to meet their own needs.

This classical definition from the WCED has obtained widespread consensus (although sometimes this consensus is purely formal and hides serious misunderstandings, such as *interpreting "sustainable development" as "sustained growth"*, which is, of course, the opposite), as well as many criticisms. Therefore, the meaning of the concept of sustainability that, as Bybee (1991) affirms, constitutes "*the central unifying idea society most needs at this point in human history*" must be studied in further depth.

One initial criticism of the WCED definition is that *sustainability simply expresses a common idea* which is more or less explicitly present in many primitive cultures: preserving our resources for future generations.

This criticism must be rejected and it must be made clear that sustainability is an absolutely new concept: it implies accepting that the Earth is not as large and limitless as we believed and cannot indefinitely bear the effects of many human predator activities. This knowledge about the unsustainability of the current socioeconomic development is recent and has been a real surprise even for experts (Meadows et al. 1972 and 1992).

This knowledge is also new in another even deeper sense: sustainability has been understood to only be possible on a planetary scale and demands the consideration of the host of Earth's interconnected problems: a sustainable city or country has no meaning because problems have a planetary dimension: there are no borders for pollution or ocean degradation. A country or a city can (and must!) *contribute* to sustainability, but cannot aspire to be sustainable by itself without considering the rest of the world. As Brazilian theologist Leonardo Boff says, this time there will be no Noah's Ark to preserve some human beings: this time salvation will be for all of us or for none of us. People will not be "chosen".

Sustainability undoubtedly implies that new knowledge is accepted only with great difficulty, because the signs of degradation have not been sufficiently visible until very recently and because, in certain parts of the world, human beings have notoriously improved their quality of life. Subordination of nature to the necessities and wishes of human beings has always been considered a distinctive sign of advanced societies, explains the former General Director of UNESCO Mayor Zaragoza (2000) in *The World Ahead: Our Future in the Making*. In fact, it was not even seen as subordination: nature was considered practically unlimited and people could focus their attention on their own necessities without bothering with potential environmental consequences. Although experts' alarm and international studies (Meadows et al. 1972; World Commission on Environment and Development 1987; Worldwatch Institute 1984-2010; Meadows, Randers and Meadows 2004) are abundant, most citizens, including scientific, economic and political leaders – and educators – are not yet aware.

We have to recognize the fact that the recent preoccupation with preserving our planet's conditions of life constitutes an authentic and difficult mental revolution (Mayor Zaragoza 2000): emerging in just one or two generations, this cultural, scientific and social metamorphosis breaks a long tradition of indifference towards the environment. It is not a question of considering development and environment as contradictory (the first "attacking" the second and the second "limiting" the first), but of recognizing that both are intimately connected and cannot be handled separately: after the Copernican revolution that unified Heaven and Earth and after the Theory of Evolution, which shows the relationship between humankind and the rest of living beings, we are now witnessing the integration of environment and development in a new paradigm of ecologic economy, based on the idea of sustainable development (Vilches and Gil 2003; Sachs 2008).

We are now faced with another, quite serious obstacle for the acceptation of the new paradigm: *to consider that the expression "sustainable development" unites two contradictory terms* and that the new concept is a stratagem of supporters of limitless economical growth that intend to make this compatible with ecological sustainability (Girault and Sauvé 2008). Nevertheless, the idea of sustainable development explicitly rejects the identification of development and growth. As Daly (1991) points out, we need to distinguish between growth and development: growth is a quantitative increase on a physical scale while development is the qualitative improvement or tapping into potential. In light of the fact that the human economy is a subsystem of a global ecosystem which does not grow, yet develops, it is clear that economic growth is not sustainable over a long period of time.

In other words, although growth cannot continue indefinitely in a limited world, development is possible. Possible and *necessary* because the current ways of life must undergo profound qualitative changes, both for those who live precariously (the majority) and for the minority who exercise a predator-like overconsumption. These necessary qualitative changes constitute development (not growth), which naturally needs to be carefully designed to avoid ambiguities and inappropriate use of the expression "sustainable development".

Following these clarifications, sustainability can play the central unifying role that Bybee (1991) attributes to this concept. This unifying role is based on the global study of problems, their causes and the adoption of suitable measures, which need to be contemplated on a global scale by moving beyond any expectations of finding simple solutions to the interconnected problems humanity faces now and in the near future.

It is essential to pay attention to these obstacles in order to overcome the poor response to the United Nations' call to contribute to citizens' awareness and understanding of the state of the world.

Keeping these obstacles in mind, we are convinced that if we promote a global discussion in some depth, which is based on founded documentation, we may obtain more correct perceptions and more favourable attitudes from teachers for the inclusion of this issue in education. Based on our experience, this can be an effective way to convey that there is a situation of planetary emergency and that each of us has the ethical commitment *and privilege* to contribute to reverting it.

What can be done to Incorporate Education for Sustainability as an Ethical Commitment?

As previously mentioned, one of the obstacles to incorporating education for sustainabilitystems from *paying more attention to the problems than to possible solutions*: therefore, it is particularly important to explore alternative approaches and to participate in actions aimed at favouring a sustainable future. With this goal in mind, we propose the following questions:

- What measures would need to be adopted to confront the situation of planetary emergency?
- Discuss what can be understood as sustainable development.
- What characteristics should technological measures have in order to favour sustainable development?
- What educational proposals may contribute to sustainable development?
- What political measures should be promoted?

With the help of founded information, this discussion makes it clear that the current process of planetary degradation can be reversed. However, this would require the *immediate joint* implementation of a variety of scientific, educational and political measures. It is this understanding that encourages *knowledge-based activism*, i.e. a positive answer to the United Nations' call to incorporate sustainability education into our teaching *and research* as a collective ethical commitment. We shall discuss this ensemble of measures in some detail.

Technical Measures

There is general agreement on the need for technologies that favour sustainable development (Gore 1992; Daly and Cobb 1989; Daly 1991; Lubchenko 1998; Flavin and Dunn 1999; United Nations Development Programme 2001; Sachs 2008). The proposed measures range from the search for new energy resources, the improvement of efficiency in food production, the prevention of illnesses and catastrophes, the reduction and recycling of waste and the attainment of responsible family planning (Vilches and Gil 2003, chapter 12).

However, a careful analysis of the technologies conceived is required, because an apparent solution may create more problems than it solves. Daly (1991) suggests two principles which must characterize technology in order to be compatible with sustainable development:

- The rate at which resources are gathered must not exceed regeneration rates (or, for resources that are not renewable, the creation of renewable substitutes).
- Waste production rates must be lower than the assimilation capacities of the planet's ecosystems.

Additionally, Daly insists on the fact that we are moving from an *empty world* economy (where technology was the limiting factor for achieving profit from the exploitation of natural resources) towards a *full world* economy, where natural capital will increasingly become the limiting factor. This imposes a third principle or characteristic:

- The aim of technology for sustainable development must be to increase the efficiency of resources, rather than raise their extraction rate. This means, for instance, that we need to produce more efficient lamps instead of constructing more electrical power stations.

These essential technical criteria must be accompanied by ethical criteria (Vilches and Gil 2003, chapter 12) such as:

- Priority must be given to technologies oriented towards satisfying basic needs and reducing social inequalities.
- A *precautionary principle* must be systematically applied to avoid the hasty application of a specific technology if possible negative effects have not been sufficiently discarded by research.

These two principles are geared towards overcoming the short-term search for topics of individual interest which has usually characterized scientific development, and they proceed to question the widespread and erroneous idea that the solutions to the serious problems faced by humanity today solely depend on better knowledge and more advanced technologies: options and dilemmas are essentially matters of ethics (Aikenhead 1985; Tilbury 1995; Delors 1996; Mayor Zaragoza 2000) and demand the consideration of educational and political measures.

Educational Measures

The importance experts place on the role of education is clearly demonstrated by the numerous appeals from the United Nations and other international institutions to educators of any subject and at any level, both formal (school curricula) and informal (press, museums, etc.) and, in particular, in the recent proclamation of the Decade of Education for a Sustainable Future (2005-2014).

The educational measures proposed to contribute to a sustainable society emphasize global analyses and the search for global and joint solutions (Delors 1996; Morin 1999; Vilches and Gil 2003). These types of measures are aimed at overcoming the usual tendency to satisfy individual interests in the short-term (or to continue habits characteristic of an "empty" world of isolated compartments). We need education which contributes to an accurate perception of the state of the world and prepares citizens for decision-making (Aikenhead 1985 and 1996) by generating responsible attitudes and behaviours (Bybee 1991; Fien 1995; Tilbury 1995; Mayor Zaragoza 2000; Rifkin 2009; Worldwatch Institute 2010) oriented towards achieving development which is both culturally plural and physically sustainable.

Questions like "What energy policy should be promoted?" or "What role should be given to genetic engineering in the food industry?" and "What controls should be placed on GM food production?" demand informed decision-making and the adoption of suitable policies. We need an education that promotes responsible behaviour, not just favourable opinions and attitudes (Vilches and Gil 2003, chapter 13).

Some authors have indicated that these responsible attitudes and behaviours cannot be attained without overcoming the usual anthropocentric stance which gives human beings priority over the rest of nature. However, in our opinion, we do not need to abandon an anthropocentric point of view in order to understand the importance of protecting the environment and biodiversity. What person could continue promoting the unsustainable exploitation of nature after learning about the serious dangers it entails for his or her *own* children?

We believe that education for a sustainable society should be based on what can be reasonably understood by most people, even if their ethical values are more or less anthropocentric. In other words, the borderline should be one that separates people who have a sound perception of problems and an inclination to contribute to the necessary decision-making processes and actions from people who lack such preparation.

This type of education must promote the analysis of concepts that are presented as "obvious" and "indisputable" without alternatives, thus eliminating the possibility of selecting a different option. This is particularly true when it comes to *competitiveness*. Everybody speaks of competitiveness as something that is absolutely necessary without realizing that this type of behaviour is incompatible with the aim of sustainable development. Indeed, the success of one person or group in, for example, a commercial competition, implies the failure of others whose future is not taken into account. We insist that this contradicts the characteristics of sustainable development which must be fundamentally global and embrace the whole planet.

Instead of promoting competitiveness, we need education that helps students and teachers analyze the efficiency of our actions *from a global viewpoint*, taking into account the short-term, medium-term and long-term repercussions, both for ourselves and for all of humanity. We need education that helps to transform the current competitive economic globalization into a democratic and sustainable project (Delors 1996) that enhances the richness of biological and cultural diversity. In fact, appeals to individual responsibility are multiplying; they include detailed lists of specific, potential actions in various fields – including water, food supply, traffic, cleaning, heating, lighting and family planning (Button and Friends of the Earth 1990; Silver and Vallely 1990; The Earth Works Group 1900; Pico 2008; Vilches, Praia and Gil-Pérez 2008).

These educational aims need to be integrated into an appropriate educational framework: research and innovation are required to conceive and incorporate sustainability education into various educational activities, both formal and informal, including teacher training.

To conclude with these considerations on the role of education in promoting sustainability, it is necessary to stress that individual contributions can, and must, extend beyond the private domain and apply to professional, social and political activities. For example, citizens can support non-governmental organizations and political parties that promote solidarity and environmental protection; they can also demand positive action on behalf of public institutions (town councils, parliaments). On the other hand, it is particularly important for these individual and collective citizen actions to avoid local or partial approaches and consider multifaceted environmental questions (pollution, depletion of resources etc.), and other related problems, such as social inequalities and conflicts, from a planetary perspective. The ecologist slogan "to think globally and to act locally" has its limitations; we now know that it is necessary to *act globally* as well (O'Connor 1992) by adopting *political measures on a planetary scale*, which are capable of avoiding the imposition of individual interests and values that are harmful to other people or to future generations. We provide commentary on these items in the next section.

Political Measures

To begin with, we have to remember that we are facing problems with a planetary dimension which cannot be tackled solely with a local approach: political measures on a local, regional *and planetary* scale are required.

The discussion about political measures which could promote sustainable development usually leads to heated debates, and demands careful analysis. Most science teachers and citizens view the adoption of planetary political measures with scepticism and a certain degree of reluctance because there is a strong tendency, as we have already mentioned, to limit our perspective to "ourselves" (our country, etc.) and to forget "the others" or even see them as a danger. Nevertheless, climate change, which knows no borders, reminds us that, for the first time in human history, we are living in an interconnected civilisation which encompasses the entire planet. Therefore, we can understand the absolute necessity for political integration to put the environment, as the common substratum of life, above the individual interests of any country, region or transnational enterprise, for the first time in human history.

We might think that the risk of employing only local approaches is becoming obsolete because of the spiraling process of economic *globalization*. Paradoxically, this process is not global at all when it comes to the survival of life on our planet. Despite the broad discussion about globalization, most approaches continue to be partial, sectorial and one-dimensional. They do not consider environmental destruction as a specific phenomenon. Rather, they take it into account but not in order to prevent it: economic globalisation promotes the displacement of production centres towards countries where ecological norms are less restrictive (Sachs 2008).

Thus, economic globalization appears to be quite one-dimensional. Planetary norms, therefore, are necessary in order to avoid the general degradation of the environment and its tremendous economic cost, which has only just begun to be evaluated (Constanza et al. 1997; Sachs 2008). In this sense, political integration on a planetary scale is deemed absolutely necessary and *urgent*; this integration must be capable of promoting and controlling the measures to protect our social and physical environments before the degradation process becomes irreversible.

In short, a new world order based on co-operation and solidarity is required, with institutions capable of avoiding short-sighted individual interests that are harmful to other people, future generations and, in fact, to the predator's future as well (French 1992; Aikenhead 1996; Renner 1999; Folch 1998; Giddens 1999; Sen 1999; United Nations Development Programme 2002; Vilches and Gil 2003, chapter 14; Sachs 2008; Worldwatch Institute 2008 and 2010; Rifkin 2009).

However, this planetary political integration, which our survival seems to depend on, also generates the fear of cultural homogenizationthat is already underway: *the fear of cultural impoverishment*. However, this destruction of cultures cannot be attributed to a process of political integration which has not yet occurred. It is just another consequence of a purely commercial integration. A democratic order on a planetary scale could incorporate the protection of the environment and defence of biological and cultural diversity, while promoting intercultural exchanges to take advantage of this diversity (Vilches and Gil 2003, chapter 14).

Therefore, a fully democratic political integration on a global scale is a prerequisite to:

- Stop current physical and cultural planetary degradation
- Put an end to unsustainable social inequalities
- Stop unilateralist actions, world terrorism and the trafficking of arms, drugs, capital and people
- Attain world security and sustainability.

Our survival, our basic *right to life*, depends on this type of integration. Recently, the hypothesis of an imminent "sixth mass extinction of species", with humankind as the main agent and the victim, has been seriously advanced and justified (Leakey and Lewin 1996). This concept is connected to the fundamental question of human rights, which is also closely linked to the attainment of sustainability. The next section is dedicated to clarifying this relationship.

Sustainability and Human Rights

It may seem strange to establish such a direct relationship between human rights and sustainability. For this reason, we shall try to clarify what is currently understood as *human rights* – a concept that has developed and now encompasses three "generations" of rights (Vercher 1998) – and how they are related to achieving a sustainable future.

First, we refer to *democratic civil rights (opinion, association) for all, without social, gender or ethnic limitations*. They constitute a *sine qua non* condition for citizens' decision-making about current and future environmental and social problems (Folch 1998). These rights are now known as "first generation human rights", because they were the first rights to be demanded and obtained (not without conflict) in a growing number of countries. In this respect, we must not forget that the *droits de l'homme* from the French Revolution (to quote a well-known example) explicitly excluded women; women only achieved the right to vote in France after the Second World War. Futhermore, we must remember that such basic rights are systematically violated every day in many countries.

Amartya Sen (1999) concluded that the expansion of liberties is, at the same time, a basic aim of social development and its principal instrument to make sustainability a reality. Nevertheless, a sustainable future demands the recognition of other rights in addition to these civil rights. We are referring to *economic, social and cultural rights* or "second generation human rights" (Vercher 1998; United Nations Development Programme 2000) such as:

- The universal right to a satisfying job, overcoming unstable situations to which hundreds of millions of human beings (including more than 250 million children) are exposed
- The universal right to an adequate dwelling in an appropriate physical and cultural milieu
- The universal right to appropriate nourishment, both quantitatively (avoiding the under-nourishment of billions of fellow humans) and qualitatively (avoiding unbalanced diets)
- The universal right to health. This requires resources, research and education to fight infectious illnesses (e.g. cholera and malaria, which are still ravaging many third-world countries) and the new "industrial" and behavioural illnesses (such as tumours, depression, AIDS). Above all, healthy milieus and habits, as well as solidarity with handicapped people, must be promoted.
- The universal right to family planning and free enjoyment of sexuality (while respecting the freedom of others) by overcoming the cultural and religious barriers that condemn millions of women to submission
- The universal right to an education of quality *throughout* one's life, without social, gender or ethnic limitations
- The universal right to culture, in the broadest sense, as a supporting axis for personal and collective enrichment and development
- The universal right to investigate any kind of subject (the origin of life, genetic manipulation, etc.) without ideological limitations, but with a suitable degree of social control that takes into account social and environmental consequences and prevents the hasty application of insufficiently tested technologies.

Finally, we have third-generation human rights, known as *solidarity rights* because they tend to preserve the integrity of the entire population (Vercher 1998). They incorporate the right to life in a suitable environment, the right to peace and the right to sustainable development for all people and future generations:

- *The right of all human beings to an environment appropriate to their health and welfare.* As Vercher (1998) states, the inclusion of this right as a fundamental human right derives from an irrefutable fact: if degradation of the environment continues at the current rate, maintaining the environment will soon be the most fundamental issue of survival for mankind. The longer it takes us to recognize this situation, the greater the sacrifices and difficulties will be on the path to appropriate recovery.

- *The right to peace,* which implies the capacity of the International Community to prevent the imposition of specific interests (economic, ethnic, cultural, etc.) over general interests and values.
- *The right to sustainable economic and cultural development* of all peoples. This involves, on the one hand, questioning the current, marked economic inequalities among different human groups and, on the other hand, defending cultural diversity and cultural crossbreeding (against racism and ethnic or social barriers).

Vercher insists on the fact that these third generation rights can only be achieved via the coordinated effort of all actors of the social scene. Therefore, we can understand the link we have established between sustainable development and universal human rights. And we can also understand the need to move towards real globalization, with democratic institutions on a planetary scale that are capable of guaranteeing this ensemble of rights.

The *ensemble* of these rights appears to be a prerequisite (and, at the same time, an objective) of a sustainable society, as *they are all interconnected*. We cannot expect, for example, that some people do not contribute to the depletion of a fishing bank … if this is their only resource to nourish their family. And we cannot imagine that the demographic explosion could be interrupted without recognizing the right to family planning and free enjoyment of sexuality, which is also connected to the right to education: as Mayor Zaragoza (2000) states, only universal education can reduce the continuous growth of population in any religious or ideological context.

In short, achieving sustainable development is synonymous with making human rights universal in the broadest sense (Vilches and Gil 2003, chapter 15). This requires:

- Creating democratic institutions on a *planetary scale*, which are capable of preventing the imposition of individual interests which are harmful to other people or future generations
- Orientating scientific development towards the creation of technologies that favour sustainable development
- Promoting education which is capable of countering the usual tendency to behave according to individual interests in the short term.

Promoting the Decade of Education for a Sustainable Future: an Ethical Commitment

We are in the middle of a decade which will be decisive in one sense or another:

Decisive on a sad note if we remain anchored in our routines and do not become aware of the need to reverse a degradation process that is constantly sending us irrefutable signals in the form of global heating, unnatural catastrophes, loss of biological and cultural diversity, millions of deaths from starvation and wars due to suicidal, short-sighted interests and fundamentalism, dramatic migrations … and a long etcetera.

Happily decisive if we are capable of generating a universal trend for a sustainable future which has to begin right now.

This is the aim that we can, and must, incorporate into science education, teaching and research – conscious of the difficulties, but determined to contribute, as educators, scientists and citizens, to create the conditions for a sustainable future (Rifkin 2009; Worldwatch Institute 2010).

And even though research in science education has shown that the state of the world and education in sustainability are absent in most curricula, there are, in fact, many opportunities to introduce them into scientific and technological education. For instance, studying energy is an excellent opportunity to deal with the world's situation and contribute to a better understanding of problems and possible actions to be taken in light of the current situation of planetary emergency (Cachapuz et al. 2005; Furió et al. 2005). Innovation and research are needed: innovation to implement changes in the curricula, research to analyze and overcome obstacles faced by the current planetary emergency.

As educators and researchers, this should be a very important ethical commitment for every one of us.

References

Aikenhead, G.S. (1985), Collective Decision Making in the Social Context of Science. Science Education 69 (4): 453-475.

– (1996), Towards a First Nations Cross-Cultural Science and Technology Curriculum for Economic Development, Environmental Responsibility, and Cultural Survival, IOSTE 1996 (http://www.usask.ca/education/people/aikenhead/iost1996.htm).

Anderson, B. (1999), Evaluating students' knowledge understanding and viewpoints concerning the state of the world in the spirit of developmental validity, pp. 149-151. Second International Conference of the ESERA, Kiel, Germany.

Broswimmer, F.J. (2002), Ecocide: A Short History of the Mass Extinction of Species. Pluto Press, London.

Brown, L.R. (1998), The Future of Growth. In The State of the World 1998, Worldwatch Institute. W.W. Norton, New York.

Brown, L.R. and Mitchell, J. (1998), Building a New Economy. In The State of the World 1998, Worldwatch Institute. W.W. Norton, New York.

Burdett, R. and Sudjic, D. (2008), The Endless City. Phaidon, London.

Button, J. and Friends of the Earth (1990), Do it Green. Century Hutchinson, London.

Bybee, R.W. (1991), Planet Earth in Crisis: How Should Science Educators Respond? The American Biology Teacher 53(3), pp. 146-153.

Cachapuz, A., Gil-Pérez, D., Pessoa, A.M., Praia, J. and Vilches, A. (2005), A necessária renovaçâo do Ensino das Ciências. Cortez Editores, São Paulo.

Colborn, T., Myers, J.P. and Dumanoski, D. (1997), Our Stolen Future. Dutton, New York.

Constanza, R., d'Arge, R., de Groot, R., Farber, S., Grasso, M., Hannon, B., Limburg, K., Naeem, S., O'Neill, R., Paruelo, J., Raskin, R., Sutton, P. and van den Belt, M. (1997), The value of the world's ecosystem services and natural capital. Nature, 387, pp. 253-260.

Daly, H. (1991), Steady-State Economics. Island Press, Washington D.C.

Daly, H.E. and Cobb, J.B. (1989), For the Common Good: Redirecting the Economy Toward Community, the Environment, and a Sustainable Future. Beacon Press, Boston.

Delors, J. (1996), L'éducation: un trésor est caché dedans, UNESCO, Commission internationale sur l'education pour le vingt et unième siècle. Jacob O., Paris.

Diamond, J. (2005), Collapse: How Societies Choose to Fail or Succeed. Viking, New York.

Duarte, C. (2006), Cambio Global. Impacto de la actividad humana sobre el sistema Tierra. CSIC, Madrid.

Duarte Santos, F. (2007), Que Futuro? Ciência, Tecnología, Desenvolvimento e Ambiente. Gradiva, Lisboa.

Ehrlich, P.R. and Ehrlich A.H. (1990), The Population Explosion. Touchstone, London.

– (1993), Why isn't everyone as scared as we are?, pp. 55-112. In Valuing the Earth: Economics, Ecology, Ethics. The MIT Press, Cambridge.

Fien, J. (1995), Teaching for a Sustainable World: the Environmental and Development Education Project for Teacher Education. Environmental Education Research 1(1), pp. 21-33.

Flavin, C. and Dunn, S. (1999), Reinventing the Energy System. In The State of the World 1999, Worldwatch Institute. W.W. Norton, New York.

Folch, R. (1998), Ambiente, Emoción y Ética, Ariel, Barcelona.
French, H.F. (1992), After the Earth Summit: The Future of Environmental Governance, Worldwatch Institute. W.W. Norton, New York.
Furió, C., Carrascosa, J., Gil-Pérez, D. and Vilches, A. (2005), ¿Qué problemas plantean la obtención y el consumo de recursos energéticos? In ¿Cómo promover el interés por la cultura científica? Una propuesta didáctica fundamentada para la educación científica de jóvenes de 15 a 18 años. OREALC/UNESCO, Santiago de Chile.
Giddens, A. (1999), Runaway World: How Globalisation is Reshaping Our Lives. Profile Books, London.
Gil-Pérez, D., Vilches, A., Edwards, M., Praia, J., Marques, L. and Oliveira, T. (2003), A proposal to enrich teachers' perception of the state of the world: First Results. Environmental Education Research 9(1), pp. 67-90.
Girardert, H. (2001), Creating Sustainable Cities. Green Books Ltd., Totnes (UK).
Girault, Y. and Sauvé, L. (2008), L'éducation scientifique, l'éducation à l'environnement et l'éducation pour le développement durable. Aster, 46, pp. 7-30.
Gore, A. (1992), Earth in the Balance: Ecology and The Human Spirit. Houghton Mifflin, Boston.
– (2006), An Inconvenient Truth. Rodale, New York.
Hicks, D. and Holden, C. (1995), Exploring the Future: a Missing Dimension in Environmental Education. Environmental Education Research, 1(2), pp. 185-193.
Hubbert, M.K. (1993), Exponential Growth as a Transient Phenomenon in Human History. In Valuing the Earth: Economics, Ecology, Ethics. The MIT Press, Cambridge.
IPCC (Intergovernmental Panel on Climate Change) (2007), IPCC 4th Assessment Report (http://www.ipcc.ch).
Leakey, R. and Lewin, R. (1996), The Sixth Extinction: Biodiversity and its Survival. Weidenfeld and Nicolson, London.
López Alcantud, J., Gil Pérez, D., Vilches, A. and González, E. (2004), El estudio de la energía en la educación tecnológica: una ocasión privilegiada para analizar la situación del mundo. Didáctica de las Ciencias Experimentales y Sociales, 18, pp. 81-104.
Lubchenko, J. (1998), Entering the Century of the Environment: A New Social Contract for Science. Science, 279, pp. 491-497.
Lynas, M. (2004), High Tide: News from a Warming World. Flamingo, London.
Maaluf, A. (1998), Les Identités meurtrières. Éditions Grasset and Fasquelle, Paris.

Mayor Zaragoza, F. (2000), The World Ahead: Our Future in the Making. Published in French with the title 'Un monde nouveau'. Editions UNESCO, Paris.

McNeill, J.R. (2003), Something New Under the Sun: An Environmental History of the Twentieth-Century World. W.W. Norton, New York.

Meadows, D. H., Randers, J. and Meadows, D.L. (2004), Limits to Growth: The 30-Year Update. Chelsea Green, London.

Meadows, D.H., Meadows, D.L. and Randers, J. (1992), Beyond the Limits. Chelsea Green, White River Junction, Vermont (USA).

Meadows, D.H., Meadows, D.L., Randers, J. and Behrens, W.W. (1972), The Limits to Growth. Universe Books, New York.

Morin, E. (1999), Les sept savoirs nécessaires à l'éducation du future. UNESCO, Paris.

O'Connor, J. (1992), Think globally, act locally? Towards an international red green movement. Capitalism, Nature, Socialism 3 (4), pp. 1-7.

Orr, D.W. (1994), Earth in Mind: On Education, Environment, and the Human Prospect. Island Press, Washington.

– (1995), Educating for the Environment: Higher Education's Challenge of the Next Century. Change, May/June, pp. 43-46.

Pearce, F. (2006), The Last Generation. Eden Project Books, Cornwall.

Pico, M.J. (2008), El planeta i tu. Bromera, Alzira (Spain).

Renner, M. (1999), Ending Violent Conflict. In The State of the World 1999, Worldwatch Institute. W.W. Norton, New York.

Rifkin, J. (2009), The Empathic Civilization: The Race to Global Consciousness in a World in Crisis. Polity Press, Cambridge.

Sachs, J.D. (2005), The End of Poverty. Penguin Press, New York.

– (2008), Common Wealth: Economics for a Crowded Planet. Penguin Press HC.

Sartori, G. and Mazzoleni, G. (2004), La Terra Scoppia. Sovrappopolazione e Sviluppo. Rizzoli, Milano.

Sen, A. (1999), Development as Freedom. Alfred A. Knopf, Inc., New York.

Silver, D. and Vallely, B. (1990), The Young Person's Guide to Saving the Planet. Virago Press, United Kingdom.

The Earth Works Group (1990), The Recycler's Handbook: Simple Things You Can Do. Greenleaf Publishers, New York.

Tilbury, D. (1995), Environmental Education for Sustainability: Defining the New Focus of Environmental Education in the 1990s. Environmental Education Research 1(2), pp. 195-212.

United Nations (1992), UN Conference on Environmental and Development. Agenda 2, Rio Declaration, Forest Principles. UNESCO, Paris.
– (1997), Earth summit +5. Programme for the Further Implementation of Agenda 21 (http://www.un.org/esa/earthsummit/index.html).
United Nations Development Programme (1998), Consumption for Human Development. Oxford University Press, New York (http://hdr.undp.org/reports/global/1998/en/).
– (2000), Human Rights and Human Development for Freedom and Solidarity. Oxford University Press, New York (http://www.undp.org/hdr2000/home.html).
– (2001), Making New Technologies Work for Human Development. Oxford University Press, New York (http://hdr.undp.org/reports/global/2001/).
– (2002), Deepening Democracy in a Fragmented World. Oxford University Press, New York (http://hdr.undp.org/ reports/global/2002/).
– (2003), Millennium Development Goals: A Compact Among Nations to End Human Poverty. Oxford University Press, New York (http://hdr.undp.org/reports/global/2003/).
– (2004), Cultural Liberty in Today's Diverse World. Oxford University Press, New York (http://hdr.undp.org/reports/global/2004/).
– (2005), International cooperation at a crossroads: Aid, trade and security in an unequal world. Oxford University Press, New York (http://hdr.undp.org/reports/global/2005/).
Vercher, A. (1998), Derechos Humanos y Medio Ambiente. Claves de Razón Práctica 84, pp. 14-21.
Vilches, A. and Gil-Pérez, D. (2003), Construyamos un futuro sostenible. Diálogos de supervivencia. Cambridge University Press, Madrid.
Vilches, A., Gil-Pérez, D., Edwards, M. and Praia, J. (2003), Science Teachers' Perceptions of the Current Situation of Planetary Emergency. In Science Education Research in the Knowledge-Based Society. Kluwer, Dordrecht.
Vilches, A., Gil-Pérez, D., Toscano, J.C. and Macías, O. (2008), Obstáculos que pueden estar impidiendo la implicación de la ciudadanía y, en particular, de los educadores, en la construcción de un futuro sostenible. Formas de superarlos, Revista CTS, no. 11, vol. 4, pp. 139-162 (http://www.revistacts.net/4/11/009/file).
Vilches, A., Praia, J. and Gil-Pérez, D. (2008), O antropoceno: entre o risco e a oportunidade. Educação: Temas e Problemas, 5, pp. 41-66.
World Commission on Environment and Development (1987), Our Common Future. Oxford University Press, Oxford.
Worldwatch Institute (1984-2010), State of the World. W.W. Norton, New York.

– (2007), State of the World 2007: Our Urban Future. W.W. Norton, New York.
– (2007), State of the World 2008: Innovations for a Sustainable Economy. W.W. Norton, New York.
– (2010), State of the World 2010: Transforming Cultures: From Consumerism to Sustainability. W.W. Norton, New York.

Chapter 2

Evolutionary Foundations for Sustainability and Environmental Ethics

–

Some Reflections on the Origin of Goodness and the Future of Environmental Education

Marina Prieto Afonso Lencastre

Abstract

The biological origin of goodness is discussed in relation to the evolution of kin selection and cooperation. Genetic and reciprocal altruism are appreciated, considering the social consequences of empathy and the theory of mind, in primates. Benevolent feelings in parental animals are presented as the precursors of love and goodness in humans. In a globalized world, the ethological and psychological conditions for the experience of goodness in expanded contexts are discussed. Signified by language, goodness and compassion are vital both in defining environment ethical principles and in education for environmental values and sustainability.

Introduction

Modern-day environmental problems call for urgent political and institutional changes. Just as important as these structural changes is the will of individuals to act in favour of the natural world under threat: endangered species, the deplorable conditions in which animals are kept for human consumption or entertainment, threatened ecosystems, polluted water and air, the stress of over-consumption, global warming etc., all of these phenomena awake in us moral feelings concerning the fragile state of the natural world. This sentiment often comes hand in hand with a feeling of impotence when we are confronted with the magnitude of the problems involved. Nonetheless, the evidence of a link between the environment and individuals has never been clearer, and never has it been more important for each of us to intervene to reduce the global impact on the Earth's ecosystem.

Goodness, as a moral sentiment, is the general willingness to do good. It is closely linked to compassion, which is the need to relieve the suffering of others, and to altruism, which is the social emotion derived from it. These feelings are at the basis of pro-social and pro-environmental behaviour which imply cooperation and confidence in others, together with a renewed vision of the importance of life on Earth. How did these feelings and behaviours evolve and what are the ethological and psychological conditions which cause them to be manifested in modern-day expanded contexts? In what ways may they be involved in the basic principles of environmental ethics and the expansion of moral considerations to include the natural world? These and other issues will be looked at in this paper in an attempt to illustrate their evolutionary origins and the way in which, signified by language, they give rise to cooperative norms and cultures.

The Biological Origin of Goodness

Kinship and Cooperation: the Evolutionary Origin of Altruism

Cooperation is a common phenomenon in the animal world but it has only recently found a satisfactory biological explanation. In an apparent contradiction of the Darwinian idea of selfish competition for food and sexual resources, cooperation and biological «altruism»[1], which is generally defined as the sacrifice of one animal for the good of another, confronts biology with a serious dilemma: how can it be evolutionarily adaptive to sacrifice oneself for the good of others, incurring the risk of condemning one's own reproductive potential? Modern studies on cooperation and «altruism» solved this problem in 1963 with W.D. Hamilton's idea of inclusive fitness. This author, who instigated contemporary research on cooperation and «altruism» and also on behavioural ecology and sociobiology, argued that it was more likely that cooperative and altruistic behaviours had evolved among related animals than among those with no kinship ties. Related animals share more similar genes than unrelated animals, meaning that an altruistic action benefiting one animal also benefits the perpetrator, inasmuch as the chances of reproduction of common genes is increased by the altruistic act. Thus biological «altruism» is, in fact, genetic selfishness. Hamilton's theory is based on a complex mathematical model which predicts that cooperation will be more frequent among related animals than among those with no kinship, and that the degree of «altruism» will depend on the degree of the genetic relationship. Together with

1 Biological altruism will appear in inverted commas following the recommendation of Ruse (1996) with regard to the distinction between the ethology of altruism in socially less complex animals and true altruism motivated by emotion, as in higher primates, or by obligation, as in the case of humans.

the work of E.O.Wilson on social insects, his theory provided a favourable context for the development of sociobiology and its theories on the evolutionary origin of «altruism».

One of the best known cases of genetic «altruism» and animal cooperation which serves to validate this theory is the case of bees – insects which live in society. In bees, altruistic behaviour was explained as having evolved from the genetic relationship among the worker bees themselves and between them and the queen bee. This species has a particular form of reproduction called haploid-diploid. The male possesses only half of the genes normally present in the gametes, and thus transmits all of his genes to his descendants; this means that worker bees inherit the totality of paternal genes. In this form of reproduction sister-worker bees share on average 75% of the genes among them but only 50% with their offspring if they reproduce in the normal way. According to the theory of inclusive fitness, it is more advantageous for the group to evolve maintaining the worker bees sterile and available to care for the sister-larvae and for the queen bee. The latter is the only bee in the hive which produces eggs after having been fertilized by various males successively during the reproduction period. When the queen gets old and dies, some of her fertile eggs will be fed with royal jelly and they will produce fertile females who will compete for the genetic possession of the hive. Males derive from sterile eggs which develop until they produce males which are haploid and fertile.

The model used by Hamilton to explain genetic «altruism» did not succeed in solving the problem of "altruistic" behaviour in animals with no kinship. In the light of the model of inclusive fitness, how is it possible to explain that an animal might come to the aid of another with which it has no genetic link? In 1971, Trivers offered an explanation for these behaviours: genes for altruistic behaviour could be selected if the individuals were differentially altruistic with those who had shown altruism towards them. The vampire bat, for example, donates food collected during the night to other bats that have not had the opportunity to collect their share of blood. Part of the blood collected is regurgitated in order to avoid the death of the animal who receives the benefit. As this behaviour is common to the group, each individual can count on the fact that in the same situation it will be fed, and will thus survive. The receiver may be kin, but it may also be an unrelated adult. It was noted that bat colonies are stable and that the members of the group recognize each individual animal, which means that the receiver will be identified and later, in turn, requested to give blood.

This type of «altruism» was called reciprocal «altruism» and, applied to humans, it was conceived by Trivers in terms of the theory of games and the prisoner's dilemma. The prisoner's dilemma presents the following scenario: two suspects of a crime are questioned by the police in separate rooms. They can cooperate with each other or they can defect. Defecting would mean telling the police that the other man is guilty, and cooperating would imply not accusing the other. Even with no confession, the police have enough evidence to hold both of the men for one year. If they accuse each other they will be imprisoned for 3 years; if only one accuses the other and the other keeps quiet, he will go free and the accused will go to prison for 5 years. The dilemma lies in the fact that although selfish behaviour offers the chance of freedom, ignorance concerning the behaviour of the other prisoner makes it more in the interest of both prisoners to cooperate than to defect. The prisoner's dilemma was developed by Hamilton and Axelrod in the 1980s and was applied to the prediction of cooperative behaviour among populations of individuals: when the probability of one individual meeting another in the near future passes a critical threshold, then the probability of cooperative behaviour is greater. According to Axelrod, this conditional cooperative strategy, called Tit for Tat, owes its evolutionary success to three factors: being nice – never being the first one to defect – being rapid to retaliate, and knowing how to forgive your partner if he/she has only defected once.

Recently, modelling of human cooperation has evolved in order to include more complex social factors such as social reputation, pro-social behaviour, the influence of creeds and religions, the social benefits of conformity, etc. They also included theories of strong altruism, which consists of cooperating with others even if the costs of the cooperation may not always be recovered. This model is successful when the exchange contexts are solid and predictably stable (Gintis et al. 2007). The co-evolutionary view of altruism (Boyd 2003) suggests, in turn, that the human capacity to cooperate in large groups derives from the co-evolution of altruistic genes and a culture of cooperation which tends to punish selfish behaviour. Selfish individuals will have no opportunity to disseminate their survival strategy and the group as a whole will tend to be altruistic and cooperative. However, the majority of these models are based on the assumption that the deep motivation for altruism is the self-interest of the individual or of the group – calculations for altruistic investment are based on an assessment of the costs and benefits in which the donor becomes the potential winner of the evolutionary race.

Together with these models, ethological observations have shown that within parental animals (birds and mammals) inclusive fitness and reciprocal «altruism» became more complex with the evolutionary development of emotions concerned with attachment and abandonment – affect and anguish. In these animals, the emotions of attachment and abandonment became autonomous and achieved their own motivational circuitry. Appropriate signal stimuli with an open beak and insistent vocalisations make it possible for birds in flight to feed the young of other birds. Recent evidence from mammal and primate ethology shows that truly altruistic behaviour, based on empathy and cooperation, occurs frequently in groups. In higher primates this behaviour is observed in sophisticated social contexts in which the effects of social and cognitive empathy intervene, the latter consisting of the elaboration of a theory of mind of the other animal. The animals behave in a pro-social way (de Waal 2006); it is unlikely that in difficult, urgent situations they make complex assessment of costs and benefits.

It is likely that some of the difficulties encountered by the evolutionary models of altruism and cooperation in the analysis of these complex behaviours lie in the neurophysiological emancipation of the cognitive and emotional responses of these animals. Indeed, both in anthropoids and human beings, pro-social motivation would appear to depend more on neurophysiological and ethological circuits of social emotions than on the selective logic of genes. Since selection is made on phenotypes and not directly on genes, it could be expected that a complex organism would present more stabilized neurophysiological and epigenetic levels for the expression of behaviour than a simple organism. This type of organization shifts original motivation to other motivational levels through processes of motor and emotional emancipation, which were described by Tinbergen (1951) for ritualization. For some primatologists, the pro-social behaviour of primates is accompanied by genuine emotional experiences which are precursors of the emotional experience of goodness and compassion in humans.

Altruism, Social Empathy and Theory of Mind

One of the most recent outcomes of ethological observation of primates is the discovery of social empathy and a highly complex system of communication. Behaviour, including reciprocal help in situations of danger, special treatment of injured or handicapped animals, aggressive behaviour among families following action which was harmful to other family members, are all examples of empathy and sympathy, which have a recognized affective value at a human level.

Sympathy with the suffering of another animal can also be observed among rhesus monkeys. De Waal (1996) recounts the story of a young female called Fawl, born in the Wisconsin primatology centre, who was attacked by adult members of strong matrilineal lineage; during the attack, the aggressors regularly stared at Fawl's mother and sisters in a particular way attempting to prevent any action on their part to come to the aid of their kin. At the end of the fight, Fawl withdrew to a corner of the cage looking very miserable. Two minutes later her sister came up, put her arm around her, touched her several times to make her react and hugged her again. The two sisters ended up playing together while the mother groomed the dominant female who had started the attack in order to make peace.

In the same group of rhesus monkeys a small female called Azalea was born with a rare congenital disease, autosomal trisomy, which is similar to Down's syndrome. Azalea's mother was a female who had reached the end of her fertile years and who had previously had eleven perfectly normal offspring. Azalea was given special attention by the group: she was more protected, she was groomed twice as often as the others, even by unrelated monkeys, though she was not spared in the conflicts which pitted her mother against other members of the group. Further studies on the same species have shown that young blind monkeys are never left alone and are always accompanied by specific members of the group. Other observations show that monkeys tolerate bizarre, out-of-context behaviour by animals with neurological disorders, which they would normally never tolerate.

The type of observation described above indicates that monkeys are better than the majority of mammals at learning and expressing social skills, which imply not only representation of relevant information about members of the group but also ways to use this information according to the empathetic knowledge of the other's condition. This fact does not necessarily imply that animals are able to represent to themselves the "mental state" of other animals as an abstract entity, or in other words, that they understand the mind beyond a social and emotional understanding. However, recent work has shown that anthropoid primates (orang-utans, gorillas and chimpanzees) are capable of attributing intentions and developing complex representations of experiences of social empathy, and to behave accordingly. These animals possess what authors call a theory of mind of their conspecifics (S. Savage-Rumbaugh et al. 1998).

Consolation behaviour is very common among chimpanzees. After a fight involving a considerable number of animals, organized according to their family, social and political affinities, they manifest an irresistible desire for reconciliation, hugging and stroking each other, and grooming each other at length. Their desire for contact and for a feeling of security is accompanied by a number of typical expressions which are homologous with human expressions in the same circumstances.

Among chimpanzees, the loss of an offspring is clearly a moment of grief for the mother who moans and shrieks inconsolably for some time. She often carries the dead baby around with her, sometimes for days, trying to make it hold on to her or to climb on to her back, until she finally abandons the corpse literally in pieces. The death of the mother has been known to cause the death of her young from sadness and apathy, as was the case of Flint, observed by Jane Goodall. Flint was a young male chimpanzee who lost his aged mother, Flo, to whom he was very attached. He stopped feeding, and gradually stopped moving about, and three weeks after the death of his mother he, too, died.

The death of an adult animal disturbs the whole group of chimpanzees, as was observed by de Waal (1996) in animals in captivity. As a result of the cold weather during the winter months, the colony at the Arnhem Zoo was divided into two groups and kept in separate compounds where the animals could hear but not see each other. One day, one of the females started to shriek hysterically, staring at another female who had been ill for some time. The cries were different from the usual cries of anger and fear; the second female replied with weak vocalisations and suddenly collapsed and lay motionless. All attempts by the researchers to reanimate her were fruitless, and the animal died. In the other compound, a different female shrieked in a similar way and then the whole colony of chimps became silent. In another case recounted by the same researcher, the death of an animal from injuries incurred during successive fights caused absolute silence among the group of chimpanzees, even during feeding time the following morning – usually the occasion of great excitement – and only after the body had been removed did vocalisations begin again.

Response to the need for help and empathy are shown when chimpanzees lick the blood from the wounds of other animals, delicately removing dirt and keeping flies away. Although this type of behaviour is common in mammals and is not therefore specific to chimpanzees, it is the manner and expressions which accompany this type of behaviour which make us intuitively recognize in these animals a deeper understanding of what they are doing.

Altruism, Goodness and Language

Empathy and Goodness

The point concerning social empathy in primates shows that there is a very strong social predisposition in higher animals for reciprocal help, consolation, compassion, and cooperation. It also demonstrates that these behaviours are observed as part of the social dynamics of the group and are expressed in specific situations amid other types of action (aggressive, reinforcing status, territorial, sexual …). In evolution, biological traits are rarely lost; structures are transformed, modified, co-opted for other functions (de Waal 2006). Thus, it is to be expected that the behaviour of primates obeys the oldest laws of kin selection and reciprocal «altruism». However, as they are more socially sophisticated animals living in individualized groups based on the existence of family nuclei and «political» dynamics, it is also to be expected that they manifest more elaborate and more diversified pro-social behaviour than other mammals. Humans inherited many of these behavioural tendencies and added others which are specific to them. Goodness and compassion and the moral pre-disposition which derives from them are specifically human traits; but it is important to understand that they have an evolutionary root which links goodness and morality to nature and not only, as is usually thought, to cultural norms. Goodness, compassion and morality are as deep-rooted in us as other everyday spontaneous, intuitive behaviours.

Empathy, which is the capacity to experience the emotional situation of others through neural and organic representations, is an automatic mechanism which enables us to identify with the emotions of others and act accordingly. De Waal (2006) argues that empathy is part of a relatively simple mechanism which immediately links representation to action. This mechanism is probably related to mirror neurons, identified by Rizzolatti in the 1990s in the prefrontal cortex of monkeys. Pre-motor neurons are generally triggered when the animal is preparing to act. However, Rizzolatti and his colleagues demonstrated that they are also triggered when the animal observes an action performed by others, or when the animal observes the intention to act. These results indicate that empathy has a neurophysiological basis, and that it is at the origin of the motivation to help others. Work published in 2006 (Langford et al.) showed that the same mechanism is active in mice, which have been observed to empathize with the pain of fellow rats but not with that of unfamiliar rats. These researchers consider that empathy is a widespread emotional response among parental animals. The closer the ties with the object of empathy, the easier it is to activate the motor and autonomic responses of the individual. Conversely, distant and different objects do not evoke empathetic response. This would appear to be one of the mechanisms used to deal with in-group and out-group members.

Cognitive empathy, which is the capacity to evaluate the reasons for the emotions of others, adds to emotional empathy. This type of empathy is found in primates which are able to build a theory of mind; it is more advanced than emotional empathy and presupposes its existence. It allows the individual to evaluate the situation and provide suitable help. It is one of the sources of human goodness and with language it acquires the possibility to develop through norms and moral narratives. New evidence has been found to answer the old question of whether affect or reason determine moral choices; genuinely generous impulses motivate us to act in favour of others and only later does reason organize the action as a narrative.

In human beings, positive behaviours based on empathy appear very early (Montagner 1988, 1993) and empathetic emotions appear to be one of the main bases for understanding and concern with others (Goldman 1999). Research on attachment and early affective synchrony (Stern 1985) points to their importance in the development of goodness and moral emotions. Sann (1993) showed that in the newborn child, bonding has a beneficial effect on the adult and that it participates in the emergence of positive feelings. Premack (1996) also illustrated that newborns already present some expectations as to what should happen in specific interactions and these studies would seem to consolidate the idea that from a very early ontogenetic stage the human being presents expectations and assessments of what later might be called the "fairness" of the social situation. The work of Hoffman (1999) suggests that early emotional experience may, through empathetic imagination, permeate moral judgement and the development of more abstract reasoning, thus demonstrating the link between conceptualization and early emotional experience. Moral narratives reveal traces of old emotional experiences and show how they inspire discourse and action (M. Johnson 1993, 1999).

Expanded Community and Language: Altruism and Moral Narratives

In humans, empathy and concern with others develop at an early stage. This depends not only on evolutionary inheritance but also, to a large extent, on ontogenetic conditions which imply proximity and constancy of the attachment figure, access to affect and adult goodwill, and the translation of positive emotions into language. It is important to understand that the evolutionary pressure which selected the capacity for empathy and goodness was not necessarily benevolent (de Waal 2006). In evolutionary terms, what matters is the successful passing on of genes from one generation to the next and many ways of doing this were stabilized by selection. Social cooperation and altruism are some of these ways, and in humans they have acquired increasing psychological and cultural importance. However, empathy, altruism and cooperation are essentially family or in-group phenomena. Human development requires a solid ontogenetic basis in order to mature, which is only possible in a small family group, and humans tend to develop a group morality, which often excludes others. Human ethology and psychology have shown that humans tend to treat out-group members much worse than they do the members of their own group. Eibl-Eibesfeldt (1989) showed that affiliative sociality or the feeling of belonging to a group, which governs a large part of group relationships in humans, decreases as the size of the group grows. There would seem to be an optimum size for spontaneous bonding and empathy, reciprocity and benevolent altruism or goodness. Recent studies have shown that a natural human group tends to be around 100-150 individuals and that above this number anonymity is likely to emerge.

However, studies of strong altruism have shown, in turn, that confidence in the contractual stability of extended groups encourages cooperation, as can be seen in studies of market altruism and cooperation (Gintis et al. 2007). Humans behave altruistically even when they seem to derive no immediate benefit from it. Strong altruism depends on the contracts entered into by humans, based on the capacity to create reciprocal responsibilities and obligations. The essential instrument to these contracts is language.

Indeed, human evolution added something essential to social emotions and to cognitive understanding. Language and the ability to develop rational arguments about our actions and the actions of others make us the only truly moral species. As the human group grew and became more complex, territoriality, hierarchy and tacit social norms related to cooperation became insufficient to achieve coordination of the group. Language allowed more powerful coordination from a distance, allowing the enunciation and recognition of social and moral norms, imposed as mental objects (Gibbard 1996). Emotions associated with socially valuable behaviour started to be coordinated by concepts which, in turn, became organized in justifying narratives. Thus, social sentiments became organized in cultural trajectories which made them particular to those cultures and different from those of others. Some cultures encourage feelings such as jealously and love because they contribute to the stabilization of conjugal ties. Other cultures encourage values such as community life, kinship and solidarity because these values discourage the desire for autonomy and possession of scarce resources. The system of behavioural norms imposed by one set of institutions to regulate social relations (law) is the result of the historical trajectory of concepts and the justifications found for them, and they tend to define what is desirable for life in society.

There is some evidence that our justification for moral action is often *post hoc* (de Waal 2006): we react emotionally and then we use language to justify ourselves. But it is also known that moral narratives are dynamic and allow us to modify our action. According to Johnson (1993), narrative is not merely linguistic or textual but characterizes the synthetic nature of experience and is prefigured in our everyday activities and projects. As we are imaginative and narrative creatures, we are able to configure our lives in new ways. The moral subject emerges inside the narrative and from its means of justification.

Some current research suggests that part of the conceptual activity present in narratives uses images arising from our emotional experience to create more abstract and less structured domains. In the case of environmental ethics, for example, basic experiences such as pain, pleasure, harm, well-being, empathy, suffering, happiness, goodness and compassion could provide a universal basis for the construction of the abstract idea of human and animal rights (M. Johnson 1993). As humans share cognitive mechanisms and also homologous physical experiences, it is likely that in cases where linguistic translation is insufficient (such as cases in which different languages do not contain equivalent expressions), experience is understood and communicated (Lakoff 1987). The understanding of the concepts shows that they contain the cognitive and emotional dimensions of the experience, which allows for imaginative expansion to contexts outside the ones in which they originated, supporting more abstract collective deliberations.

Conclusion: Evolutionary Implications for Environmental Ethics and Education for Sustainability

We live in a world in which ethical questions have become global. However, the ethological and psychological conditions for the experience of goodness and altruism are not much different today from the conditions which characterized a reduced human world. We continue to choose to belong to small groups and our altruistic tendencies tend to benefit our families, then the group, and only afterwards the broader community. In evolutionary terms, our pro-social behaviours were closely related to our aggressive tendencies. The evolutionary pressure which selected cooperation and altruism for the survival of the group is similar to the pressure which led us to discriminate against out-group members, who therefore become potential enemies (Eibl-Eibesfeldt 1989). However, humanity's current situation has drastically changed the evolutionary scenario of previous centuries. Today we are faced with a globalized world, presented to us daily by the media. The group we belong to has become enlarged and we identify with those living in the antipodes almost as much as with our next-door neighbours. All of them have become part of the extended group of humanity. The frequent solidarity campaigns organized on a global scale and contemporary reflection on the global village are proof of this.

It is obvious that in these extended contexts altruism, the exercise of goodness and moral obligation are all diluted. The individual recognition, the stability and the potential reciprocity of small groups are all missing. This is one serious problem of ethics and environmental education. However, an understanding of the ethological and psychological conditions for altruism and cooperation could help to create the conditions for pro-social motivation in expanded communities. As we have seen, this includes the understanding of cooperation and altruism as biological forces for the promotion of social groups, the ontogenesis of affect and the early experience of compassion and goodness, the exercise of emotional and cognitive empathy which is active mostly in individualized groups, the capacity for identification with others whom we consider morally similar to ourselves, the possibility of instituting a system of norms and penalties resulting from them, the capacity to imagine narrative arguments to justify a better global culture.

Contemporary situations confront us with problems which were unheard of in the history of mankind: from bioethics to environmental problems – a whole new range of issues call for conscious reflection and wise choices. Most of these situations raise problems of perception and comprehension at many levels. They break away from behavioural habits and do not permit direct awareness of their degree of practical necessity. They occur in scenarios which are distant in time and space, and are the result of uncertain predictions. These new issues call for new educational responses and new sustainable values: a new literacy able to deal with the complexity and unpredictability of contemporary social and environmental situations and a new ethics sensitive to the condition of others, both human and non-human, present and future. Goodness associated with a deeper understanding of the collective meaning of life on Earth is fundamental to the understanding of global issues and their sustainability. Emotional intelligence may then find the cooperative solutions which our collective future requires of all of us.

References

Axelrod, R. and Hamilton, W.D. (1981), The evolution of cooperation. Science, 211, pp. 1390-1396.

Boyd, R. (2003), Cultural evolution of human cooperation. In Hammerstein, P. (ed.), Genetic and Cultural Evolution of Cooperation. MIT Press.

Eibl-Eibesfeldt, I. (1989), Human Ethology. A. de Gruyter, N.Y.

Gibbard, A. (1996), Moralidade e evolução humana. In Changeux, J.P. (ed.), Fundamentos naturais da ética. Instº Piaget, Lisboa.

Gintis, H. et al. (2007), Explaining altruistic behavior in humans. In Dunbar, R.I.M. and Barrett, L., The Oxford Handbook of Evolutionary Psychology. Oxford Un. Press, USA.

Goldman, A. (1999), Ethics and cognitive science, Ethics, p. 103, citado por Johnson, M. Ethics. In Bechtel, W. and Graham, G. (ed.), A companion to cognitive science. Blackwell Publ.

Hamilton, W.D. (1964), The evolution of altruistic behavior. Am. Nat., 97, 354-356.

Hoffman, M. (1999), The contribution of empathy to justice and moral judgement, citado Johnson, M., Ethics. In Bechtel, W. and Graham, G. (ed.), A companion to cognitive science. Blackwell Publ.

Johnson, M. (1993), Moral imagination. Implications of cognitive science for ethics. University Chicago Press, USA.

– (1999), Ethics. In Bechtel, W. and Graham, G. (ed.), A companion to cognitive science. Blackwell Publ.

Langford, D.J. et al. (2006), Social modulation of pain as evidence for empathy in mice. Science, 1967-1970.

Lakoff, G. (1987), Women, fire, and dangerous things. What categories reveal about the mind. Chicago Un. Press, USA.

Montagner, H. (1988), L'attachement. Les débuts de la tendresse. Odile Jacob, Paris.

– (1993), La précocité de l'emergence des conduites sociales complexes, fondement de la vie relationnelle et des valeurs morales. In Thinès, G. (ed.), Évolution biologique et comportement éthique. Acadámie Royale de Belgique, Brux.

Premack, D. (1996), "Conhecimento" moral do recém-nascido. In Changeux, J.P. (dir), Fundamentos naturais da ética. Inst° Piaget, Lisboa.

Rizzolatti, G. (1996), Premotor cortex and the recognition of motor actions. Cognitive Brain Research, vol. 3, 2, pp. 131-141.

Sann, L. (1993), Aspects éthologiques de l'emergence éthique chez l'enfant premature. In Thinès, G. (ed.), Évolution biologique et comportement éthique. Acadámie Royale de Belgique, Brux.

Savage-Rumbaugh, S., Shanker, St.G. and Taylor, T.J. (1998), Apes, language and the human mind.Ox.Un.Press., USA.

Stern, D. (1985), The interpersonal world of the infant: a view from psychoanalysis and developmental psychology. Basic Books, N.Y.

Tinbergen, N. (1951), The study of instinct. London, Oxford University Press, N.Y.

Trivers, R. (1971), The evolution of reciprocal altruism. Q.Rev.Biol. 46, pp. 35-57.

Waal, F. de (1996), Good natured: the origins of right and wrong in humans and other animals. Harv.Un.Press., USA.

– (2006), Primates and philosophers. How morality evolved. Princeton Un. Press.

Chapter 3

Implementing Sustainable Development in Higher Education

Walter Leal Filho, Evangelos Manolas

Abstract

The implementation of sustainable development in higher education is a global trend. However, the intensity and the depth with which higher education institutions are taking on the challenge of sustainability differs significantly.

This paper presents a global analysis on how higher education institutions are including sustainability issues in different areas (e.g. teaching, research, outreach and institutional management) and describes the progress made in this field during the last 10 years. A special emphasis is given to the problems and barriers which prevent developments in respect of the integration of sustainability issues in higher education. Finally, some concrete steps are outlined which may be undertaken in order to enable universities to integrate sustainable development in their activities along with regional development.

Introduction: Higher Education and Sustainability

Sustainability is today one of the most widely used words in the scientific field as a whole and in environmental sciences in particular, the analysis of the evolution of such a concept is a difficult exercise. This is because the records of the systematic use of such an expression, whose reference in the current vocabulary and political discourse is nowadays so popular, are scattered around. Until the late 1970s, the "word" sustainability was only used occasionally, in most cases to refer to ways in which forest resources should be used. It has, in other words, strong connections with the forestry sector from where, some believe, it originated.

A key question at this stage – and one which is posed time and again, every day, by millions of people all over the world – is *what does sustainable development really mean*? Depending on the ways it is looked at, it may have many meanings such as:

- the systematic, long-term use of natural resources – as defined in the Brundtland Report described elsewhere in this chapter – so that these are available for future generations (here referring to country and local policies);

- the modality of development that enables countries to progress, economically and socially, without destroying their environmental resources (here referring to country policies);
- the type of development which is socially just, ethically acceptable, morally fair and economically sound (here referring to the social ramifications of development);
- the type of development where environmental indicators are as important as economic indicators (here referring to the close links it bears with economic growth).

Many other variants may be listed and are indeed used by different organizations, taking into account their political perspectives and institutional aims. IUCN, which in liaison with UNEP and WWF produced "Caring for the Earth" (IUCN, WWF, UNEP 1991), suggested at the time that the expression "sustainable development" be replaced in some contexts by "sustainable living", since although "development" is associated with governments and refers to a government's responsibilities, the word "living" is closer to an individual's life.

Even though over the past ten years or so significant achievements have been made in respect of sustainability (Singh et al. 2009; et al. 2007) and sustainability science (e.g. Kates et al. 2001), there is still considerable differences among countries. This state of affairs can be partly explained by the different levels of emphasis given to sustainability in countries as a whole and by institutions in particular.

The authors defend the view that there is unlikely to be a consensus – at least a total one – on the meaning of sustainable development, although most people would agree on what it is all about. The reason for this is rather simple: one's own definition will be influenced by one's training, working experience and one's own political and economic setting. There is nothing negative about that, but, equally, there is the need to establish some ground rules so that the search for a consensus on what it is and what it means may not be made hopelessly impossible due to individual differences in opinion and perspectives. Another way to overcome the problem is by looking at approaches to *sustainability* – this meaning the processes which may ultimately lead to sustainable development. This paper will thus refer to sustainability, as opposed to the broad spectrum of sustainable development, having universities as a focal point.

A significant step forward in the international efforts towards recognizing the value and usefulness of sustainability was the momentum generated by the publication of "Our Common Future", also known as the "Brundtland Report" (WCED 1987), which reported on the deliberations of the *World Commission on Environment and Development* (WCED), an official group set up by the UN to investigate ways by which environmental conservation could be systematically pursued internationally, paying due attention to economic, social and political considerations. After nearly three years of work, a panel comprising specialists from various disciplines and from different geographical regions provided an in-depth examination of what the matter of sustainability is, what is required to pursue it and the measures that countries should adopt in order to put it into practice. The definition of sustainability proposed by the WCED and referred to above has also contributed towards some degree of acceptance on what it means. This is an important achievement since, before that, emphasis to it was piecemeal.

The process leading to the UN Conference on Environment and Development (UNCED) in Rio de Janeiro in 1992 – set in motion partly thanks to the Brundtland Report – the national meetings which preceded the conference and the preparation of the so-called "National Environment Reports" by many of the countries represented at Rio gave a new impulse to the debate on sustainability. The publication of Agenda 21 (UN 1992) as the ultimate framework of action in the environment sector and its subsequent endorsement by over one hundred governments, reiterates the value the international community afford – at least in principle – to sustainability. Agenda 21 emphasizes the importance of sustainable development time and time again in its forty chapters. The Johannesburg conference, held in 2002, revised progress made since Rio, and it was seen that comparatively little had been achieved despite promising prospects, and many of the problems raised in Rio in 1992 had still not been sufficiently addressed ten years later.

Things were a little different in the university sector. Nearly 15 years ago, a pioneering book outlined how sustainable development could be implemented in universities (Leal Filho, MacDermitt and Padgham 1996). Over ten years ago, in a set of volumes on sustainability in the context of university education (Leal Filho 1998; 1999a; 1996b; 1996c), the author also discussed the fact that, since Rio, a lot had been said, written and published on the broad subject of sustainability at various levels, from curriculum (Creighton 1996) to planning (Blowers 1993) to the broad field of policy (e.g. Selman 1996; Baker, Kansis, Richardson and Young 1997; Brown 1997) and local environmental initiatives (Whittaker 1995; ICLEI 1997) at the university level too. Within the tertiary sector, there have been various landmarks in the process of design of approaches and mechanisms to bring in environmental concerns to university policies – some of which set in motion well before UNCED – which include many important documents such as:

- the Magna Charta of European Universities (1988)
- the Talloires Declaration of University Presidents for a Sustainable Future (1990)

- the Halifax document "Creating a Common Future: an Action Plan for Universities" (1991) (Lester Pearson Institute for International Development 1992)
- the "Urgent Appeal from the CRE" to the Preparatory Committee of UNCED (1991)
- the COPERNICUS "Universities Charter or Sustainable Development" (1994)
- the Lüneburg Declaration on Higher Education for *Sustainable* Development (2001)
- The Ubuntu Declaration on Education and Science and Technology for Sustainable Development (2002)
- Graz *Declaration* on Committing *Universities to Sustainable* Development (2005)
- G8 University Summit Sapporo Sustainability Declaration (2008)
- G8 University Summit: Statement of Action (2010).

Moreover, in a further reflection on the maturity of the topic at a scientific level, a specialist journal was launched in 2000. Entitled "*The International Journal of Sustainability in Higher Education*" (IJSHE), this unique periodical has established a prestigious international network of practitioners, academics and legislators working on sustainable development in higher education. IJSHE is the world's first journal to specifically focus on the subject of sustainability and sustainable development at universities, especially the 600 or so of them worldwide which have committed themselves towards sustainability by signing international agreements and conventions such as the Bologna Charter, The Halifax Declaration, the Talloires Declaration and the Copernicus Charter for Sustainable Development, among others. IJSHE differs from other journals in the field (e.g. Journal of Cleaner Production, Journal of Sustainable Development, etc.) in the sense that:

- it provides up-to-date information on new developments and trends on sustainable development at universities;
- it reports on international experiences in an academic context, especially from Europe and North America, but also from other parts of the world (e.g. Africa, Latin America, Asia, Oceania);
- it is a tool for the dissemination of reliable information on matters related to campus greening or institutional change;
- it enables continuous networking and information exchange on sustainability at universities on a global basis.

Furthermore, IJSHE represents a step forward in the provision of a peer-reviewed journal, which may promote the work of the people working in the field and disseminate their work and their own institutions on an international basis. Now in its 12th vol. (2011), IJSHE has already provided a significant contribution towards addressing the lack of specific scientific articles on sustainability in higher education, addressing matters such as misconceptions, comparison of policy and practice, overcoming barriers to campus greening and promoting approaches such as back-casting or a university's role in promoting sustainability in healthcare. IJSHE has already received many awards for its innovative focus, including the Aurelio Puccine Award, conferred on it during a special ceremony held in Rome, Italy, in the summer of 2001. The medal states "the challenge of current generations is to make sure that a better world will be here waiting for the new ones".

Moreover, over the course of the past ten years, literature has registered various works which have attempted to throw some light onto the ways by which sustainability – seen as both a process and as a goal – may be effectively implemented, emphasizing what can be done at university level. For example, a publication outlining *practical* examples of the introduction of sustainability components into university activities entitled "Sustainability and University Life" (Leal Filho 1999a) provides concrete examples of action on various fronts such as administration, planning, teaching, extension and research. The ultimate aim of the book is to show the "hows" of sustainability and the practical problems experienced as part of the process at the time. Since then, much progress has been made and the following publications have provided a solid basis for the development of sustainability as a research topic and as an area of importance in university programmes:

- Leal Filho, W. (ed.) (2000), Communicating Sustainability. Peter Lang Scientific Publishers, Frankfurt
- Leal Filho, W. (ed.) (2002a), Teaching Sustainability – Towards Curriculum Greening. Peter Lang Scientific Publishers, Frankfurt
- Leal Filho, W. (ed.) (2002b), International Experiences on Sustainability. Peter Lang Scientific Publishers, Frankfurt
- Leal Filho, W. and Ubelis, A. (eds.) (2004), Integrative Approaches towards Sustainability in the Baltic Sea Region. Peter Lang Scientific Publishers, Frankfurt
- Leal Filho, W. (ed.) (2005), Handbook of Sustainability Research. Peter Lang Scientific Publishers, Frankfurt
- Leal Filho, W., Ubelis, A., Berzina, D. (eds.) (2006), Sustainable Development in the Baltic and Beyond. Peter Lang Scientific Publishers, Frankfurt
- Leal Filho, W. (ed.) (2006), Innovation, Education and Communication for Sustainable Development. Peter Lang Scientific Publishers, Frankfurt
- Leal Filho, W., Salomone, M. (eds.) (2006), Innovative Approaches to Education for Sustainable Development. Peter Lang Scientific Publishers, Frankfurt
- Leal Filho, W., Manolas, E., Sotirakou, M., Boutakis, G. (eds.) (2007), Higher Education and the Challenge of Sustainability: Problems, Promises and Good Practice. Evrographics, Orestiada

- Leal Filho, W. (ed.) (2010), Sustainability at Universities: Opportunities, Challenges and Trends. Peter Lang Scientific Publishers, Frankfurt.

They illustrate the fact that the topic has become mature and is the subject of much scholarly attention on a worldwide scale.

Sustainability at Universities: Some Case Studies

This section introduces some current trends on sustainability in higher education, drawing on various case studies from around the world. Starting with Bould (2009), who focused on what is happening in New Zealand's tertiary education system with regard to design education. Bould's principal research data consisted of four perspectives collected from the country's higher education institutes between February 2008 and March 2009. The data comprise graduate attributes for six institutes with a design department, course guides, prospectuses and websites from the design departments and interview transcripts involving 15 design staff and 23 students attending their second year of studies.

Regarding the manner in which sustainability is incorporated into design education, an exploration of the attributes of the six national institutes that have a design degree programme revealed that although only one explicitly mentions sustainability as its objective, all attributes of the six institutes nevertheless indicate a capacity for learning about issues of sustainability.

In their prospectuses, course guides or online information all six institutes emphasize social or environmental sustainability. Two describe key experiences, another two assert that students will learn some form of environmental responsibility after graduation whilst the last two indicate that students will acquire an understanding of relationships between technology and society. From the six design departments two offer an optional specialized course on sustainability issues within design and, in addition, one of these two also offers a general design course that includes elements of sustainability. Two other design departments offer compulsory, general design courses with elements of sustainability whilst the last two offer no courses at all. In order to examine what was actually taught Bould interviewed 15 staff members from three of the six design departments named as case studies A, B, and C. Although between the two cases A and B, B specifically teaches sustainable design, there is nevertheless little difference between them with regard to how the design team in each case study discusses sustainability. Do teachers in case study A have similar knowledge regarding sustainable design when compared to teachers in case study B? Bould found a variety of opinions held by design team members in case A, and this design team emphasized the need for a "sustainability champion" and "better collegiality amongst staff". The design team from case B shows a level of collegiality amongst themselves, and all indicate that they play a leading role in promoting sustainability.

In order to answer the question whether it makes a difference to the design team if sustainable design is integrated or separated, Bould interviewed 23 second-year students from case studies A and B with the aim of identifying their level of general awareness of sustainability issues with regard to design. Bould divided the level of awareness in both cases into three categories, low, medium and high. Although the students, who have a specific course in sustainable design (case B), have relatively more awareness regarding sustainability in the high level category, the findings nevertheless show a different distribution of students for all three levels with a considerable spike for the medium level category in case A. In order to determine what or who influences students in regards to sustainability within design, Bould analyzed their responses in order to identify motivating and demotivating factors. The findings show that an integrated approach to sustainable design promotes the consideration of a greater variety of motivations. When sustainability is taught separately, as in case B, students seem to be motivated through the course and the staff associated with the institute and have increased motivation to locate others associated with sustainability.

According to Wallis et al. (2005), universities can contribute to the enhancement of regional sustainability in four main ways: through best practice in sustainability, through producing graduates who are skilled and committed to sustainability, through research and through partnerships.

Regarding best practice in sustainability, university campuses may themselves serve as an example in sustainable practices. The *University of British Columbia* was the first university in Canada to have a sustainable development policy and a Campus Sustainability Office, e.g. it is worth noting the running score of paper, the use of electricity and water used on campus plus the fact that students are encouraged to sign a sustainability pledge that requires them to consider the social and environmental consequences of all their actions. *Michigan State University* has produced a sustainability report describing social and economic as well as environmental indicators of performance. *Hong Kong University* is also committed to sustainability, and this is evident from its website and booklets promoting environmentally sustainable practices.

As regards the creation of knowledgeable and skilled graduates who are practitioners in sustainability, most existing programmes are postgraduate, even leading to doctorate qualifications such as *Colombia's* PhD in Sustainable Development. Others have a particular focus such as the MSc in Environmental Sustainability, which is offered by the *University of Edinburgh*, and the MBA in Corporate

Social Responsibility, which is offered by *Nottingham University*. Undergraduate degrees in sustainability are less common although many universities aim to produce graduates who are familiar with the principles of sustainable development. For example, *Deakin University* in Australia has a suite of generic attributes graduates should have called "The Deakin Advantage", where attributes such as "an understanding of the principles and applications of sustainable development" are meant to be applied to the graduates' own disciplinary fields and work situations.

Regarding the dimension of research, Wallis et al. make a distinction between traditional Type 1 and Type 2 research. The latter "is problem-driven, applied, multi-disciplinary, iterative and partnership-based in which collaboration is established from the outset through the problem definition phase, data collection and interpretation" (Wallis et al. 2005). For example, *Deakin University* has led the realization of the project Regional to Catchment Scale Indicators of Sustainability in Southwest Victoria. The project has 17 local partners, and it is worth noting the 19 indicators of sustainability which have been produced and are considered most important.

In terms of *partnerships*, it is worth mentioning the South West Sustainability partnerships established in 1998 and comprising 15 public sector organizations, local government bodies and tertiary education institutions in Victoria, Australia. The Sustainability Blueprint which was produced aimed at building capacity for regional action, at conserving and enhancing the natural resource base, at creating greater prosperity through sustainability and at achieving effective implementation and monitoring.

Witham (2007) presents and discusses a research project undertaken by the Economics Network (EN) which, under the auspices of the Higher Education Academy (HEA) in the UK, focused on how the subject of sustainable development was to be embedded in the higher education economics curriculum. In the first phase of the project, a total of 133 economics lecturers from 62 UK departments completed an online questionnaire. The second phase was a series of focus groups for economics lecturers and other interested academics with the purpose of identifying the barriers embedding ESD into the curriculum as well as exploring ways of overcoming those barriers. The respondents felt the discipline of economics should become heavily involved with ESD and identified several concepts which were important in linking ESD and economics. However, there was disagreement on what those key concepts were. In addition, several barriers to the embedding of ESD in economics curricula were revealed.

The economists thought that their subject was important for ESD for a number of reasons, ranging from offering a way to thinking about the world to adding to the understanding of policy instruments such as Environmental Impact Assessments. Although there was no consensus on what the most important concepts were, the respondents nevertheless identified several key concepts for linking ESD with economics curricula:

> market failure (externalities); inter/intra-generational issues (particularly equity); utility; discounting; value theory, and methods of valuation; non-use value; tradeoffs; renewable and non-renewable resources; resource endowment; circular flow models; dematerialization issues; substitutability; scarcity; trade; long-term equilibrium; steady state; growth; weak versus strong sustainability; natural capital; marginality; and the general concept of limits, whether this applies to growth or to economics itself (Witham 2007).

The respondents identified several barriers to embedding sustainable development in the higher education economics curriculum. These were:

> the lack of staff awareness and expertise, and the associated need to acquire new knowledge; a lack of time to update courses; the lack of relevant course examples; misunderstanding or lack of academic rigour; problems of internal accreditation and systems of validation, including benchmarks; a lack of institutional drive and commitment; perceived irrelevance by students, perhaps manifested in a lack of student demand; and the financial restrictions which might be placed on any ESD project, particularly if it is a new initiative (Witham 2007).

What follows identifies the three main barriers and explores ways in which such barriers could be overcome: 1. Economics lecturers are not convinced of the relevance of embedding sustainable development in the curriculum mainly because it seems to conflict with the employability of students. Such an obstacle could be overcome by developing sustainable products and markets, and / or through social marketing. 2. Those willing to embed sustainable development are uncertain as to what the key concepts are as well as how these concepts should be embedded in the curriculum. The way to overcome this barrier would be firstly to determine the concepts to be included, then survey the students to see what they know, followed by a survey at the end of the term to see what they were taught. Concepts not taught will be embedded during the following academic year. 3. If the problem of sustainable development involves multiple disciplines, its solution should be multi-disciplinary if not interdisciplinary. Barriers to interdisciplinary work for economists could be that economics is not welcoming to other disciplines, the belief that economics has no limitations when answering questions of sustainable development and the feeling that some economists feel insufficiently 'trained' in communicating with other disciplines. Such problems could be overcome by dividing the economists into groups and giving them the same problem to solve, e.g. discuss the London Olympics viz., to identify factors which would make this project sustainable. This topic is suitable because it embodies a multiplicity of economic, social and environmental effects.

Murphy et al. (2009) evaluated the degree to which sustainability concepts have been incorporated into engineering education and research in the U.S. As regards sustainable engineering education, the authors sent out two questionnaires. The first focused on the development of sustainable engineering at the programme level and was sent to the heads of all academic units, while the second was sent to engineering faculty identified as sustainable engineering champions. The first questionnaire was sent to 1300 programme heads and answered by nearly 300 (a 21% response rate), while the second was sent to 327 engineering faculty and answered by 137 (a 43% response rate). Although the overall representation of engineering programmes is <30%, more than 80% of the respondents nevertheless include 73% of schools with PhD programmes that rank in the top 100. From these responses it was made clear that most of the top 100 engineering programmes in the U.S. support teaching and research in sustainable engineering.

The findings of the above effort are presented in two categories: (1) courses, course modules and curricula, and (2) research. As this effort was not a census of activities, the findings should be considered directional. In addition, the terms "sustainability" or "sustainable engineering" were not defined during the research in order to have engineering programmes identify themselves the content that constitutes sustainable engineering. Indicatively, the section below discusses curricula and research.

It is worth noting that most of the top 100 programmes offer courses, while a much smaller number of programmes offer degrees. 23% of respondents (65 departments and 53 institutions) offer Bachelor's or Master's degree programmes.

In terms of the curricula and the graduate level, it is worth mentioning two Master degree programmes. The programme offered by the *University of Michigan (Ann Arbor)*, launched in 2007, is inter-departmental and offers three tracks: sustainable energy systems, sustainable design and manufacturing, and sustainable water resources. The *Yale* programme also involves core courses from both the School of Engineering (water resources, industrial ecology, and sustainable design) and the School of Forestry and Environmental Studies (environmental science, social ecology, economics, and policy and law). The *Yale* programme was also launched in 2007.

At the undergraduate level, the Department of Engineering and Public Policy at *Carnegie Mellon University (Pittsburgh, PA)* has been offering a Bachelor of Science (B.S.) programme for more than 30 years. The programme basically consists of selected courses in decision analysis, economics, statistics, and technical topics, plus two project courses where students work in teams on a current problem for an outside client. The engineering faculty at the *University of Texas (Austin)* has developed "Signature Courses" such as the course entitled "Sustaining a Planet", which describes material and energy cycles in the natural world, how natural systems interact and are modified by engineered systems and how students' lives fit into these systems.

As regards research, there has been substantial funding in sustainable engineering. For the purposes of their paper, Murphy et al. identified roughly a quarter of a billion dollars in funding which is concentrated in top-tier institutions and engages more than 500 graduates and roughly 400 undergraduates. The most topical area for research has been energy and power generation systems. Other important areas include industrial processes, materials, building and construction, water, transportation, pollution prevention, climate change.

One of the most challenging areas of university education is to provide students with opportunities to put sustainability principles into practice. Steinemann (2003) first presents university initiatives to promote research, education, and application of sustainable development in the U.S. Secondly, she examines problem-based learning (PBL) as a practice- based pedagogy and as an approach for learning how to implement sustainability, then she brings the above together by means of presenting her course "Sustainable Urban Development". Fourthly, she summarizes experiences and lessons drawn from the course, and finally offers recommendations.

According to Steinemann, universities promote sustainability in five main ways – policy, research, education, infrastructure and outreach activities. Regarding policy, one example would be the *California Institute of Technology Environmental Task Force* which encourages students, faculty, and staff to work on campus environmental initiatives such as recycling, bicycling, and energy conservation. The *University of California at Davis* pursues research on strategies and technologies for sustainable management of urban forests. In terms of education, one may mention *Harvard University's Center for the Environment, Brandeis University's* Master's degree in Sustainable International Development and *Brown University's* student projects on campus sustainability. As for infrastructure, the green buildings of the *University of California in Santa Barbara* and the *University of Texas* in *Houston* are well-known. As far as outreach is concerned, *Bates College* has established responsible purchasing, composting, and recycling initiatives, and at the *University of South Carolina* students and campus recycling staff work with charitable organizations to donate food, clothes, building materials and furniture in order to reduce the university's disposal costs.

PBL emphasizes learning by doing and students are asked to solve real-world problems. PBL has many advantages. It can make knowledge more accessible and applicable. It helps students develop skills for solving practical and real world problems. In PBL it is the students, not teachers, who are responsible for solving the problem. PBL provides motivation for learning and for solving a problem. Finally, PBL helps students cultivate professional skills such as cooperative and interdisciplinary problem-solving.

In Steinemann's course "Sustainable Urban Development", taught at the *Georgia Institute of Technology, Atlanta,* students identify a problem and then develop a sustainability plan to address it. The course attracts students from several disciplines. The course emphasizes interactive implementation, i.e. communicating and working with university staff, students, companies and the wider community. Besides instructor feedback, students make oral presentations of their work every two weeks, both in terms of content and in terms of form.

Some thematic areas for student projects in the four years of teaching the course were: framework for sustainability, sustainable buildings, energy efficiency, water conservation, storm water management, sustainable landscaping, recycling and composting, transportation, hazards reduction and pollution prevention, education and outreach.

Finally, besides the potential contributions and strengths of PBL Steinemann examines the challenges of her course. First, she notes the importance of finding a balance between giving students feedback and giving them freedom in project design and implementation. Second, she mentions the importance of helping students identify and overcome barriers. A third challenge is the design of suitable and feasible projects. Fourth, she mentions time, pointing out the use of some projects even several years later. Fifth, is maintaining student enthusiasm. Sixth, she points out that a PBL course is very demanding in terms of preparation, delivery and evaluation.

Boyle et al. (2007) has summarized *MIT's* environmental activities in education and research.

Besides the wide array of courses and programmes at both undergraduate and graduate levels, there are striking successes in education which should be used as models for the future:

> For example, the Terrascope freshman experience provides out-of-the-box thinking about environmental problem-solving from the first day students arrive at **MIT.** Another is D-Lab, an interdisciplinary offering, which uses case studies and projects enabling students to understand challenges to sustainable development. A third is Sloan's Laboratory for Sustainable Business (S-Lab), a subject in which student teams collaborate with volunteer clients to meet environmental and social challenges in business models. The Department of Urban Studies and Planning offers an international environmental practicum that takes students (after a semester of problem-focused study in collaboration with overseas partners) into the field for a month-long problem-solving effort Boyle et al. (2007).

All of *MIT's* five schools engage in the study of the environment and sustainability. In the School of Science, for example, ground-breaking work is being done with regard to predicting an increase in the frequency and intensity of tropical cyclones due to global warming and in addressing how to use reaction chemistry (catalysis) to replenish/remove fundamental elements to maintain natural chemical cycles.

Many faculty members in the School of Engineering are pursuing research driven by environmental concerns such as "the need for new materials, substitutions for existing materials, changes in new devices for increasing product efficiency and reducing materials use, new energy storage, sources and demand reduction, and better, more holistic design methods." A new field in *MIT* is environmental genomics. Scientists in this field study the diversity and abundance of microbes – the most significant agents of biochemical cycles.

Research staff at the Sloan School of Management study how business can operate in a sustainable fashion and how it can play a positive role in the regeneration of the planet and support of human health and welfare. Scholars there also study "the organizational dynamics that often thwart or delay the adoption of innovative technologies, emphasizing how firms and other organizations can successfully implement large-scale changes in technology and practices that determine the impact of business on the environment."

The School of Architecture and Planning has worked through the Media Lab and the Departments of Urban Studies and Planning and Architecture to improve visualization, urban design and public participation tools to promote more sustainable patterns of urban development. Academic staff from across the school have engaged in projects aimed "at enhancing the energy efficiency and environmental sustainability of buildings, cities, and regions, including a special focus on the extreme challenges facing China and other rapidly urbanizing countries."

The School of Humanities, Arts, and Social Science has emphasized the study of the costs and benefits associated with various environmental protection strategies through a variety of departments. Examples include the impacts of regulation, economic costs of climate change and the influence of organizational structure on sustainability.

Some campus interdisciplinary efforts should also be mentioned. Of particular importance is the multi-school Center for Global Change Science that implements the joint programme of the Science and Policy of Global Change. CGCS concentrates on "climatic processes with the goal of identifying critical thresholds in the climate system and accurately predicting changes in the global environment. Modelling and data analysis from this effort are figuring significantly in policy discussions, and the programme is expanding to address directly influences on the environment from current and future energy sources."

Columbia University's Earth Institute was established in 1995. The research institute's stated mission is to address complex issues facing the planet and its inhabitants, with particular focus on sustainable development and the needs of the world's poor. The Earth Institute's activities are guided by the idea that science and technological tools that already exist could be applied to greatly improve conditions for the world's poor, while preserving the natural systems that support life on Earth. The Institute comprises 16 research units. The units, which have a strong environmental focus, are The Columbia Water Center, The Center for Rivers and Estuaries, The Center for Climate Systems Research, The Center for Environmental Research and Conservation, The Earth Engineering Center, The Lenfest Center for Sustainable Energy, The Center on Globalization and Sustainable Development, The Center for Sustainable Urban Development, The Center for International Earth Science Information Network, The International Research Institute for Climate and Society, The Columbia Climate Center, The Lamont-Doherty Earth Observatory, and The Center for Hazards and Risk Research.

Indicatively, this paper will present some of the work done in Columbia Water Center (CWC) and in the Center for Environmental Research and Conservation (CERC). Columbia Water Center (CWC) was established in January 2008. The centre focuses on water-related issues, especially issues of water scarcity around the globe. It concentrates on finding strategies to tackle the current and impending levels of global water scarcity. With a team of hydrologists, climatologists, engineers, and water policy analysts, CWC aims at designing water management models that are both dependable and sustainable as long-term global solutions.

According to Columbia Water Center, it has undertaken projects in four countries since its establishment: India, China, Mali, and Brazil.

In Brazil, CWC has partnered with the *Federal University of Ceará* in the northeast to focus on sophisticated climate-based forecasting systems used for water-allocation decisions across diverse use sectors. Recently, ground was broken in Brazil on a centre that will provide water to 500 people living in a rural area. In India, the project focuses on reducing water consumption in the agricultural sector by encouraging sustainable crop choice patterns, as well as working with corporations to deploy better irrigation technologies (and increase reliability) throughout their supply chains. CWC has partnered with *Punjab Agricultural University* to develop innovative solutions to eliminate overuse of water. Some of the methods developed are currently being used in the 2010 planting season with over 500 farmers participating. The work in Mali coincides with the Millennium Villages Project. CWC is focusing on designing irrigation and cropping systems that can be operated and maintained locally in order to provide farmers with higher-value cash crops. The centre has installed large pumps which improve the amount of water available. Projects emphasize bringing together local partners, market forces and public-private partnerships (Columbia Water Center 2010).

The Center for Environmental Research and Conservation aims to stem the loss of biological diversity and cultivate leadership to address the ecological challenges of the 21st century. CERC is a consortium of five world renowned scientific institutions based in New York City: *Columbia University*, the American Museum of Natural History, the New York Botanical Garden, the Wildlife Conservation Society and Wildlife Trust. Each member of the consortium contributes to developing environmental leadership through a variety of education and research programmes. Collectively, this work is conducted in over 60 countries.

Throughout CERC's 13-year history, consortium researchers, volunteers, interns, students, faculty and staff have been involved in:

- Finding new species of plants and animals in biodiversity hotspots
- Mapping the movement of wildlife and zoonotic diseases that pass from animals to humans
- Studying the evolution of primate behaviour
- Examining how forests respond to disturbance
- Studying ecosystem processes and ecosystem services like carbon storage by tropical trees and grasslands
- Understanding how to develop participatory conservation programmes
- Working on the restoration of damaged habitats
- Exploring models for sustainable development through a balance of good economics, governance and conservation (Center for Environmental Research and Conservation 2010).

One of the most important tasks of higher education institutions is to transform learners into inquisitive, reflective, experience and critical-thinking individuals. Transformative pedagogies dictate assessment processes which move away from traditional formats and address a variety of learning styles at the cognitive, psychomotor and affective levels.

Pace (2010) explores the impact of transformative pedagogies on pre-service teachers who participated in an environmental education programme (EEP) of the *University of Malta* by means of an action project. A total of 22 pre-service teachers participated in the programme. Enrolment in the EEP was on a voluntary basis. They were all prospective primary school teachers. This particular programme was designed not to target content in particular but rather to encourage the cultivation of pro-environmental values by engaging students in their own learning as well as providing them with opportunities for reflection and self-evaluation. Pace relies on qualitative methods to analyze his data. In particular, he analyzed evaluation questionnaires on the programme's study units, reflective questionnaires drawing upon the students' reflective journals, focus group interviews and one-to-one interviews.

According to Pace, the major lessons which can be drawn from his paper are three:

- The true change factor that influenced student behaviour was the methodology adopted during the programme rather than what content matter was provided.

Experiencing methodologies that acknowledged different learning needs and provided opportunities for active engagement in learning enabled students to develop skills, attitudes and values necessary for the promotion of sustainable development principles.

- Responding to the different learning needs of students was only possible through a novel way of designing, implementing and evaluating an educational programme. Educational institutions (particularly universities) need to adapt their administrative structures to promote and be more open to new and alternative approaches.
- The EEP experience evidences that change – both institutional and personal – occurs gradually and in small steps, provided that an institution and an individual learner is given space to grow.

True environmental education can only happen if educators and institutions disengage from traditional methods of teaching and use transformative and learner-centred pedagogies.

A final example of what is happening today comes from Hamburg, Germany, where the Hamburg University of Applied Sciences organized the "World Sustainable Development Teach-In Day", held in December 2010. The aims of the "World Sustainable Development Teach-In Day" (fig. 1) are:

- to disseminate information on the concept, aims and purposes of sustainable development so that it can be understood by a broad public, and will include elements relating to its environmental, social, economic and policy aspects;
- to raise awareness among university students on the complexity of matters related to sustainable development and the need for personal engagement and action;
- to provide an opportunity to introduce projects and other initiatives on sustainable development that are being undertaken at international, regional and local level by schools, universities, government bodies, NGOs and other stakeholders;
- to discuss the problems, barriers, challenges, opportunities and the potential achievements with regard to implementing sustainable development globally, regionally and locally.

Fig. 1: Website of the "World Sustainable Development Teach-In Day"

A set of lectures in English prepared by some of the leading experts on sustainable development, combined with videos on the scientific, economic and social aspects of sustainable development, and a discussion forum ensured a wide participation.

Problems in Implementing Sustainability in Higher Education

Sustainability as a whole and education for sustainable development in particular have greatly benefitted from the international momentum created by the UN Decade of Education for Sustainable Development (DESD). Running from 2005 to 2014, DESD, for which UNESCO is the lead agency, seeks to integrate the principles, values, and practices of sustainable development into all aspects of education and learning in order to address the social, economic, cultural and environmental problems faced in the 21st century.

Despite DESD and even though there have been important cultural changes in the general perception of what sustainability is or in relation to what it means to people, the wide implementation of the concept of sustainability still faces major problems and challenges. These are also seen in the higher education context, where a number of barriers still prevent further advancements both in the use of sustainable development techniques or in the execution of sustainable development programmes. A key question to be posed at this stage is:

If sustainability in higher education is so important, why is it not more widely implemented?

This is a question whose answer is not easy to provide since there are many social, political and economic considerations which influence it. However, part of the answer lies in the fact that there have been – and there still are – many structural problems which prevent the wider dissemination of the principles of sustainable development in higher education. The wide range of problems in relation to the implementation of sustainable development in higher education can be classified into four categories as outlined in Table 1.

Table 1: Categories of problems faced in implementing sustainable development in higher education

Category	Nature of the problem	Consequence
Perception	Low value attached to sustainability	Little or no interest
Attitude	Indifference to sustainability	No engagement
Perspective	No acknowledgment of the value of sustainability	Resistance to include it in university programmes
Focus	Sustainability not perceived as relevant	Emphasis given to other themes or subjects
Cost-effectiveness	Sustainability seen as costly area of work	Little or no investment in curricular or in-campus activities

The categories outlined above have some real implications and lead, in turn, to a number of serious problems. Some of them are as follows:

Problem 1: sustainable development is deemed as a subject per se. Principles of sustainable development are across the board and, as such, do not always fit in a single subject. Since it is not classified as being of the domain of any given science – rather being a component which may be incorporated into all disciplines– there tends to be a trend towards perceiving sustainable development as a concept which needs to be "packed" under a single subject or discipline.

Problem 2: the basis of sustainability is often very theoretical. Here, part of the difficulty lies in the fact that sustainability and sustainable approaches are seen as theoretical matters. To make things worse, sustainability, as a term, has been hijacked by the political discourse and, to many people, it is merely a theoretical expression or vague jargon.

Problem 3: sustainability is perceived as too broad. This feeling is strongly felt in some contexts (e.g. in engineering programmes), where sustainability is seen as too broad and, by default, as difficult to handle.

Problem 4: *expertise on sustainability is not easily available.* This is observed in some universities which regard sustainable development as a new issue and a new field of action in which they have no specific expertise. This trend has been changing since there has been many appointments of professors in areas related to sustainable development, but it is still an issue.

Problem 5: sustainability is a fashion. This is unfortunately still observed in many universities, where the sceptical approach used by some vice chancellors and presidents do not really encourage more developments.

Other problems may also be added to the list. Indeed, a recent study performed by Leal Filho (in press) has identified the following issues as further problems in the implementation of education for sustainable development at universities, but especially at school level:

1. lack of training of teachers to handle matters related to education for sustainable development;
2. lack of specific materials which may facilitate its handling in classrooms;
3. limited provisions for continuous education in the field of education for sustainable development;
4. limited availability of institutions which specialize in education for sustainable development and can provide advice on the ground.
5. difficulties in evaluating education for sustainable development;
6. lack of coordination between countries and institutions on education for sustainable development;
7. lack of continuous political support to education for sustainable development beyond the political discourse at international events.

It is noticeable that, during the research outlined above, lack of training was the problem most frequently mentioned. And that was also the case across Europe. It was closely followed by a lack of teaching materials. Insufficient ongoing education programmes (i.e. in-service training) are also deemed as an obstacle. A lack of sustained financial support is acknowledged as a barrier too, albeit to a lesser extent.

Perhaps one of the most worrying features in relation to the above state of affairs is that one or more of the above points are used, every day, as an excuse for not taking action. It is crucial that these are duly addressed.

Some Concrete Steps

In trying to solve or at least minimize the impacts of the problems outlined here, a number of objective and concrete measures are needed. Some of them may be:

- more political support for sustainability programmes in higher education
- more coordination between the individual initiatives
- more funds for sustainability projects
- better provision of suitable infrastructure
- more specific literature and appropriate teaching materials.

In addition, further action is needed in respect of ensuring sustainability is embedded into university programmes, as opposed to being a marginal part of it which has largely been the case so far. Figure 2 outlines some elements which need to be considered in order to catalyze a greater integration of sustainability in university programmes.

Fig. 2: Some elements to catalyze a greater integration of sustainability in university programmes.

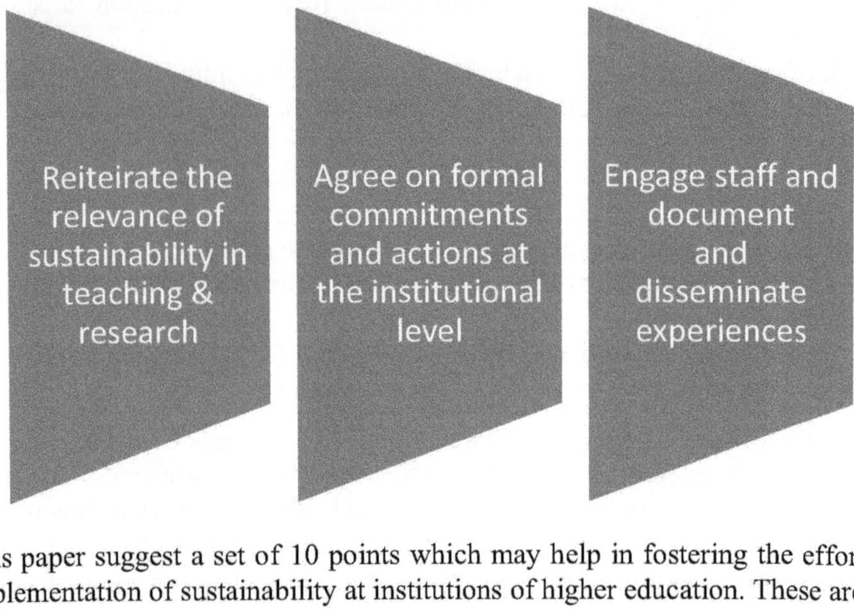

This paper suggest a set of 10 points which may help in fostering the efforts of implementation of sustainability at institutions of higher education. These are:

1. Encourage and promote the development of initiatives and projects on sustainable development at universities, not only in respect of subjects (e.g. curriculum) and campus greening, but also in terms of research and extension (e.g. training events to an external audience).

2. Ensure universities take part in local, regional and national initiatives related to sustainable development in order to put the principles of sustainability into practice and support initiatives in respect of sustainable consumption and production and the promotion of ecological, social and economic development.
3. Identify and use tools, tookits, practical measures and monitoring mechanisms, which show how much progress can be achieved as a result of implementing sustainability efforts.
4. Provide examples of approaches and methods which show how sustainability principles may lead to improvements in efficiency and in reduction of costs, as well as in reducing environmental degradation, pollution and waste.
5. Mobilize sources for financial and technical assistance and capacity building for works in industrialized but also in developing countries.
6. Develop materials and instruments which show or illustrate environmental and health impacts, using case studies to implement them when possible and deemed appropriate.
7. Run regular awareness-raising initiatives illustrating the importance of sustainable development and take part in information and dissemination schemes using the media.
8. Establish and seek peer support for sustainability programmes. If appropriate, set up centres where university professors from different disciplines (faculties) may gather.
9. Collect and disseminate information on the activities taking place at the university and promote the exchange of best practices and know-how on sustainability methods and processes.
10. Involve the university administration (e.g. rector, vice rector, dean, etc.) on sustainability activities so that they can see it is worth it.

It may be the case that not all 10 measures can be implemented at once. However, in trying to get them implemented universities may realize they have a potential which has been dormant and which can be put into concrete terms if they have the political will to do so.

Conclusions

As shown in this paper, higher education institutions worldwide are including sustainability issues in different areas such as in teaching, research, outreach and institutional management, but it has to be said that this is happening at different levels of depth and with different degrees of success.

Even though much progress has been made over the past years, there are still many problems and barriers which prevent developments in respect of the integration of sustainability issues in higher education. These barriers and problems can be addressed if the political will needed is available. When applied to various contexts such as water, energy, industry or transport, sustainable development has proven to be an important tool in fostering life quality and in reducing the human impact on the environment.

There are many universities which have excelled in putting the principles of sustainable development into practice, as outlined in this paper. Part of the success of some of them goes back to the fact that they have not only chosen innovative approaches and have pursued innovative ideas, but also because they have established partnerships and were therefore able to provide added value to their own works. Cooperation therefore may be one effective means by which universities may consolidate their own good work and, at the same time, open the door for new opportunities.

References

Baker, S., Kansis, M., Richardson, D. and Young, S. (eds.) (1997), The Politics of Sustainable Development. Routledge, London.

Blowers, A. (ed.) (1993), Planning for a Sustainable Environment. Earthscan, London.

Bould, N. (2009), Sustainable Design Education in New Zealand. http://www.iasdr2009.org/ap/Papers/Special%20Session/Design%20for%20 Social%20Inclusion%20and%20Social%20Sustainability/Sustainable%20Design%20Education%20in%20New%20Zealand.pdf, accessed 20. August 2010.

Boyle, E. A., Bras, R. L., Chisholm, S. (Penny), Eltahir, E., Entekhabi, D., Green, W.H., Jaillet, P., Marks, D.H., Nocera, D.G., Oye, K.A., Rizzoli, P.M., Steinfeld, J.I., Sterman, J., Susskind, L.E., Tester, J.W., Vale, L.J., Zuber, M.T. (Chair) (2007), Creating a Sustainable Earth: An MIT Research, Teaching, and Public Service Initiative for Understanding, Restoring and Managing the Environment A Report by the Committee to Assess Environmental Activities at MIT (available: http://web.mit.edu/provost/reports/environ. assessment.pdf, accessed 20 August 2010).

Brown, V.A. (ed.) (1997), Managing for Local Sustainability. Policies, Problem-Solving, Practice and Place. National Office of Local Government, Canberra.

Center for Environmental Research and Conservation (2010) (http://www.cerc.columbia.edu, accessed on 10th December 2010).

Columbia Water Center (2010), Will Climate Change Exacerbate or Mitigate Water Stress in Central Asia? Working Paper. Columbia Water Center, New York.

Creighton, S.H. (1996), Greening the Ivory Tower. Improving the Environmental Track Record of Universities, Colleges, and Other Institutions. John Wiley, West Sussex.

International Council for Local Environmental Initiatives/ICLEI (1997), Local Agenda 21 Survey: a Study of Responses by Local Authorities and their National and International Associations to Agenda 21. ICLEI, Toronto.

IUCN, WWF, UNEP (1991), Caring for the Earth. Oxford University Press, Oxford.

Kates, R.W., Clark, W.C, Corell, R., Hall, M.J., Jaeger, C.C. and Lowe, I. (2001), Environment and Development: Sustainability Science. Science, 292 (5517), pp. 641-642.

Leal Filho, W. (ed.) (1998), Umweltschutz und Nachhaltigkeit an Hochschulen: Konzepte-Umsetzung. Verlag Peter Lang, Frankfurt.

– (ed.) (1999a), Sustainability and University Life. Verlag Peter Lang, Frankfurt.

– (1999b), Getting people involved. In Buckingham-Hatfield, S. and Percy, S. (eds.), Constructing Local Environmental Agendas. Routledge, London.

– (ed.) (2000a), Communicating Sustainability. Peter Lang Scientific Publishers, Frankfurt.

– (ed.) (2000b), Dealing with misconceptions on the concept of sustainability. In International Journal of Sustainability in Higher Education, vol. 1, issue 1, pp. 9-19.

– (ed.) (2002a), Teaching Sustainability – Towards Curriculum Greening. Peter Lang Scientific Publishers, Frankfurt.

– (ed.) (2002b), International Experiences on Sustainability. Peter Lang Scientific Publishers, Frankfurt.

– (ed.) (2005), Handbook of Sustainability Research. Peter Lang Scientific Publishers, Frankfurt.

– (ed.) (2006), Innovation, Education and Communication for Sustainable Development. Peter Lang Scientific Publishers, Frankfurt.

– (ed.) (2010), Sustainability at Universities: Opportunities, Challenges and Trends. Peter Lang Scientific Publishers, Frankfurt.

– (In Press), An Overview of Education for Sustainable Development in European Countries. In Global Environmental Research vol. 14 no. 2/2010, Special Issue on "Education for Sustainable Development (ESD): Creating Synergies for Learning for a Sustainable World".

Leal Filho, W., Manolas, E., Sotirakou, M. and Boutakis, G. (eds.) (2007), Higher Education and the Challenge of Sustainability: Problems, Promises and Good Practice. Evrographics, Orestiada.

Leal Filho, W., MacDermott, F.D.J. and Padgham, J. (eds.) (1996), Implementing Sustainable Development at University Level. CRE, Geneva.

Leal Filho, W. and Salomone, M. (eds.) (2006), Innovative Approaches to Education for Sustainable Development. Peter Lang Scientific Publishers, Frankfurt.

Leal Filho, W. and Ubelis, A. (eds.) (2004), Integrative Approaches towards Sustainability in the Baltic Sea Region. Peter Lang Scientific Publishers, Frankfurt.

Leal Filho, W., Ubelis, A. and Berzina, D. (eds.) (2006), Sustainable Development in the Baltic and Beyond. Peter Lang Scientific Publishers, Frankfurt.

Murphy, C.F., Allen, D., Allenby, B., Crittenden, J., Davidson, C.I., Hendrickson, C. and Matthews, H.S. (2009), Sustainability in Engineering Education and Research in U.S. Universities. Environmental Science and Technology, 43, pp. 5558-5564.

Pace, P. (2010), Self-Evaluation as a Tool in Developing Environmental Responsibility, Journal of Teacher Education for Sustainability, vol. 12, no.1, pp. 5-26.

Selman, P. (1996), Sustainable Development: Managing and Planning Ecological Sound Places. Paul Chapman, London.

Singh, R.K., Murty, H.R., Gupta, S.K. and Dikshit, A.K. (2009), An overview of sustainability assessment methodologies. Ecological Indicators 9, pp. 189-212.

Steinemann, A. (2003), Implementing Sustainable Development through Problem-based Learning: Pedagogy and Practice. Journal of Professional Issues in Engineering Education and Practice, vol. 129, no. 4, pp. 216-224.

United Nations (1992), The UN Conference on Environment and Development: a Guide to Agenda 21. UN Publications Service, Geneva.

Wallis, R.L., Wallis, A.M., Harris, C.M. (2005), How Universities Can Enhance Sustainable Development through Successful Engagement with their Regional Communities (available: http://www.engagingcommunities2005.org/abstracts/Wallis-Robert-final.pdf, accessed 18 August 2010)

Whittaker, S. (1995), Case Studies of LA 21. Local Government Management Board, London.

Wilson, J., Tyedmers, P. and Pelot, R. (2007), Contrasting and comparing sustainable development indicator metrics. Ecological Indicators 7, pp. 299-314.

Witham, H. (2007), Embedding Sustainable Development in the Higher Education Economics Curriculum (available: http://www.economicsnetwork.ac.uk/ projects/esd/ESDreport.pdf, accessed 18 August 2010)

World Commission on Environment and Development/WCED (1987), Our Common Future. Oxford University Press, Oxford.

Chapter 4

An Assessment of Requirements to Reframe Science Education towards Education for Sustainable Development

Maria Arminda Pedrosa

Abstract

Documents published by United Nations agencies, or in cooperation with them, along with relevant research literature, namely on science education and on environmental education, are analyzed to articulate concepts involved in sustainable development and in education to help promote it. The goals of the Decade of Education for Sustainable Development are outlined and the importance of this initiative to move towards the millennium development goals is highlighted. Strategic actions for education on sustainable development, educational policies and curricular orientations are briefly reviewed in the framework of the second half of this decade, at policy level regarding member states and at practice level as defined in the Bonn Declaration. Some key ideas on educational policies and reform in the European Union are presented, focussing on the formal education and curricular orientations of the national curriculum for sciences in Portugal. Research-based recommendations on science and technology education are highlighted, emphasizing the STS movement in science education and good practices, at different levels, especially regarding science teachers' education, to help reorient education towards education for sustainable development. Concluding remarks are presented focussing on the requirements for reshaping science education towards this end, within the United Nations Decade and beyond, on a practical level in schools, classrooms and laboratories. Ways forward are suggested to promote and sustain educational innovations, based on pertinent research-based knowledge. Synergies and partnerships are advocated to stimulate science education research involving practising teachers with the aim of helping to define and expand zones of feasible innovation suitable for the integration of education for sustainable development in curricular sciences at all levels.

Introduction

Education for sustainable development (ESD) aims to help pursue sustainable development (SD) and UNESCO, as the lead agency of the "United Nations Decade of Education for Sustainable Development (2005-2014)", is willing to help pursue its objectives, namely "help countries make progress towards and attain the millennium development goals through ESD efforts" and "provide countries with new opportunities to incorporate ESD into education reform efforts" (UNESCO 2005a: 1). As an important international strategy to catalyze ESD, the Decade of Education for Sustainable Development (DESD) aims to encourage changes in people's behaviour that are required for more just and sustainable societies for everybody; these are becoming increasingly necessary.

Indeed, recent financial and economic global crises have had a negative impact on the millennium development goals (MDG) and slowed down their progress, especially in developing countries, although in different ways (UNDP 2010). Furthermore, climate change impacts on ecosystems, affecting the range and distribution of flora and fauna, and "on migration, particularly rural-urban migration", which is already significant "and will have major implications for social service provisions, as well as natural resource conflicts and changing livelihoods" (UNDP 2010: 18). Therefore, "adaptation is urgently required to ensure sustainability of all MDG achievements" for various reasons such as "variability in agricultural production as a result of increased unpredictability of weather" (UNDP 2010: 17).

The overall goal of DESD is to integrate values encompassed in SD into all aspects of learning[1] in order to help face the ESD challenge: "Educating people to deal with complex issues that threaten planetary sustainability" (UNESCO 2005a, Annex I: 3). Therefore, through DESD it is sought to stimulate international organizations, governments, civil society and stakeholders, in general and at different levels, to develop efforts required to help people understand and address the social, economic, cultural and environmental problems of the 21st century. Poverty, migration and climate change are among today's problems, of which many were identified at the Earth Summit in 1992 and reaffirmed at the World Summit on Sustainable Development in 2002.

Within DESD, it is sought to stimulate institutions, organizations and stakeholders, in general, to face "global issues of sustainability that affect individual nations and communities" (UNESCO 2005a, Annex I: 3). Understanding and addressing these highly complex issues requires education reforms and "broad and sophisticated educational strategies for this and the next generation of leaders

---

1 *http://www.unesco.org/en/education/decades-days-prizes/un-decades/* – retrieved 23/07/2010).

and citizens to find solutions", as well as concerted actions involving "a broad and deep effort from many sectors of society" (UNESCO 2005a, Annex I: 3).

The overall goal for the second half of the DESD, "to support member states and other stakeholders in addressing global sustainable development challenges through ESD, thus addressing the challenges of learning for bringing about a more sustainable world" (UNESCO 2010: 3), was set drawing on the DESD monitoring and evaluation processes and building on the UNESCO World Conference on ESD and the Bonn Declaration, as well as on the outcomes of other key international ESD meetings. The "UNESCO Strategy for the Second Half of the United Nations Decade of Education for Sustainable Development" (UNESCO 2010) reaffirms the importance of the International Implementation Scheme (UNESCO 2005b) and acknowledges that the global framework for the second half of the decade "is also shaped by the shared vision of education emerging from the four 2008–2009 UNESCO international education conferences [...]: one of education systems that encourage equity and inclusion, quality learning, flexibility and innovation" (UNESCO 2010: 3). While achieving "education for all" is highly regarded, it is stressed that the education actions of UNESCO are extended "to pay explicit attention to inclusion, social cohesion and social justice, to lifelong learning and to the crucial role of higher education and research" (UNESCO 2010: 3).

Concerning Europe, it is included (with North America) in one of the five regions considered in the UNESCO strategy for the second half of the DESD (UNESCO 2010) in which "key regional challenges and opportunities for the 2nd half of the DESD" are outlined. Among others, key challenges in this region concern a lack of expertise "to address the interdisciplinary and holistic nature of ESD, which was recognized as a persistent bottleneck in advancing ESD by Ministers of Education and of the Environment in Belgrade in 2007", along with "the lack of appropriate ESD teaching tools and research" (UNESCO 2010: 7).

Education nowadays should take these challenges into account, responding to worldwide calls for innovative practices to implement "successful negotiations agreed upon after 2000 within the United Nations" (Pedrosa and Moreno 2008: 284), especially with regard to SD issues. These should encompass various levels, from local to global, and with the aim of "eduction for all", to help enable active and responsible citizenship. This perspective means and implies taking into account past, present and future scenarios that are appropriate to approaching complex problems whilst acknowledging the "interdependence of environment, economy, society, and cultural diversity from local to global levels" in order to develop "creative and critical approaches, long-term thinking, innovation and empowerment for dealing with uncertainty, and for solving complex problems" (UNESCO 2010: 18).

Therefore, some of today's envisaged complex problems and solutions, when they are available and appropriate, should shape "curriculum orientations to help students develop an informed citizenship" that is required to figure out "ways to actively and responsibly exercise it, nowadays and in the future" (Pedrosa and Moreno 2008: 283). While calling upon the integration of ESD in formal education, this perspective on curricular orientation gives rise to several questions such as:

In formal education systems structured in curriculum fields and disciplines, what is required to help students develop an informed citizenship in a perspective of SD? Regarding science education, what is required to do so: i) at different policy levels? ii) to implement policies set up at different levels? iii) on a practical level in schools, classrooms and laboratories? How to incorporate the best knowledge available to integrate ESD across curriculum fields and disciplines, especially with the aim of "education for all" and physical sciences?

The questions above emerged from several assumptions regarding science, science education and ESD, in particular:

i) scientific knowledge is essential for understanding complex problems that humankind faces today and helping to solve them whilst being aware that they will not be solved "with the same values and approaches that created them" (UNESCO 2010: 4);
ii) science and technology should be oriented towards developing expertise and making the required knowledge available, including that needed to reorient education according to principles of ESD and guidelines agreed upon for the second half of the DESD (UNESCO 2010);
iii) "it is the obligation of the scientific and technological community to accept major responsibility for the advancement of science education", for various reasons, namely because "access to scientific knowledge is part of the right to education and the right to information belonging to all people" and "science education is essential for human development and for creating endogenous scientific capacity" (Lederman 2000: 162).

Answering the above-mentioned questions, which are complex and difficult ones, requires the participation of social actors belonging to different institutions and with various responsibilities and decision-making levels regarding research undertakings. Therefore, this paper is not intended to provide answers to the above questions; it is intended to stimulate reflection and discussion among those involved in science education research and teaching and related scientific and professional fields, and to help develop innovations required to construct answers. Specifically, the objectives of this paper are to make explicit some relevant information on international contexts and knowledge made available through research with the aim of reshaping science education and, taking the above into account, clarifying what is required in practice and in school contexts.

In the next section, *Strategic actions for ESD, educational policies and curricular orientations*, the UNESCO strategy for the second half of the DESD (UNESCO 2010) is briefly reviewed, paying particular attention to Europe. Key ideas are briefly presented, emphasizing formal education both at policy level regarding member states and at practice level of the Bonn Declaration.

Strategic Actions for ESD, Educational Policies and Curricular Orientations

For the second half of the DESD, UNESCO efforts will focus on areas of strategic action: "1) enhancing synergies with different education and development initiatives and strengthening ESD partnerships; 2) developing and strengthening capacities for ESD; 3) building, sharing and applying ESD-related knowledge; 4) advocating ESD, and increasing awareness and understanding of sustainability" (UNESCO 2010: 7). Strategic actions concerning enhancing synergies with different education and development initiatives and strengthening ESD partnerships are based on the premise that education is vital to achieve SD. The main idea underlying these strategic actions is that it is essential to achieve SD, a complex and difficult endeavour requiring "a global change of mindsets, beliefs and behaviours" through "the building of partnerships at all levels – local, national, regional and global – and this certainly applies to education" (UNESCO 2010: 8).

Furthermore, it is crucial to create synergies between different education initiatives, building up partnerships for SD, as "ESD has a demonstrable relevance to the wider development agenda, particularly the MDG", and the "integrated and systemic ESD approach is especially well suited to creating and enhancing such synergies" (UNESCO 2010: 8).

By developing and strengthening capacities for ESD it is envisaged to support "member states to reorient teacher education and training programmes towards sustainability" and "educators' communities of practice for adapting tools to their contexts, for enhanced exchange on ESD approaches, on innovative experiences in ESD teaching and learning, including through electronic networking", as well as to "assist member states in the further development of monitoring and evaluation frameworks, tools and indicators to assess ESD progress" (UNESCO 2010: 10-11).

Since knowledge is regarded as an essential component of ESD, a strategic action concerning building, sharing and applying ESD-related knowledge has been set up in which support is envisaged to develop "national research agendas and assist in building the capacity of higher education institutions to engage in ESD-related research and innovation" and where "central activities within the DESD" are identified: "Promoting research, supporting the development of scientific understanding, and sharing and disseminating the vast amount of available knowledge, including traditional and indigenous knowledge" (UNESCO 2010: 11).

Strategic action that concerns advocating ESD and increasing awareness and understanding of sustainability has been set up to help mobilize support for the DESD and develop a shared and common vision of ESD. It identifies partners to achieve "large-scale mobilization", e.g. civil society and the media, and considers necessary to increase "the visibility" and improve "the profile of ESD, especially aiming at educators, policy-makers and young people" (UNESCO 2010: 14) so that knowledge on the SD agenda may be improved and its acceptance may increase at targeted and general levels.

To promote the DESD goals, it is necessary to integrate ESD into all settings of education at all levels, aiming at contributing to accomplish other United Nations (UN) goals, specifically the "Education for All (EFA)" and the "Millennium Development Goals" (UNESCO 2010: 14). However, various differences between EFA and ESD have been identified, for instance: "ESD questions development and the assumption that current dominant forms of development are necessarily inevitable and desirable" implying that, in this respect, "ESD is more contentious than EFA"; "ESD is arguably aimed not just at poverty reduction but also at poverty prevention", meaning that ESD must challenge education that supports "trends towards social and environmental degradation" (Wade and Parker 2008: 6).

In addition, while EFA tends to simply view basic education as a means to development, without considering that "the picture is made more complex by the fact that education is also viewed as an integral part of development", "ESD involves a spectrum of models of education and change from the purely instrumental, whereby educators are charged to 'deliver' policy on SD and to change views and behaviour in accordance with the current understanding of policy-makers, to the development of skills and capacities that enable a range of unpredictable and creative responses to the challenges to meet sustainability" (Wade and Parker 2008: 6). By contrast, the range of the desired scope of change, from the individual to the wider society, represents a point of similarity between EFA and ESD, though "ESD also emphasises the need to change personal/individual and social relations to the local and global ecosystems that we inhabit, including relations of consumption and production" (Wade and Parker 2008: 6).

To enhance "synergies with different education and development initiatives (EFA, MDGs, UNLD, EDUCAIDS) and strengthen partnerships among ESD stakeholders", it is necessary to "continue and further extend dialogue with governments and other key stakeholders in order to facilitate global and regional partnerships" through different priority actions, for instance to "advocate the incorporation of ESD into school curricula and national educational plans as well as in informal and non-formal educational methods" (UNESCO 2010: 9).

The Bonn Declaration, agreed upon by the participants gathered at the UNESCO World Conference on Education for Sustainable Development, is meant to "provide the global strategic framework for UNESCO and its DESD partners during the second half of the Decade" (UNESCO 2010: 3) calling for actions at different levels, specifically at "policy level in member states" and at "practice level". In fourteen numbered paragraphs it pays due attention to the progress achieved until 2009, and outlines progress in scientific knowledge in specific domains (see UNESCO 2010: 17-19) to emphasize "the importance of education for the global development agenda and underlines the essential contribution of ESD to shaping the purpose, content and quality of all education" (UNESCO 2010: 3).

The Bonn Declaration considers that progress in scientific knowledge is focussed on "a better knowledge of climate change and of the Earth's life-support systems", "significant knowledge about HIV and AIDS, malaria, tuberculosis, heart diseases, and other serious health challenges", to stress the relevance of making more knowledge available "about natural systems, and human impacts on them, and the ways that biodiversity supports our well-being". It emphasizes "that current economic thinking has to change, and that there is a need to avoid unsustainable production and consumption and promote and support the emergence of 'sustainably developed' countries". It values insights provided by social science "into ethical, cultural, cognitive and affective aspects of human development, as well as sociologies of change", while regarding education as a potential "significant factor in improving human well-being". Moreover, it recognizes that there are "knowledge and experience available to significantly improve the contents, methods and purposes of education" and "to begin reorienting education systems to emphasise lifelong learning". In the Bonn Declaration it is valued and highlighted "learning how to improve links between formal, non-formal and informal education" through ESD and the importance of "strengthening and sharing knowledge of educational change processes" (UNESCO 2010: 18).

The call for action in the Bonn Declaration relies on the recognition that the "progress of ESD remains unevenly distributed and requires different approaches in different contexts" and that "there is a clear need for both developed and developing countries, civil society and international organisations to make significant efforts" (UNESCO 2010: 19). It considers the policy level in member states, regardless of their human development indexes, comprising five paragraphs, and the practice level, comprising thirteen paragraphs (UNESCO 2010: 19-21). The main ideas conveyed at policy level in member states are summarized in Figure 1 and those concerning the practice level in Figure 2.

Fig. 1: Graphic representation of main ideas conveyed in the five paragraphs calling for action at policy level in the Bonn Declaration (UNESCO 2010: 19).

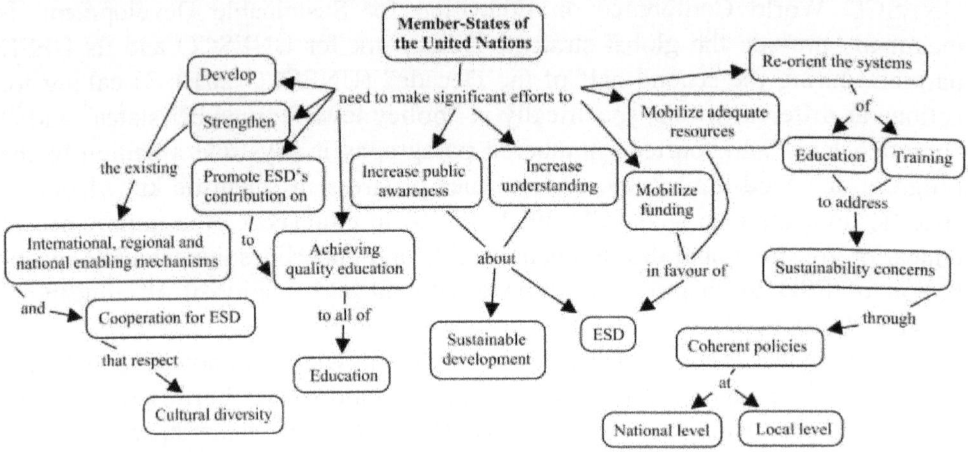

These diagrams were created by the author using "Cmap Tools"[2] to show outcomes of the DESD mid-term Bonn Conference, namely making explicit significant efforts that need to be undertaken in specific domains and/or for particular purposes to ensure implementing ESD within the DESD and beyond.

Concerning formal education, the need is reaffirmed to re-orient "education and training systems to address sustainability concerns through coherent policies at national and local levels" (UNESCO 2010: 19), that have been recognized in other former UN initiatives, notably in the International Implementation Scheme for DESD (UNESCO 2005b). It is stressed that educational reforms are required worldwide in which challenges not only concern the need for equitable access to knowledge, but also encompass issues on how to characterize socially valid knowledge and how to organize society to make possible lifelong learning (Aguerrondo 2009).

2 http://cmap.ihmc.us/download/ – retrieved 29/01/2011).

Fig. 2: Graphic representation of main ideas conveyed in the thirteen paragraphs at practice level in the Bonn Declaration (UNESCO 2010, p. 19-21).

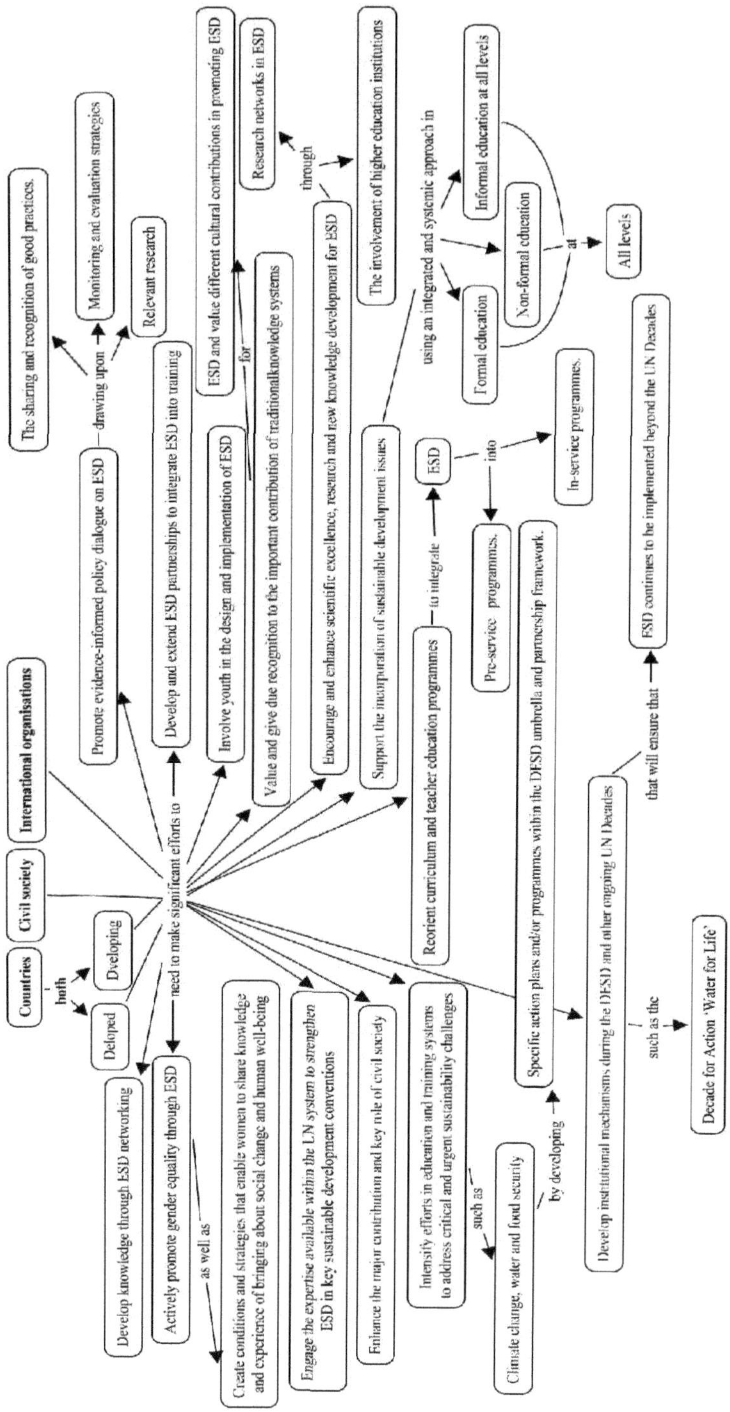

A brief review of literature on educational policies and reforms in the European Union (EU) is presented in the next section, in particular, on formal education and curricular orientation for sciences in Portugal. Research-based recommendations on science and technology education at policy and practice levels are discussed and highlighted to help identify needs to reorient curricular disciplines and teacher programmes concerned, regarding in particular sciences in the Portuguese national curriculum.

Policies, Guidelines, Education Reforms and Curricular Orientations in the EU

Educational skills required to help solve problems that matter to today's society appears desirable for every country, regardless of its geographical location or human development index. However, particular emphasis is noticeable in the EU within a knowledge-based economy framework and focussing on strategies that emphasize competition. For example, in a document whose publication was recently requested in the Official Journal of the European Union, the "Joint Progress Report of the Council and the Commission on the implementation of the "Education & Training 2010" work programme", it is stated: "More than ever, Europe's success in global competition is dependent on its skills and innovation capacities and a swift transition to a low-carbon, knowledge-based economy" (CEU 2010, Annex: 19).

The eight key competences regarded as "necessary for personal fulfilment, active citizenship, social inclusion and employability in a knowledge society" include "mathematical competence and basic competences in science and technology" (CEU 2010, Annex: 5) – a key competence that concerns specifically science education. To develop it in order to help solve problems that matter to today's society, learners are required to articulate dimensions of both performance and ethical commitment. Science education should also be concerned with helping develop cross-curricular competences, e.g. "Social and civic competences", another key competence defined in "The European Framework for Key Competences for Lifelong Learning" (CEU 2010, Annex: 5).

Apparently, there is an effective reform movement across the EU towards "the Key Competences framework", as reforms that explicitly use it as a reference are taking place in a "large number of countries", and progress in adapting school curricula is considered good. Simultaneously, it is recognized that efforts must be made in teacher education and to ensure access equity: "But there is still much to be done to support teachers' competence development, to update assessment methods, and to introduce new ways of organising learning in an innovative school environment. There is a major challenge to ensure that all learners benefit from innovative methodologies, including the disadvantaged" (CEU 2010, Annex: 5).

In some countries the "European Framework of Key Competences" has been the "key in policy reform" and has considerably contributed to a "clear trend across the EU towards competence-based teaching and learning, and a learning outcomes approach", so that progress achieved is considered significant, particularly in school curricula. Regarding these reform dimensions, the "transversal key competences are becoming more prominent and more explicit" and science, like one's own language, foreign languages or mathematics "are being treated in a more cross-curricular way, with more emphasis on developing skills and positive attitudes alongside knowledge, and with more 'real-life' applications" (CEU 2010, Annex: 6). In addition, as a result of poor performance by member states in the 2006 PISA survey, several countries "have created strategies or action plans to raise basic skills levels, particularly in reading, maths and science" (CEU 2010, Annex: 6).

Curricular changes required to implement competence-based approaches are generally occurring, though they are not enough on their own. Competence-based approaches, involving "the skills and attitudes to apply knowledge appropriately, and the development of positive attitudes towards further learning, critical thinking and creativity competence" (CEU 2010, Annex: 6), require that school leaders and teachers face the real challenges they represent and are provided with the means and support needed. For instance, learning to acquire competences, which are present in many curricula, "require more support to embed them systematically in teaching and learning processes and to promote the learning ethos across the whole school" (CEU 2010, Annex: 7).

Furthermore, support involving other stakeholders is required, in particular researchers with research experience and agendas relevant to specific curricular areas and disciplines, notably science-related ones. As devising and implementing inquiry-based learning is regarded as "especially useful for those whose previous experiences in school have been unsuccessful or negative" (CEU 2010, Annex: 7), partnerships involving researchers with these profiles are essential. It is also essential to open schools to "outreach activities with employers, youth groups, cultural actors, and civil society", especially "when seeking to impart social and civic competences, a sense of initiative and of entrepreneurship and cultural awareness" to provide students with "opportunities to take initiatives and learn in schools which are open to the worlds of work, voluntary action, sport and culture" (CEU 2010, Annex: 7).

It should be stressed that these innovations, and others that key competences for lifelong learning of the EU reference framework (OJEU 2006) call for, are essential to progress and succeed in implementing competence-based approaches. In this EU reference framework and for each of the eight key competences considered (one of which is "mathematical competence and basic competences in science and technology"), a definition, essential knowledge, skills and related attitudes are presented.

Concerning competences in science and technology, the definition states that it "involves an understanding of the changes caused by human activity and responsibility as an individual citizen" and related attitudes comprise "critical appreciation and curiosity, an interest in ethical issues and respect for both safety and sustainability, in particular as regards scientific and technological progress in relation to oneself, family, community and global issues" (OJEU 2006: 15). The focus of essential knowledge is on the natural world and on the impact of science and technology on it – a technological and scientific perspective that "should enable individuals to better understand the advances, limitations and risks of scientific theories, applications and technology in societies at large (in relation to decision-making, values, moral questions, culture, etc.)" (OJEU 2006: 15). The "essential knowledge comprises the basic principles of the natural world, fundamental scientific concepts, principles and methods, technology and technological products and processes" (OJEU 2006: 15). Similarly, what is considered to be skills indicates a technological and scientific perspective, as they "include the ability to use and handle technological tools and machines as well as scientific data to achieve a goal or to reach an evidence-based decision or conclusion", that should enable learners "to recognize the essential features of scientific inquiry and have the ability to communicate the conclusions and reasoning that led to them".

The definition, essential knowledge, skills and attitudes presented in the EU reference framework (OJEU 2006) regarding competences in science and technology refer and are applicable to curricular areas and disciplines generally designated by sciences within the Portuguese education system. Sciences comprise physics, chemistry, biology and geology disciplines in secondary education, while in the third cycle of basic education (that is meant to be for all) they include two disciplines, natural sciences and physics-chemistry sciences. They are considered together in the national curriculum, designated by natural and physical sciences, focus on competences and are organized in four general themes: "Earth in Space", "Earth in Transformation", "Sustainability on Earth" and "Living Better on Earth" (DEB 2001a).

The ideas included are explored and deepened in specific teaching guidelines published by the Ministry of Education (DEB 2001b), in which "educational experiences" are presented for each discipline separately, e.g. on "energy sources and forms" for physics-chemistry sciences and on "energy flux and matter cycle" for natural sciences, as well as for both, e.g. on "risks of scientific and technological innovations for the individual, society and environment". Pedrosa and Loureiro (2008) found that innovative and integrative approaches are suggested for "educational experiences" regarding both disciplines considered as a whole, in contrast with guidelines for each discipline alone. These appear more traditionally oriented with no suggestions or incentives to approach energy in a perspective of ecologic sustainability and education for citizenship that call for suitable changes in behaviour.

Competences in science and technology presented in the EU reference framework (OJEU 2006) and referred to before are considered in tune with research-based recommendations for science education, especially with regard to three essential dimensions to re-orient science teaching and learning: education in sciences, about sciences and through sciences. To interlink and balance these dimensions appears essential to develop competences required to understand human dimensions of global environmental changes in a perspective of ESD (Loureiro et al. 2008).

Furthermore, in the science education thematic meeting of the World Conference on Science it was recognized that science and science education worldwide have contributed to the perception of development "as progress in desired directions which includes improvements in the material welfare of a people as well as the eradication of mass poverty, illiteracy and disease". Moreover, to promote countries' development it was defended that interlinks between science, technology and society (STS) should be explored "in innovative, challenging and pragmatic science education programmes" (Bajah 2000: 156).

In the report on this thematic meeting the role of science education for human development and science education as a human right are stressed to call upon the international science community "to accept the moral obligation to fully participate in radical improvement of science education at all levels for all students. Scientists must contribute some of their time to collaborating with teachers and educators on advancing universal science literacy" (Lederman 2000: 162). A learner-centred science education focussing on local issues is reported to have been defended by panellists and participants, considering that local "experiences are most effective, especially when applied using the 'hands-on' inquiry method of instruction" (Lederman 2000: 162).

In the next section, *Science Education, STS Interrelations and Good Practices and ESD*, some literature on STS interrelations in science education and on good practices relevant to reorient education towards ESD are briefly reviewed at different levels, from global to geopolitical regions, integrating Portugal and ending on local intervention in teacher education programmes.

Science Education, STS Interrelations and ESD

Science education required within a framework of knowledge societies should be integrated in educational systems envisaging non-traditional thinking operations that could be translated into complex competences in action that comprise dimensions of performance and ethical commitment. These dimensions are involved in developing problem-solving competences that should be useful at personal and societal levels, including those required to run everyday personal and work life and to develop broader citizenship competences (Aguerrondo 2009). This perspective means politicizing sciences' curricula, once "such an education is to enable young citizens to look critically at the society we have, and the values that sustain it, and ask what can and should be changed in order to achieve a more socially just democracy and to ensure more environmentally sustainable lifestyles" (Hodson 2003: 654). To achieve these goals requires grounding traditional science content in socially and personally relevant contexts integrated in issues-based approaches. These "can provide the motivation that is absent from current abstract, de-contextualized approaches and can form a base for students to construct understanding that is personally relevant, meaningful and important" (Hodson 2003: 654).

Much of the STS curricula of sciences recognize negative impacts of scientific and technological development on the environment but scarcely recognize that often their impacts on society are differentiated and that "scientific and technological decisions are taken in pursuit of particular interests, justified by particular values and sometimes implemented by those with sufficient economic and political power to override the needs and interests of others" (Hodson 2003: 656). Few science curricula appear to be concerned with stimulating students to develop critical views to enable them to: i) recognize that wealth and power are linked with scientific and technological development and ii) "begin to see the ways in which problems of environmental degradation are rooted in societal practices and in the values and interests that sustain them" (Hodson 2003: 656), for instance addressing the symptoms of poverty, including socio-political and historical analyses of their causes.

The STS movement, which was originally inspired by environmentalism and the sociology of science, is rooted on research, showing that for the great majority of students canonical science content is not achievable in the context of traditional school science (Aikenhead 2005). Environmental issues have been incorporated in science education curricula within educational reforms that are happening at the beginning of the 21st century in many countries, including Portugal. Environment-oriented science education, without paying due attention to social and economic dimensions, will not ensure a level of critical scientific and technological literacy for all – an essential means to bring about social reconstruction – while mistaking the purpose of the dimension of education for citizenship. In Hodson's words it is necessary to politicize science and technology curricula, which requires addressing four major curricular components, "*Learning Science and Technology*", "*Learning About Science and Technology*", "*Doing Science and Technology*" and "*Engaging in Sociopolitical Action*". This can be achieved by using socially and personally relevant contexts and issues-based approaches in which students are given opportunities "to confront real world issues that have scientific, technological or environmental dimension" (Hodson 2003: 654).

Some of the arguments that Hodson (2003) uses to call for action for an alternative and sustainable future, defending that "*Engaging in Sociopolitical Action*" should be a curricular dimension of science education, resemble ideas defended by Saramago in the closing ceremony of the 2002 World Social Forum: "entre tantas outras discussões necessárias ou indispensáveis, é urgente, antes que se nos torne demasiado tarde, promover um debate mundial sobre a democracia e as causas da sua decadência, sobre a intervenção dos cidadãos na vida política e social, sobre as relações entre os Estados e o poder económico e financeiro mundial, sobre aquilo que afirma e aquilo que nega a democracia, sobre o direito à felicidade e a uma existência digna, sobre as misérias e as esperanças da humanidade, ou, falando com menos retórica, dos simples seres humanos que a compõem, um por um e todos juntos" (Saramago 2008: 9). José Saramago, a Portuguese writer "*who with parables sustained by imagination, compassion and irony continually enables us once again to apprehend an elusory reality*", was awarded the 1998 Nobel Prize in Literature[3].

For science education to help meet the eight MDM (UNDP 2010), it is essential to change key features of the status quo based on which Hodson (2003) argues that a fourth curricular component, "*Engaging in Sociopolitical Action*" is needed. Developing synergies with other disciplines is essential to this end, as is defended in the UNESCO strategy for the second half of the DESD, at practice level (see fig. 2), e.g. stating that significant efforts are required to support the incorporation of SD issues using integrated and systemic approaches in all education types and at all levels (UNESCO 2010).

As far as the national curriculum for sciences in Portugal (DEB 2001a) is concerned, it comprises environmental issues, e.g. global change or water at local and global levels, but does not consider social and economic dimensions

3 http://nobelprize.org/nobel_prizes/literature/laureates/1998/ (retrieved 31/08/2010).

involved in the Hodson's critical perspective outlined above. This requires addressing other dimensions involved in issues approached in sciences' curriculum suitable for citizenship education purposes. Besides arguments rooted in traditional views of science education that help explain resistance and barriers to consider socio-political dimensions on issues-based approaches (Hodson 2003), environmental dimensions of issues at any level, from global, e.g. climate change, to local, e.g. tap water quality, appear easier to relate to canonical science content.

In addition, among the reasons claimed to explain the rapid growth of environmental education research and associated activities, Hart (2007) suggests global environmental issues that are increasingly more pervasive, changing societal expectations and educational reforms, focussing on "new narratives that give guidance and inspiration to people and purpose to schooling" (Hart 2007: 689). These narratives "range from teaching young people to accept the world as it is, to those that encourage critical thinking more distanced from conventional wisdom and intended to change what is wrong" and include one that stresses human beings' moral obligation to the planet – "a story of human beings as stewards of the Earth, caretakers with a global consciousness and a sense of educational responsibility"(Hart 2007: 689).

Concerning geopolitical contexts comprising Portugal and Spain, Vilches and Gil-Pérez (2010) analyzed communications presented in what became the Iberian CTS Seminars on science teaching that started in 2000, take place every two years and became the First Iberian-American Seminar in 2008. Based on specific papers presented in those meetings or published in the proceedings concerned, these authors concluded that socio-environmental degradation issues have been addressed in all five editions of these seminars, from 2000 to 2008. The third edition of the Iberian CTS Seminar is highlighted for various reasons, due to the increased number of papers focussed on these issues, the appearance of a specific reference to the DESD and especially due to the opportunities provided during the seminar to elaborate a statement to support the DESD, which was the departing point of the OEI (acronym for Organización de Estados Iberoamericanos para la Educación, la Ciencia y la Cultura) website www.oei.es/decada. Referring to the number of educators and educational institutions supporting the statement on this website, Vilches and Gil-Pérez (2010) consider it as an effective instrument to promote the objectives of DESD. In addition, they stress that socio-environmental degradation issues have been increasingly addressed in all editions of these seminars, from 2000 to 2010, contributing to generate a climate of increased attention and commitment for a sustainable future.

Recognizing the importance of policies to reframe science education towards ESD, namely those reviewed before from the UN and its agencies to the EU and country governmental bodies concerned, it should also be recognized that the distance between these policies and the different dimensions considered in the strategy set up for the second half of the DESD (UNESCO 2010) may be enormous, in particular regarding science education practices in school contexts. Fortunately, there are good practices towards ESD reviewed in the next section.

Good Practices Towards ESD

To stimulate good practices in broader educational contexts and at a global level, UNESCO made available several publications on the ESD website, including "education for sustainable development in action", a series to reflect ESD actions in different fields and countries[4] that has five titles, one of which is "Good Practices". This heading comprises the two publications referred to in the previous two paragraphs and information on "ESD Good Practices using the Earth Charter". This is a compilation of "27 educational experiences from non-formal, primary and secondary education and higher education", carried out in 17 countries and co-published with the Earth Charter International[5].

The item "ESD Good Practices using the Earth Charter" provides a link to the website "The Earth Charter Initiative®", which emphasizes ethical dimensions by adding "values and principles for a sustainable future" and is presented as "an extraordinarily diverse, global network of people, organizations, and institutions that participate in promoting and implementing the values and principles of the Earth Charter"[6]. Among the many and diverse resources available, "Good Practices using the Earth Charter" (ECI 2007) stands out for offering case studies from around the world "that highlight how groups and individuals are using the Earth Charter as a tool for education for sustainable development"[7]. Although most resources offered on this website are written in English, some are available in other languages including Portuguese, e.g. "The Earth Charter Text" and "The Earth Charter Guidebook for Teachers", which are among the 10 items considered as "key resources".

Regarding Europe, more precisely in the UNECE (acronym for UN Economic Commission for Europe) region, which comprises 56 countries including Portugal, there are reports illustrating good practices (UNESCO/UNECE 2007). Furthermore, the "Good Practices in Education for Sustainable Development"

4 *http://www.unesco.org/en/education-for-sustainable-development/resources/publications/* (retrieved 31/08/2010).

5 *http://www.unesco.org/en/education-for-sustainable-development/resources/publications/good-practices/* (retrieved 31/08/2010).

6 *http://www.earthcharterinaction.org/content/* (retrieved 31/08/2010).

7 *http://www.earthcharterinaction.org/invent/details.php?id=248* (retrieved 31/08/2010).

web page[8] provides information concerning activities undertaken by "various governments, international organizations, research institutions, NGOs and other stakeholders", and asks for sharing other "good practices in ESD" in addition to announcing a new collection of "good practices in addressing sustainable consumption and production (SCP), including sustainable transportation, through ESD in the context of climate change". This indicates that in spite of the cultural and socio-economic differences among the member states, addressing these issues is considered: i) important to face today and future challenges, and ii) relevant to promote informed decision-making and democracy through ESD: "ESD is crucial for good governance, informed public decision-making and the promotion of democracy, and plays a fundamental role in overcoming social, economic and environmental challenges" (UNECE-SCESD 2009: 2).

At local level, Pedrosa (2009) refers to one in-service chemistry and physics teacher education programme focussed on STS themes in chemistry teaching and on approaches using laboratory work. The Earth Charter was one of the resources used and as part of activities developed in small groups, teachers reflected on the potential of the Earth Charter to help integrate STS interrelations in chemistry teaching in basic and secondary education levels and expressed their views. Although most teachers manifested agreement with principles of "ecological integrity" as expressed in the Earth Charter without providing any elaboration, some elaborated on their thinking. They explicitly affirmed that chemical education should help change behaviour towards exerting consciously and actively one's citizenship, including questioning and discussing everyday behaviour and their interrelations with present-day problems, for instance climate change.

Another in-service teacher education programme for science teachers that specifically addressed ESD and focussed on climate change is briefly presented in Pedrosa (2010). The activities carried out by teacher trainees in their schools during the programme were further developed through projects involving students. Pedrosa et al. (2010) describe the main features of this in-service education programme and further developments of projects that led to the organization of a one-day symposium. It was decided to organize it on a Saturday to allow teachers and students to participate without disrupting school activities that they were supposed to undertake according to previously agreed planning of schools and classes concerned.

Some of the projects led by teachers in their schools were presented and discussed in the symposium that took place on 5th June 2010 – the World Environment Day – and included other pertinent communications. As regards projects developed following on from the in-service programme, the communications,

8 *http://www.unece.org/env/esd/GoodPractices/index.html* (retrieved 31/08/2010).

oral and in poster, were presented by leading teachers together with some of the students who participated in the projects, and abstracts were elaborated for publication and are available in Janicas (2010).

Cooperation was a key for these undertakings. Indeed, organizing the symposium required cooperation among teachers from different education levels and institutions, planning the in-service science teacher education programme required cooperation between researchers with different profiles and interests though complimentary, as did setting it up. Planning and setting it up also required stakeholder cooperation of a different kind, namely the "Projecto Climántica" and the "Nova Ágora". The former is an "environmental education project produced by the *General Directorate of Sustainable Development* (Regional Government of Galicia, Spain) in cooperation with the *General* Directorate *of Educational Innovation*"[9], while the latter is an education centre[10] accredited with providing in-service teacher education programmes in Portugal. The UNESCO National Commission recognized the in-service programme and the symposium as contributions to implement the DESD.

Case studies on good practices at teacher training institutions, pre-service and in-service, to address sustainability show a broad diversity of efforts and deep impacts and cover different aspects: "professional development for in-service teachers, curriculum revision at the pre-service level, research with students in a local school, greening of a building and its garden, creating a network of universities, starting a journal, and creating new undergraduate and graduate programs" (Hopkins and McKeown 2007: 5).

Science Teacher Education and ESD

To restructure curricular sciences towards ESD, in practice and at levels involved in school contexts, innovative interventions are required that are difficult to devise and implement. Taking into account that teachers are inclined to reproduce teaching models that they were exposed to, special care and attention need to be taken to move from traditional approaches focussed on canonical science contents to innovative approaches. In particular, to adequately integrate ESD in formal science education requires devising and implementing innovations that should be grounded in research-based knowledge, beginning with teacher education, pre-service and in-service.

9 *http://www.climantica.org/proyectoPrensa/en/page/proyecto* (retrieved 30/08/2010).

10 *http://www.cfagora.net/* (retrieved 30/08/2010).

Rogan (2007), drawing on relevant research, defends the statement that implementing innovations should be ongoing processes "in which teachers and other members of a school are given a say in where they begin and how fast they feel they are able to go" (Rogan 2007: 443). As criteria to assist in developmental planning processes at school level, this author suggests manageability, coherence and consonance, comments on them and considers that it is not clear how they might be practically realized. Nevertheless, manageability, coherence and consonance are considered useful pointers regarding the nature and boundaries of a zone of feasible innovation that "consists of a collection of teaching strategies that go beyond current practice, but are feasible given the existing resources available to that teacher, or group of teachers, and the prevailing environment of the school in terms of its ability to foster and sustain innovation. The strategies in the ZFI are regarded as stepping stones to some defined goal, which is not yet attainable given the prevailing context"(Rogan 2007: 441).

Exploring issues, e.g. climate change, in the framework of "ecological integrity" as expressed in the Earth Charter, might constitute an initial zone of feasible innovation, as the first step towards more critical perspectives required in ESD. In spite of difficulties encountered in addressing social and economic dimensions involved in Hodson's critical perspective outlined above, these dimensions should also be addressed, discussed and articulated with environmental ones in coherence with balances envisaged by ESD. This idea of exploring climate change based on more critical perspectives of ESD takes into account the apparent willingness of teachers who participated in an in-service programme referred to in the section "good practices towards ESD" to explore the Earth Charter.

However, whatever issues are chosen, these should be explored using critical perspectives required in ESD and strategies in the zone of feasible innovation, as first steps to help move towards sustainability, building up synergies through partnerships to promote competence building and raise science teachers' awareness in various issues such as:

- The great and "unprecedented scale of human modification of geological and ecological processes" and the Anthropocene (Adams and Jeanrenaud 2008: 14);
- Conceptions of SD, including mainstream ones "built on the idea of market-driven approaches and strategies based on technology and intense regulation (termed ecological modernisation)" and promising "to steer the world towards sustainability in ways that do not demand too many dramatic changes, and that do not upset the comfortable, the rich or the powerful" (Adams and Jeanrenaud 2008: 12);
- Changes in the environmental movement that lead to broadening its scope evolving to third generation environmentalism – it "recognises that current organizations, institutions and political processes are part of the sustainability

problem, and seeks to mainstream the environment within the existing matrices of power, and influence domestic and international public policy agendas" (Adams and Jeanrenaud 2008: 83);
- Integration of different forms of knowledge "to consider the culture and values of the research community itself, as this affects the choices that are made regarding what to observe and study, and what types of information are considered to be legitimate knowledge" (ICSU 2005: 19);
- Lifestyle patterns related to sustainability, how they are determined and changed (ICSU 2005), how they are related to today's problems, at local and global levels, and what can, and should, be done to change the situation in a perspective of critical humanism and glocal responsibility, that is articulating local and global needs (Murga-Menoyo 2009);
- Articulation between canonical science content (Aikenhead 2005) and issues that matter to integrate ESD in science curricula in all education levels to adequately balance them, on the one hand, designing and implementing assessment procedures and instruments congruent with the nature of learning activities undertaken and their relative weight, on the other.

Moreover, the selection of activities and resources to be used in each teacher education programme should be such that these allow the target teachers viewing that the innovations suggested fulfil the three dimensions of "praticality ethic" (Rogan 2007: 444):

- "Instrumentality" – requiring the presentation of "classroom examples rather than abstract principles";
- "Congruence" – requiring compatibility "with current practice and with the limitation imposed by prevailing conditions";
- "Cost" – requiring expected benefits to "outweigh the perceived cost of the required effort to bring them to fruition".

In such programmes, to help develop teachers' competences to integrate ESD in science curricula at schools, in classrooms and laboratories, issues should be approached in ways that help clarify environmental, economical and social dimensions involved in them. They should also make clear interrelations among these dimensions, as well as between these, ethical factors and curricular requirements, including those concerned with canonical science content. Imbernón (2007), warning that the latest reforms in the last century generated contexts that presuppose training generations of teachers more concerned with the control of classes, competences and timetable than with learning itself, defends that what matters most in in-service teacher education is to value and raise individual and collective teachers' self-esteem.

The innovations suggested to integrate ESD in school science contexts require teachers strongly motivated to help produce actual sustainability and aware that the "voice of economics must be joined by the strong voices of social and natural science, principled political leadership, idealistic citizen activism, cultural questioning of consumerist habits and values, and much more" (AtKisson 2009: 7). Then they will be able to contribute to cultural and structural changes that are likely to affect the power and position of citizens, both as individuals and as institutional faces, thus originating tensions (between the needs for bureaucratic certainties and for flexibility towards change) and resistance to them manifested in insufficient innovative power and rigid disciplinary organization of sciences (Jansen 2010).

Concluding Remarks and Ways Forward

Restructuring science education in school contexts towards ESD nowadays should take into account SD problems and related challenges, as suggested by strategic actions envisaged for the second half of the DESD (UNESCO 2010). Therefore, besides curriculum guidelines delivered by the educational boards concerned and in that context, it is essential to reorient science teacher education programmes to integrate ESD, intensifying efforts to address critical and urgent sustainability challenges, e.g. climate change and water. Such efforts should be made within the DESD umbrella, building on suitable partnerships and developing institutional mechanisms to ensure that ESD continues to be implemented beyond DESD and related UN initiatives, as is schematically depicted in parts concerned in Figure 2.

The innovations required to restructure science education in everyday practices in schools and in science classrooms and laboratories towards ESD are paramount and should be based on research on science education, in particular in curriculum development and evaluation seeking "to encourage and support teachers in becoming critically literate about their own educational practice" (Hodson 2003: 665). Curriculum and resources development should: i) be articulated with teacher education and professional development to help develop science education practices that "go beyond imparting scientific conceptual knowledge and skills" and are consistent with "critical participation in a world dominated by science conceptualized its sociocultural and political interests" (Carter 2008: 171); ii) take into account and interrelate the dimensions involved in SD: economic, environmental and social ones.

Developing action research programmes involving science teachers as partners will help expand zones of feasible innovation – a requirement to progress in integrating ESD in science education. Simultaneously, feedback mechanisms must be developed to ensure that evidence gathered and knowledge built in such programmes are used to make adjustments at higher policy levels. These are among the efforts recommended to be made by international organizations, civil society and countries to promote evidence-informed policy dialogue on ESD (see fig. 2), drawing upon feedback provided by relevant research – a fundamental feedback source.

Moving towards sustainability requires cooperation at various and different levels to raise public awareness and enhance engagement to help solve sustainability problems. All are necessary, according to each one's particular contexts and personal and social responsibilities, to overcome difficulties and barriers to move towards sustainability. Moreover, higher education institutions are required to make and intensify efforts to address critical and urgent sustainability challenges (see fig. 2), in particular taking into account their responsibilities regarding teacher education and professional development.

Many barriers against moving towards sustainability have been identified[11], as have many challenges. One fundamental "barrier is the lack of willingness to accept a fair global distribution of environmental capabilities as a condition for common survival". The main challenge concerns "evolutionary change in an unsustainable world", which calls upon integrating ESD in all education levels and curricular disciplines to develop sustainability oriented to "visions in civil society, integrating a future orientation in public policy making and in public debate, transforming structures of society for SD, and breaking through barriers" (Jansen 2010).

Last but not least, the contexts, institutional commitments and partners to work with science teachers are very important to ensure that issues to promote competence building and raise science teachers' awareness (see former section) are approached, focussing on concrete practices in schools, classrooms and laboratories, i.e. in contexts that matter to teachers. Furthermore, to integrate ESD in science education, as in other curricula, requires flexibility, is incompatible with disciplinary rigidity, and should imply challenging education practices that support social and environmental degradation trends (Wade and Parker 2008). Developing synergies and partnerships to successfully integrate ESD in science education is also essential to question overemphasis on competition – whose main goal is generating winners, which goes hand in hand with generating losers militating against cooperation, appreciate cooperation and, in so doing, show that development concerns all human beings in their place and level in society (Jansen 2010).

---

11 *http://www.unesco.org/en/esd/publications/technical-papers/* – retrieved 30/08/2010).

Cooperation and partnerships are at the heart of ESD and are essential to build up public awareness and understanding about complex problems that humankind faces today. Consequently, it becomes clear that maintaining the values and approaches that originated them are incompatible with progress to solve them – today's problems cannot be solved based on the same values and using the same approaches that originated them (UNESCO 2010).

References

Adams, W. M. and Jeanrenaud, S. J. (2008), Transition to Sustainability: Towards a Humane and Diverse World. 108 p. IUCN, Gland, Switzerland (ISBN: 978-2-8317-1072-3).

Aguerrondo, I. (2009), Conocimiento complejo y competências educativas. IBE Working Papers on Curriculum Issues. 13 p. UNESCO Oficina Internacional de Educación, Ginebra (http://www.ibe.unesco.org/fileadmin/user_upload/Publications/Working_Papers/knowledge_compet_ibewpci_8.pdf – retrieved 23/07/2010).

Aikenhead, G.S. (2005), Research into STS Science Education. Educación Química, 16, pp. 384-397 (http://www.usask.ca/education/people/aiken head/research_sts_ed.pdf – retrieved 27/08/2010).

AtKisson, A. (2009), Pushing "Reset" on Sustainable Development. Sustainable Development Insights. University of Boston, Boston (http://www.bu.edu/pardee/publications/sdi-001-reset-sustainable-development/ – retrieved 04/03/2010).

Bajah, S.T. (2000): Science education for development: case study of Project 2000+, ELSSA. In Cetto, A.M., Schneegans, S. and Moore, H. (eds.), Science for the Twenty-First Century – A New Commitment, pp. 156. UNESCO, Paris (ISBN: 1 903 598 001).

Carter, L. (2008), Sociocultural Influences on Science Education: Innovation for Contemporary Times. Science Education, 92(1), pp. 165-181 (doi: 10.1002/sce.20228).

CEU (Council of the European Union) (2010), Joint Progress Report of the Council and the Commission on the Implementation of the "Education & Training 2010" Work Programme. Council of the European Union. 19 p. Council of the European Union, Brussels (http://register.consilium.europa.eu/pdf/en/10/st05/st05394.en10.pdf – retrieved 10/08/2010).

DEB (Departamento da Educação Básica) (2001a), Currículo Nacional do Ensino Básico: Competências Essenciais. 240 p. Ministério da Educação, Departamento do Ensino Básico, Lisboa.

– (2001b), Ensino Básico: Ciências Físicas e Naturais – Orientações Curriculares para o 3° Ciclo do Ensino Básico. 42 p. Ministério da Educação, Departamento do Ensino Básico, Lisboa.

ECI (Earth Charter International) (2007), Good Practices in Education for Sustainable Development Using the Earth Charter. 162 p. UNESCO, Paris, and ECI San Jose, Costa Rica (ISBN: 978-9977-925-54-7) (http://www.earth charterinaction.org/invent/images/uploads/Good%20Practices.pdf – retrieved 31/08/2010).

Hart, P. (2007), Environmental Education. In Abel, S.K. and Lederman, N.G., Handbook of Research on Science Education, pp. 689-726. Lawrence Erlbaum Associates, Inc., Publishers, New Jersey (ISBN: 0-8058-4714-6).

Hodson, D. (2003), Time for action: science education for an alternative future. International Journal of Science Education, 25(6), pp. 645-670 (doi: 10.1080/0950069032000076643).

Hopkins, C. and McKeown, R. (2007), Good Practices in Teacher Education Institutions. 54 p. UNESCO, Paris (http://unesdoc.unesco.org/images/0015/001524/152452eo.pdf – retrieved 15/10/2009).

Imbernón, F. (2007), La formación permanente del professorado – Nuevas idas para formar en la innovación y el cambio. 147 p. Graó, Barcelona (ISBN: 978-84-7827-502-1).

ICSU (International Council for Science) (2005), Harnessing Science, Technology, and Innovation for Sustainable Development. 39 p. ICSU, Paris (ISBN: 0-930357-63-9) (http://www.icsu.org/Gestion/img/ICSU_DOC_DOWNLOAD/584_DD_FILE_Consortium_Report.pdf – retrieved 23/07/2010).

Janicas, J.P. (Coordinator) (2010), Simpósio Práticas de Educação para o Desenvolvimento Sustentável com Edublogues Climántica. 49 p. Nova Ágora – Centro de Formação de Associação de Escolas, Coimbra.

Jansen, L. (2010), Higher education's contribution to sustainable development: the way forward (http://www.guni-rmies.net/news/detail.php?id=1611 – retrieved 27/08/2010).

Lederman, L.M. (2000), Thematic meeting report. In Cetto, A.M., Schneegans, S. and Moore, H. (eds.), Science for the Twenty-First Century – A New Commitment, pp. 162-164. UNESCO, Paris (ISBN: 1 903 598 001).

Loureiro, C., Pedrosa, M.A. and Gonçalves, F. (2008), Problemas Globais e Educação Científica Formal Tripolar. In Vieira, R.M., Pedrosa, M.A., Paixão, F., Martins, I.P., Caamaño, A., Vilches, A. and Martín-Díaz, M.J. (Coord.), Educação Científica e Desenvolvimento Sustentável, pp. 200-203. Universidade de Aveiro, Aveiro (ISBN: 978-972-789-267-9).

Murga-Menoyo, Mª Á. (2009), La Carta de la Tierra: un referente de la Década por la Educación para el Desarrollo Sostenible. Revista de Educación, número extraordinario, pp. 239-262 (http://www.revistaeducacion.mec.es/re2009/re2009_11.pdf – retrieved 15/10/2009).

OJEU (Official Journal of the European Union) (2006), Recommendation of the European Parliament and of the Council of 18 December 2006 on key competences for lifelong learning. Official Journal of the European Union, L 394, pp. 10-18 (http://eur-lex.europa.eu/LexUriServ/LexUriServ.do?uri=OJ:L: 2006: 394:0010:0018:en:PDF – retrieved 24/08/2010).

Pedrosa, M.A. (2009), Formação Contínua de Professores, Temáticas CTS e Trabalho Laboratorial em Educação Formal em Química – Perspectivas de Professores Sobre a Carta da Terra. Boletín das Ciencias, 68, pp. 57-59 (http://www.enciga.org/congreso/2009/index.htm – retrieved 02/09/2010).

– (2010), Ciências, Educação Científica e Formação de Professores para Desenvolvimento Sustentável. Revista Eureka sobre Enseñanza y Divulgación de las Ciencias, 7, pp. 346-362 (http://www.apac-eureka.org/revista/Volumen7/Numero_7_extra/13_Pedrosa_2010.pdf – retrieved 19/02/2010).

Pedrosa, M.A., Gonçalves, F. and Mendes, P. (2010), Práticas Educativas para Desenvolvimento Sustentável – Necessidades, Ênfases e Iniciativas. In Janicas, J.P. (Coordinator) (2010), Simpósio Práticas de Educação para o Desenvolvimento Sustentável com Edublogues Climántica, pp. 8-21. Nova Ágora – Centro de Formação de Associação de Escolas, Coimbra.

Pedrosa, A. and Moreno, M.J. (2008), Sustainable development, problem solving and citizenship nowadays – reflections based on a study with university students. In Azeiteiro, U.M., Gonçalves, F., Pereira, R., Pereira, M., Leal Filho, W. and Morgado, F. (eds.), Science and Environmental Education – Towards the Integration of Science Education, Experimental Science Activities and Environmental Education, pp. 283-304. Peter Lang GmbH, Frankfurt am Main.

Pedrosa, M. A. and Loureiro, C. (2008), Desenvolvimento Sustentável, Energia e Recursos Energéticos em Documentos Oficiais para o Ensino Básico e Manuais Escolares de Ciências. In Rodriguez, J.M. and Domínguez M.A.F. (eds.). Educación Enerxética, Enerxías Renovables e Cambio Climático, pp. 177-186. Universidade de Santiago de Compostela.

Rogan, J.M. (2007), How Much Curriculum Change is Appropriate? Defining a Zone of Feasible Innovation. Science Education, 91(3), pp. 439-460 (doi: 10.1002/sce.20192).

Saramago, J. (2008), Este mundo da injustiça globalizada. CopIt ArXives, Ciudad de México (http://scifunam.fisica.unam.mx/mir/copit/SC0001PTES/SC0001PTES_PT.pdf – retrieved 31/08/2010).

UNDP (2010), What Will It Take to Achieve the Millennium Development Goals? An International Assessment. 58 p. United Nations Development

Programme, New York. http://content.undp.org/go/cms-service/stream/asset/?asset_id=2620072 – retrieved 22/07/2010).

UNECE-SCESD (2009), Contribution to the UNESCO World Conference on Education for Sustainable Development. 5 p. Economic Commission for Europe, Committee on Environmental Policy, UNECE Steering Committee on Education for Sustainable Development, Geneva (http://www.esd-worldconference- 2009.org/fileadmin/download/Speeches/UNECE_SteeringCommitteeOnESD.e.pdf – retrieved 16/11/2009).

UNESCO (2005a), Report by the Director-General on the United Nations Decade of Education for Sustainable Development: International Implementation Scheme – Draft International implementation scheme for the United Nations Decade of Education for Sustainable Development. 27 p. UNESCO, Paris (http://unesdoc.unesco.org/images/0014/001403/140372e.pdf – retrieved 23/07/2010).

– (2005b), United Nations Decade of Education for Sustainable Development (2005-2014): International Implementation Scheme. 31 p. UNESCO, Paris. http://unesdoc.unesco.org/images/0014/001486/148654e.pdf – retrieved 18/08/2010).

– (2010), UNESCO Strategy for the Second Half of the United Nations Decade of Education for Sustainable Development – Supporting Member States and other stakeholders in addressing global sustainable development challenges through ESD. 22 p. UNESCO, Paris (http://www.preventionweb.net/files/15341_unescostrategyfortheunitednationsde.pdf – retrieved 30/01/2011).

UNESCO/UNECE (2007), Good Practices in the UNECE region. 190 p. UNESCO, Paris (http://unesdoc.unesco.org/images/0015/001533/153319e.pdf – retrieved 15/10/2009).

Vilches, A. and Gil-Pérez, D. (2010), Educación para un nuevo orden socio-ambiental: Reflexiones acerca del futuro del Seminario Ibero-americano CTS. In Universidade de Brasília (ed.). II Seminário Ibero-americano Ciência-Tecnologia-Sociedade no Ensino das Ciências (VI Seminário Ibérico CTS no Ensino das Ciências), pp. 1-8. Universidade de Brasília, Brasília (ISBN: 978-85-62810-01-5).

Wade, R. and Parker, J. (2008), EFA-ESD Dialogue: Educating for a Sustainable World. 57 p. Education for Sustainable Development Policy Dialogue no. 1. UNESCO, Paris (http://unesdoc.unesco.org/images/0017/001780/178044e.pdf – retrieved 23/07/2010).

Chapter 5

Education for Sustainable Development Indicators, Competences and Science Education

Ana Capelo, Maria Conceição Santos, Maria Arminda Pedrosa

Abstract

Education for Sustainable Development (ESD) aims at encouraging changes required to promote sustainable development. Sustainable development includes three dimensions, i.e. social and economic developments, along with environmental protection, each one involving several interacting variables. Therefore, adequate tools are required to help monitor and evaluate those changes with regard to ESD at local, national and regional levels. Within this perspective, monitoring and assessment instruments related to ESD indicators are relevant to clarify which educational approaches are necessary (in particular, in formal education) and to evaluate the quality of these approaches in terms of their relevance to promote sustainable development. As the integration of ESD in formal education and in each geopolitical context has been recognized as important, it is vital to clarify the meaning of ESD indicators (ESDI), as well as the criteria underlying their selection. By allowing evaluation the extent to which education, in each particular subject, is aligned within ESD perspectives, the use of ESD indicators is crucial to help implement the Millennium Development Goals (MDG). In this chapter, the purposes and meanings attributed to ESDI are analyzed, the criteria adopted by some countries and regions are outlined to help select a set of ESDI for the European geopolitical context, specifically for the formal science education sector. These ESDI selected are meant to be used by educational policy makers, science curriculum developers, science teacher educators, science teachers, school leaders and researchers, specially in countries where ESDI indicators have not yet been developed. Therefore they are regarded as a starting point to assess and monitor progress in ESD that would be adjusted and improved as they are used and evaluated through research.

Introduction

In order to help people to better understand the world they live in, and to stimulate them to act appropriately, working creatively with others, in order to contribute to sustainable development (Pedrosa and Leite 2004), it is necessary to involve them in educational practices aligned with sustainable development (SD) perspectives. ESD integrates main decisions within the United Nations on social, environmental and economic dimensions required towards SD (UNESCO 2005a). It aims at promoting critical thinking about principles and values underlying education, in all forms (informal, non-formal and formal), and at all levels, helping countries make progress towards the MDG. Taking into account SD issues, ESD envisages stimulating meaningful learning that may help all citizens to become more aware of today's problems and build knowledge required to assist them in making decisions guided by SD concerns (UNESCO 2005a, b).

To assess the quality of ESD, it is necessary to develop tools for monitoring and evaluating educational practices within SD perspectives. Some of the instruments for these monitoring and evaluation processes specifically seek to evaluate i) management policies that respect the environment, ii) national and/or regional ESD strategies within legal and institutional frameworks, iii) availability of funds to ensure adoption of educational approaches that incorporate ESD perspectives, and iv) availability of qualified human resources for sustainable environmental management (Tilbury et al. 2007). Given these monitoring and evaluation foci, geographic and/or specific economic regions (e.g. UNECE [acronym for United Nations Economic Commission for Europe] and the Asia-Pacific) and countries (e.g. UK, Netherlands, Belgium, Germany) set up inter-ministerial working groups in order to establish instruments for monitoring and evaluating educational practices, using national strategies or plans for the Decade of Education for Sustainable Development (DESD). Some countries have established a set of ESD indicators (ESDI), e.g. "Anglo-German Expert Meeting on Indicators for ESD" (Reid et al. 2006) and the "Asia-Pacific Expert Group on Indicators" (Tilbury and Janousek 2006).

The criteria adopted to establish ESDI are linked with different perspectives of ESD, as Van Raaij (2007) highlighted: "Sustainable development in Europe seems to be more connected with environmental protection, (over)consumption, behaviour change, and technological development, whereas in some parts of Asia "Education for All" parallels ESD, there within the Millennium Development Goals are an explicit part of the SD and ESD strategies" (Van Raaij 2007: 11).

Tilbury and Janousek (2006) proposed a set of guidelines that are relevant for the selection of ESDI, as they help deepen meaning(s), functions, approaches and criteria for their selection. In this chapter:

a) The purposes and meanings attributed to ESDI are analyzed. This analysis is based on pertinent research (e.g. Bossel 2001; Tilbury and Janousek 2006), or research-related groups of national, regional or international organisms (e.g.

Sustainable Development Commission 2006; UNECE 2005, 2006, 2008, UN-DESA 2007);

b) The criteria adopted by some countries (United Kingdom, Germany, Netherlands, Belgium) and regions (economic region of the UNECE, Asia-Pacific region) for the ESDI selection are outlined paying particular attention to formal education contexts. Based on relevant literature (e.g. UNESCO 2005a,b, c; UNECE 2005, 2006, 2008; Tilbury et al. 2007), some criteria considered significant for the ESDI selection are highlighted, in particular those related with: *i)* sustainability issues that need to be included in educational activities (UNESCO 2005a, c); *ii)* the role of values in ESD (UNESCO 2005b, c); *iii)* linkages between the Decade and other educational initiatives (e.g. "EFA-ESD dialogue", Wade and Parker 2008); *iv)* relational processes and behavioural outcomes which should characterize learning in all circumstances (UNESCO 2005c);

c) A set of ESDI for the European geopolitical context is suggested, specifically for the formal science education sector, including the Portuguese one.

ESD Indicators – Purposes and Meanings

Prior to presenting the purposes for and meanings of ESDI, it should be made clear that these are different from SD indicators, albeit interrelated, like SD differs from and is interrelated with ESD. Indeed, to achieve SD requires that its economic, environmental and social components are integrated at all levels, which "is facilitated by continuous dialogue and action in global partnership, focusing on key *sustainable development issues*"[1]. This vision, while encompassing "populations, animal and plant species, ecosystems, natural resources" and integrating "concerns such as the fight against poverty, gender equality, human rights, education for all, health, human security, intercultural dialogue, etc." [2], is closely related with ESD. As a means to help achieve SD, ESD aims at helping "people to develop the attitudes, skills and knowledge to make informed decisions for the benefit of themselves and others, now and in the future, and to act upon these decisions" (see footnote 2). Besides conceptual differences between SD and ESD, the indicators concerned may also reflect different objectives set up and methodological approaches adopted to develop them, for instance as Ramos and Caeiro (2010) show with regard to the national sustainable development indicators system for Portugal. It is worth mentioning that the national strategy for ESD

1 *http://www.un.org/esa/dsd/index.shtml* (retrieved 20/02/2011).
2 *http://www.unesco.org/en/esd/* (retrieved 20/02/2011).

was published on 26th November 2009[3], it explicitly acknowledges the international responsibilities that Portugal has assumed, especially within the European Union, and to the best of our knowledge there is no ESDI available.

The need to develop SD indicators was already recognized in the last decade of the past century. Since the Rio Conference in 1992, the interest in developing SD indicators is largely recognized. This recognition is highlighted in some chapters of Agenda 21 (Rode 2006), for instance, Chapter 8: "Countries could develop systems for monitoring and evaluation of progress towards achieving sustainable development by adopting indicators that measure changes across economic, social and environmental dimensions"[4].

A set of indicators for MDG is also defined, allowing comparisons among countries (UN-DESA 2006). These indicators are articulated with SD indicators defined by a UN committee – United Nations Commission on Sustainable Development (UNCSD) – proposed by the United Nations – Department of Economic and Social Affairs (UN-DESA) (UN-DESA 2007). In order to monitor progress in SD, the European Union, along with the indicators developed by the UN, developed a set of SD indicators in 2000 relating to areas such as: a) economic development; b) sustainable consumption and production; c) demographic change and d) global cooperation (EU 2007).

The use of the word indicator was linked with concerns about improving the quality of education. Reinforcing intentions of international initiatives such as Education for All (EFA) and MDG (UNESCO 2005d), the PISA studies showed up a list of indicators to monitor and evaluate the quality of education (Rode 2006). These are expressed in the form of basic indicators concerning context and performance of education systems in different countries assessing their 15-year old students in reading, mathematics and science (OECD 2006, 2009). The selected indicators reflect characteristics of education systems: "their demographic and economic contexts – for example, costs, enrolments, schools and teacher characteristics, and some classroom processes", as well as their impacts on the professional market (OECD 2006: 15).

Besides indicators related to improving the quality of education in terms of teachers' performance (e.g. EC 2000, 2002; EU 2004; CEC 2004; Pigozzi 2008), skills' development (e.g. CEC 2004) and the recruitment of students to the sciences and technologies (e.g. CEC 2004), others are being suggested such as those concerning requirements for more attractive learning (e.g. CEC 2004) that focus on developing competences to enable students to act thoughtfully and responsibly as active citizens in decision-making processes.

---

3 *http://www.ipad.mne.gov.pt/images/stories/Educacao/ened.pdf* (retrieved 20/02/2011).

4 *http://www.un.org/esa/sustdev/documents/agenda21/english/Agenda21.pdf* (retrieved 27/09/2010).

The relevance of clarifying ESDI purposes and meaning is widely recognized (e.g. Tilbury and Janousek 2006). In general, these and other indicators express only specific features of a particular system or environment, providing limited information about that system/environment (Bossel 2001). Given the importance of ESD to help achieve SD, for SD or ESD evaluations, the use of complementary ESDI is crucial in order to provide a more comprehensive perspective. However, the selection of adequate ESDI is complex: "There seems to be general agreement that it is impossible to define only a single indicator of sustainable development, and that a substantial number of indicators are necessary to capture all the important aspects of sustainable development in a particular application" (Bossel 2001: 1).

The report: "Draft Guidance for Reporting – (EG on ESD Indicators)" (UNECE 2008) identifies indicators as those pointing "to an issue or condition" and having as purpose "to show how well the system is working" (UNECE 2008: 6). It further notes that the indicators only assess progress in implementing a plan on ESD and that it is a combination of answers, and not a single indicator or sub-indicator, that will allow the evaluation of the effectiveness of the implementation of the UNECE strategy for ESD.

The Australian Research Institute in Education for Sustainability (ARIES) compiled much of the existing information on ESDI involving specific proposals from countries like the UK and Germany, or groups of countries such as Austria and Germany, and regions such as the UNECE. It assigns the following ESDI meanings: a) "Identify progress in ESD rather than in the various components of SD"; b) "Determine the quality or characteristic property of a system"; c) "providing that information assists with the development of ESD" (Tilbury and Janousek 2006: 11). It also assigns them different purposes, "acting as a warning system or providing clues on the status of a system", "accounting for the effectiveness of resources that have been spent on a program or policy", "determining the performance of individuals or groups over time", "comparing progress across a region", "guiding planning", "promoting learning" (Tilbury and Janousek 2006: 11).

Within the "European Research Project Indicators for ESD", the ESDI developed were considered as those serving to guide practitioners at all education levels, adapting their practices to the objectives and methods of ESD (Tilbury and Janousek 2006). The German proposal (Rode 2006; Rode and Michelsen 2008) highlights the following functions for ESDI, "a tool for self-evaluation", "support instrument for implementation and dissemination of educational processes", "measurement for the implementation degree of innovations", and identifies three ESDI levels to support ESD: macro (at the level of federal, regional and national structures), meso (within institutions) and micro (at the classroom level) (Rode 2006). The UK proposal for ESDI, "UK: Developing an ESD Indicator for Formal Education within the UK Strategy for SD" suggests five types of ESDI indicators (Huckle 2006). Rode and Michelsen (2008) highlight that ESDI are mostly focused on the results, effects and performances of the students.

In short, due to its dynamic nature, the ESDI contribute, among other things, to disseminate, implement (Simer et al. 2006) and further clarify the intentions of ESD: "help increase the understanding of the community and to raise questions such as: what is sustainability, what does the community want in relation with sustainability and what are their limits to action" (Ortega-Cerdà 2005: 10). Therefore, they also constitute an essential learning tool for promoting ESD at all levels of education, formal, informal and non-formal, trying to guide the citizens in making decisions on facing problems concerning SD.

Criteria for ESDI Selection

Different criteria are available to assess the suitability of ESDI within specific contexts. According to Tilbury and Janousek (2007), some questions should be taken into account when selecting ESDI: Are the indicators relevant to assess national priorities and goals of ESD? How can an indicator assess intentions to ESD? What are degrees of difficulty presented by the collected data? Are the collected data reliable? Are they comparable?

When analyzing proposals concerning ESDI monitoring, evaluation and expression, Tilbury et al. (2007) concluded that the selection of the ESDI criteria is based on the aims defined in political strategies of SD or ESD by countries (e.g. UK, Germany, Netherlands, Belgium), group of countries (e.g. Nordic Minister Council) or at regional level (e.g. the UNECE and the economic region of Asia-Pacific regions). In general, the ESDI meet a defined set of requirements, i.e. they must be: a) clear, easy to interpret and serve as basis for international comparisons; b) based, whenever possible, on national responsibilities and commitments arising from overall intentions and objectives established for the DESD; c) grounded on data sources that are reliable, measurable and consistent, so that they can be used for a long period of time; and d) aligned with achieving the MDG.

Covering different geopolitical contexts, the UNECE defines ESDI (UNECE 2006, 2008), which were built on goals expressed in the UNECE's own strategy for ESD. These goals cover aspects related to the process of implementing the ESD strategy (e.g. "4.1.2. Is public (national, sub-national, local) authority money invested in this activity?", p. 6), or related with the extent to which the ESD strategy is effectively being implemented (e.g. "Is there any public support for mechanisms to share the results and examples of good practices in ESD authorities and among stakeholders?", p. 9). The proposal shows concerns about assessing: a) issues and values, inherent to SD, which may be incorporated into educational practices; b) linkages between the ESD strategy into the UNECE region and other UN initiatives being promoted and c) attitudes and behaviours, in line with SD perspectives, that may be considered to follow up the implementation of the strategy. However, the ESDI defined by UNECE do not allow a clear and specific characterization of each geopolitical context, since they do not take into account the specificity of MDG goals within each context. Moreover, they cover all forms of formal, non-formal and informal education, without taking into account the UNESCO recommendations for the second half of the DESD (Wals 2009), in particular regarding formal education.

The UNECE proposes four types of ESDI and defines them in terms of the information provided: "checklist indicators" that refer to policy, instruments and measures taken by a government to implement the strategy; "input indicators" that concern several activities taking place within the strategy; "output indicators" regarding "the direct results of these activities (e.g. performance of trained teachers, number of businesses involved in ESD projects, number of educators who received training on ESD issues)"; "outcome indicators" that refer to "the possible impact of the implementation of the strategy particularly on values, attitudes and choices in favour of SD (e.g. learning outcomes resulting from ESD partnerships, community-based projects and business involvement)" (UNECE 2008: 7).

The project ARIES (Tilbury and Janousek 2006) expresses, in general terms, a set of recommendations for defining ESDI, based on national priorities for the DESD, revealing specific concerns: a) target people who should be selected to set ESDI; b) types of ESDI that are being developed; c) how the ESDI should be developed; d) which plans are to be considered to collect data and e) who should collect these data. However, it does not address specific recommendations for the selection of ESDI to assess content, values, attitudes and behaviours that should underlie and be given priority in educational practices according to ESD perspectives. Neither does it express recommendations for the ESDI that enable an evaluation of synergies between DESD and other UN initiatives.

Regarding ESDI selection, the Asia-Pacific DESD Indicators Project team (Tilbury et al. 2007) produced a set of guidelines that assists UNESCO member states in the region with the development of national ESDI. Based on initiatives across the globe from regions in Europe, to North America and the Asia-Pacific, (e.g. Huckle 2006; UNECE-CEP 2006), this team suggested a set of key elements and procedures to follow. In particular, they recommended various ESDI under the following headings that must be used simultaneously (Tilbury et al. 2007: 31-34):

Baseline – "provide information about the initial or existing position of ESD within a country or region";

Context – "used to identify the existence of a supportive ESD setting or context";

Process	– “used to identify the existence of ESD processes, activities and opportunities”;
Learning	– “used to identify learning and knowledge built during the development, implementation and assessment of ESD efforts”;
Output	– “used to assess ESD outputs such as products, resources and immediate results”;
Outcome	– “used to assess outcomes or the mid-term results related to changes in awareness or understanding from ESD efforts”;
Impact	– “used to assess impacts that result from ESD efforts at the end of the DESD and beyond”;
Performance	– “used to assess the changing status of the overall ESD picture in a region or country”.

These indicators are grouped in three categories (Tilbury et al. 2007: 30):

Status	– indicators under this heading “determine the position or standing of ESD in a country”, e.g. baseline;
Facilitative	– indicators under this heading “assess variables that assist, support or encourage engagement with ESD” e.g. context, process, learning;
Effect	– indicators under this heading “assess variables relating to initial, medium and long-term achievements during the DESD”, e.g. output, outcome, impact, performance.

However, like ARIES, the Asia-Pacific DESD indicators project team did not provide recommendations for the selection of ESDI to assess content, values, attitudes and behaviours that should be aligned with ESD perspectives, neither did they provide recommendations for the ESDI to enable evaluating synergies between DESD and other UN initiatives.

In short, the lack of common visions for SD and ESD hampers the definition and the selection of common patterns for ESDI. Therefore, finding a common pattern for each ESDI geopolitical context would be advantageous, namely to facilitate international comparisons.

When defining criteria for ESDI selection within the European geopolitical context, besides the relevant aspects of ESDI proposed by the UNECE (2008) and countries within this region, namely Germany (Rode 2006) and the UK (Huckle 2006), it is also important to take into account ESDI proposed by Asia-Pacific DESD project team (Tilbury et al. 2007). However, it should be stressed that excepting the UNECE proposal, all the others lack comprehensive perspectives on ESDI, as it was briefly mentioned in the Introduction. For instance, a) sustainability issues that need to be included in educational activities in order to "bring relevancy to the curriculum and contribute to deeper understanding of the complexity of the issue and its impact on the community and region" (UNESCO 2005a: 36); b) ESD values that "need to be integrated into educational policies, strategies and programmes to strengthen achievement of the MDG"[5]; c) Linkages between the DESD and other educational initiatives that need to be considered (GC-UNESCO 2009; Wade and Parker 2008; Wals 2009); d) Relational processes and behavioural outcomes which should characterize learning in all circumstances (UNESCO 2005c).

Therefore, in selecting ESDI it is important to fulfil the requirements proposed and briefly reviewed. It is highlighted that in so doing, the following principles (that are intimately interconnected with each other and with the requirements referred to above) should be considered: a) Priority areas of intervention identified in Agenda 21; b) Key themes proposed by UNESCO for the DESD ("Key action themes"); c) UNESCO strategies for the DESD expressed in the International Implementation Scheme (IIS) (UNESCO 2005c); d) The links between the goals set up for DEDS and those established within other UN initiatives, e.g. MDG (GC-UNESCO 2009; Wals 2009).

Considering that the use of ESDI allows evaluating, the extent to which education is aligned within SD perspectives, it is crucial that using these indicators may help pursue the MDG, Education for All (EFA) and Decade of Literacy (DL) purposes. These UN initiatives refer to goals and purposes concerning all nations, they are global. However, achieving them does not integrate the immediate priorities of all nations. In particular, among the eight MDG for the European geopolitical context, "eradicate extreme poverty and hunger" (goal 1), "ensuring environmental sustainability" (goal 7) and "develop a global partnership for development" (goal 8) are relevant targets that, having not yet been achieved, call upon urgent actions (UN-DESA 2010). European educational policies (EC-Eurostat 2009) appear to be aligned with MDG concerns, of which fighting poverty and social inequalities are two of the goals expressed in the SD European Union's strategy (EC-Eurostat 2009). According to the Eurostat report that is based on SD indicators for the European Union (EC-Eurostat 2009), promoting cooperation and education are strategic forms to achieve these goals. UNESCO also stresses the role of education in achieving these goals: "It is widely recognized that education is fundamental for just, peaceful, adaptable societies without poverty and that none of the international development goals can be achieved without education" (UNESCO 2008: 8).

Accordingly, both cooperation and education processes should be used as MDG indicators (particularly for monitoring and evaluating MDG 1 and 8). This perspective was already stressed by the Eurostat report (EC-Eurostat 2009) that refers to 12 leading SD indicators ("headline indicators"). One of them is the

5 http://unesdoc.unesco.org/images/0017/001791/179120e.pdf (retrieved 26/09/2010).

high risk of poverty (addressed by the first MDG). As a sub-indicator of this indicator (high risk of poverty), the report proposes education and subdivides it in three items: "Public expenditure on education", "adults with low educational attainment" and "lifelong learning". This means that "investment in education", the "educated level" and "lifelong learning" are considered sub-indicators to prevent "high risk of poverty".

Overall, it is suggested that for the European geopolitical context (formal science education in particular), the ESDI should be relevant to clarify what educational approaches are necessary to enable local contexts to be taken into account too, and then assess their quality regarding their relevance to promote SD. These educational approaches may help develop students' competences, including comprehension about sustainability issues, values, attitudes and behaviours aligned with ESD perspectives. These competences should enable students to act thoughtfully and responsibly as active citizens in decision-making processes.

ESD Indicators for the European Geopolitical Context

It is important that the ESDI for the European geopolitical context, and specifically for formal science education, allow the monitoring and evaluation of ESD practices, according to the assumptions and criteria previously outlined. The list proposed by UNECE (UNECE 2008) provides a good starting point. However, it is necessary to analyze whether or not the ESDI therein reveal concerns with: a) Key action themes proposed for the DSD; b) Key themes recommended to the second half of the DESD, "which are emphasized less at present in curricula" of formal education (Wals 2009: 49), such as MDG targets, particularly the three aforementioned (to "eradicate extreme poverty and hunger", to "ensure environmental sustainability" and to "develop a global partnership for development"), disaster prevention and corporate social responsibility.

In addition, it is equally important to analyze whether the ESDI proposed by UNECE (2008) address the development of competences by students, within a perspective of inter-relating DESD and other educational initiatives, to help them build: "Broad and coherent *knowledge* of the nature and scope of the problems (e.g. health and sanitation issues), how they arose, who and what is affected by the problems and knowledge of what can be done; *Commitment and values* that motivate them to participate in contributing to changes in society; An *interest in the future*, and capacity to predict what change might be possible in a given context; *Social, critical and creative thinking skills*, why things are as they are and what needs to be done; *Experience of real-life situations* gained through participating individually or collectively in facilitating changes" (bold in the original, see footnote 5).

Last but not least, it is important to analyze the extent to which the ESDI proposed by UNECE (2008) match the "ESD learning outcomes" (Wals 2009: 48) required for formal education in the second half of the DESD, such as: "Critical reflective thinking; Understanding complexity/systemic thinking; Futures thinking; Planning and managing change; Understanding inter-relationships across disciplines; Applying learning in a variety of life-wide contexts; Decision-making, including in uncertain situations; Dealing with crises and risks; Acting with responsibility, locally and globally; Ability to identify and clarify values; Acting with respect for others; Identifying stakeholders and their interests; Participating in democratic decision-making; Negotiating and consensus building" (Wals 2009: 49).

Therefore, to further elucidate the relevance of these aspects, based on the ESDI list of UNECE region (UNECE 2006, 2008), a set of ESDI grids (including sub-indicators) focusing on formal science education (that may be adapted to different education cycles) and applicable to the European geopolitical context (see tables 1-7) is presented. Given that school curricula are structured and organized in disciplines, including those encompassing the broad area of science education, e.g. biology, chemistry, geology and physics, it is important to make available and disseminate information on ESDI, focussing on this curricular area, as would be the case concerning other curricular areas or disciplines. However, this does not mean neglecting other forms of education, namely non-formal and informal, as they are all important in ESD and clearly recommended in the UNESCO strategy for the second half of the DEDS (UNESCO 2010). Indeed, it should be viewed as a means to help implement school curricular reforms and research-based recommendations, namely those valuing STS interrelations in science education (Aikenhead 2005; Hodson 2003; Vilches and Gil-Perez 2010).

The selection of the seven ESDI grids presented below was extracted from the UNECE list of ESDI and adjusted based on the dimensions of analysis identified in former paragraphs of this section, one to three. In deciding grids and selecting ESDI for each, three levels of implementation of these indicators, macro, meso and micro levels, were also taken into account (Rode 2006) along with different categories of ESDI (e.g. "status", "facilitative" and "effect") that should underlie any ESD evaluation and monitoring process (Tilbury et al. 2007).

Table 1 is adapted from the corresponding UNECE list of indicators covering the global objective 1 for ESD strategy. It presents a set of ESDI for assessing and monitoring national programmes and education policies that support the promotion of ESD. It contains sub-indicators defined by UNECE (except for the sub-indicators 1.1.2. and 1.3.) and allows evaluating whether or not ESDI defined by UNECE are relevant to evaluate how educational policies and programmes support ESD promotion.

Overall, they constitute indicators of "status" and "facilitative", according to the categories defined by Tilbury and collaborators (Tilbury et al. 2007).

Table 1: Grid of indicators and sub-indicators, adapted from UNECE list of indicators (UNECE 2008), to be applied to evaluate and monitor national educational policies and programmes that support ESD promotion.

Objective 1 **Ensure that Policy, Regulatory and Operational Frameworks Support the Promotion Of ESD**		
Indicator 1.1. Prerequisite measures are taken to support the promotion of ESD		
Numbers	Sub-indicators	Numbers used in UNECE (2008)
1.1.1.	Is the UNECE Strategy for ESD available in your national language(s)?	1.1.1
1.1.2.	Do you have a coordinating body for the implementation of ESD?	1.1.3
1.1.3.	Do you have a national implementation plan for ESD?	1.1.4
1.1.4.	Are there any synergies at the national level between UNECE ESD process, the UNESCO global process on the United Nations Decade of ESD, and other policy processes relevant to ESD, namely: a) MDG? (yes/no); b) DL? (yes/no); c) EFA? (yes/no)	1.1.5
Indicator 1.2. Policy, regulatory and operational frameworks support the promotion of ESD		
Numbers	Sub-indicators	Numbers used in UNECE (2008)
1.2.1.	Is ESD reflected in any national policy document(s)?	1.2.1
1.2.2.	Is ESD: a) Addressed in relevant national education legislation/regulatory document(s)? (yes/no); b) Included in your national curricula and national standards at all levels of formal education, as understood by your education system in accordance with ISCED(*)? (yes/no)	1.2.2
1.2.3.	Is public awareness in relation to ESD addressed in relevant national document(s)? (yes/no)	1.2.4
1.2.4.	Does a formal structure for interdepartmental cooperation relevant to ESD exist in your Government? (yes/no)	1.2.5
1.2.5	Does a mechanism for multi-stakeholder cooperation on ESD exist with the involvement of your Government? (yes/no).	1.2.6
1.2.6.	Are public budgets and/or economic incentives available specifically to support ESD? (yes/no).	1.2.7
Indicator 1.3. National policies support synergies between processes related to SD and ESD		
1.3.1.	Is ESD part of SD policy(s) if such exist in your country? (yes/no).	1.3.1

Table 2: Grid of indicators and sub-indicators, adapted from UNECE list (UNECE 2006, 2008), to be applied to evaluate, in a complementary way, if ESD topics and skills are integrated in the formal education practices (included in the category "facilitative" according to Tilbury and collaborators 2008).

Objective 2 **Promote SD Through Formal Learning** **(adapted from objective 2 of the UNECE ESD strategy)**		
Indicator 2.1. SD key themes are addressed in formal education		
Numbers	Sub-indicators	Numbers used in UNECE (2008)
2.1.1	Are key themes of SD addressed explicitly in the curriculum/programme of study at various levels of formal education? Please specify the school years and the course units concerned for each positive answer to the following: **a)** Eradicate extreme poverty and hunger (objective 1 of MGD) (yes/no); **b)** Ensure environmental sustainability (objective 7 of MDG) related with: water (Key action theme of the DESD) (yes/no), climate change (Key action theme of the DESD) (yes/no), biodiversity (Key action theme of the DESD) (yes/no), disaster prevention (re Key action themes recommended specifically to the second half of the DESD) (yes/no), natural resources management(yes/no), environmental protection (yes/no); **c)** Develop a global partnership for development (objective 8 of MDG) (yes/no); **d)** Corporate social responsibility (Key action theme recommended specifically to the second half of the DESD); **e)** Human rights (e.g. (Key action theme of DESD); **f)** Health promotion (e.g. combat HIV/AIDS) (Key action theme of DESD) (yes/no); **g)** Rural/urban development (related with two of the key action themes of the DESD) (yes/no); **h)** Cultural diversity (Key action theme of the DESD) (yes/no); **i)** Peace and human security (Key action theme of the DESD) (yes/no); **j)** Sustainable consumption (Key action theme of the DESD) (yes/no); **l.** Ecological principles/ecosystem approach (yes/no).	2.1.1
2.1.2	Are learning outcomes (skills, attitudes and values) that support ESD addressed explicitly in the curriculum/programme of study at various levels of formal education? Please specify the school years and the course units concerned for each positive answer to the following: **a)** Broad and coherent knowledge of the nature and scope of the problems (e.g. health and sanitation issues), how they arose, who and what is affected by the problems and knowledge of what can be done (yes/no); **b)** Highlight performances with a sense of responsibility both locally and globally (yes/no); **c)** An interest in the future, and capacity to predict what change might be possible in a given context (yes/no); **d)** Apply learning outcomes to real life situations and question about them (yes/no); **e)** Understand complexity/ systemic thinking (yes/no); **f)** Plan and manage change (yes/no); **g)** Understand inter-relationships across disciplines (yes/no); **h)** Decision-making, including in uncertain situations (yes/no); **i)** Deal with crises and risks (yes/no); **j)** Ability to identify and clarify values (yes/no); **l)** Act with respect for others; (yes/no); **m)** Identify partners in concrete situations, their skills and interests (yes/no); **n)** Participate in democratic decision-making (yes/no); **o)** Participate in negotiation and consensus building (yes/no).	2.1.2

Within the UNECE general objective 2 of the ESD strategy (see table 2), all the proposed ESDI (UNECE 2008) were considered relevant, as all may contribute to evaluate, in a complementary way, if ESD topics and learning outcomes are integrated in formal education practices. However, in this adapted grid, the ESDI were readjusted (see table 2), taking into consideration the dimensions of analysis made explicit in paragraphs 1 to 3 above in this section.

Some of the sub-indicators suggested in the original UNECE list (UNECE 2006, 2008) were not included in table 2 ("Ethics and Philosophy", "Citizenship, democracy and governance" and "Economics" – UNECE (2006: 12) as they do not appear suitable to be approached as themes in science education, although they encapsulate dimensions that are very important in ESD. However, rather than presenting them as themes they should be integrated in other sub-indicators within both key themes, e.g. "Eradicate extreme poverty and hunger", "Cultural diversity", "Peace and human security" or "Sustainable consumption", and learning outcomes, e.g. "Broad and coherent knowledge of the nature and scope of the problems [...], how they arose, who and what is affected by the problems and knowledge of what can be done", "Apply learning outcomes to real life situations and question about them", "Understand inter-relationships across disciplines", "Ability to identify and clarify values" or "Participate in democratic decision-making". Other themes, such as "Peace studies", "Biological and landscape diversity", "Production or consumption patterns", "Personal and family health" "Environment health" and "Poverty alleviation" that are defined in the original UNECE list (UNECE 2006: 12) were included, directly or indirectly, in the indicator 2.1.1. (see table 2).

Some skills included in the UNECE original list (UNECE 2006) were not adopted in this grid as they were considered to be already incorporated in other items (see table 2), for instance: "overcoming obstacles/problem-solving", "applying learning in a variety of life-wide contexts", "acting with responsibility", "acting with self-respect", "acting with determination", "coping under stress", "distributing responsibilities (subsidiary)", "acting with responsibility locally and globally" (UNECE 2006: 13), "Commitment and values that motivate them to participate in contributing to changes in society", "Social, critical and creative thinking skills, why things are as they are and what needs to be done", "experience of real-life situations gained through participating individually or collectively in facilitating changes" (see footnote 5).

Regarding the ESDI defined under the UNECE general objective 2 of the ESD strategy (see table 3), in particular to evaluate and monitor strategies and institutions that promote ESD, all the proposed indicators (UNECE 2008) were considered to be relevant, given the context of formal science education (with the exception of indicator 2.5 considered not be relevant and applicable to formal education).

Table 3: Grid of indicators and sub-indicators, adapted from UNECE list (UNECE 2008), to be applied to evaluate and monitor strategies that support ESD promotion (included in the category "facilitative" according to Tilbury and collaborators 2008)

Further description of the indicators and sub-indicators underlying objective 2 of the UNECE strategy – Promote SD Through Formal Learning		
Indicator 2.2. Strategies to implement ESD are clearly identified		
Numbers	Sub-indicators	Numbers used in UNECE (2008)
2.2.1	Is ESD addressed through: a) Existing subjects only? (yes/no); b) A cross-curriculum approach? (yes/no); c) The provision of specific subject programmers and courses? (yes/no); d) A stand-alone project? (yes/no); e) Other approaches? (yes/no)	2.2.1
Indicator 2.3. A whole institution approach to SD/ESD is promoted		
2.3.1	Do educational institutions adopt a "whole institution approach" to SD/ESD? (yes/no)	2.3.1
2.3.2	Are there any incentives (guidelines, award scheme, funding, technical support) that support "a whole institution approach to SD/ESD"? (yes/no)	2.3.2
2.3.3	Do institutions/learners develop their own SD/ESD indicators for their institution/organization? (yes/no)	2.3.3
Indicator 2.4. ESD is addressed by quality assessment/enhancement systems		
2.4.1	a) Are there any education quality assessment/enhancement systems? (yes/no); b) Do they address ESD? (yes/no); c) Are there any education quality assessment/enhancement systems that address ESD in national systems? (yes/no).	2.4.1

Table 4: Grid of indicators and sub-indicators, adapted from UNECE list (UNECE 2008), to be applied to evaluate and monitor strategies that are relevant to equip educators with the competences to include SD in their teaching (included in the category "facilitative" according to Tilbury and collaborators 2008

Objective 3. Equip Educators with the Competence to Include SD in Their Teaching		
Indicator 3.1. ESD is included in the training of educators		
Numbers	Sub-indicators	Numbers used in UNECE (2008)
3.1.1	Is ESD a part of educators' initial training? (yes/no).	3.1.1
3.1.2	Is ESD a part of the educators' in-service training? (yes/no).	3.1.2
3.1.3	Is ESD a part of training of leaders and administrators of educational institutions? (yes/no).	3.1.3
Indicator 3.2. Opportunities exist for educators to cooperate on ESD		
3.2.1	Are there any networks/platforms of educators and/or leaders/administrators who are involved in ESD in your country? (yes/no).	3.2.1

Objective 3. Equip Educators with the Competence to Include SD in Their Teaching		
Indicator 4.1. Teaching tools and materials for ESD are produced		
4.1.1	Does a national strategy/mechanism for encouragement of development and production of ESD tools and materials exist? (yes/no).	4.1.1
4.1.2	Is public (national, sub-national, local) authority money invested in this activity? (yes/no).	4.1.2
Indicator 4.2. Quality control mechanisms for teaching tools and materials for ESD exist		
Numbers	Sub-indicators	Numbers used in UNECE (2008)
4.2.1	Do you have quality criteria and/or quality guidelines for ESD-related teaching tools and materials that are: a) Supported by public authorities? (yes/no); b) Approved by public authorities? (yes/no); c) Tested and recommended for selection by educational institutions? (yes/no).	4.2.1
4.2.2	Are ESD teaching tools/materials available: a) In national languages? (yes/no); b) For all levels of education according to ISCED? (yes/no).	4.2.2
Indicator 4.3 Teaching tools and materials for ESD are accessible		
Numbers	Sub-indicators	Numbers used in UNECE (2008)
4.3.1	Does a national strategy/mechanism for dissemination of ESD tools and materials exist? (yes/no).	4.3.1
4.3.2	Is public authority money invested in this activity? (yes/no).	4.3.2
4.3.3	Are approved ESD teaching materials available through the Internet? (yes/no).	4.3.3
4.3.4	Is a register or database of ESD teaching tools and materials in the national language(s): a) Accessible through the Internet? (yes/no); b) Provided through other channels? (yes/no).	4.3.4

Table 5 presents a set of ESDI required to assess and monitor the production, quality and accessibility of resources that support ESD promotion (under the UNECE general objective 4 of the ESD strategy). The ESDI and sub-indicators defined by UNECE (UNECE 2008) are considered relevant to promote educational practices aligned with ESD.

Table 5: Grid of indicators and sub-indicators, adapted from UNECE list (UNECE 2008), to be applied to evaluate and monitor the production, quality and accessibility of resources that support ESD promotion (included in the category "facilitative" according to Tilbury and collaborators 2008).

Objective 4. Ensure that Adequate Tools and Materials for ESD Are Accessible		
Indicator 4.1. Teaching tools and materials for ESD are produced		
Numbers	Sub-indicators	Numbers used in UNECE (2008)
4.1.1	Does a national strategy/mechanism for encouragement of development and production of ESD tools and materials exist? (yes/no).	4.1.1
4.1.2	Is public (national, sub-national, local) authority money invested in this activity? (yes/no).	4.1.2
Indicator 4.2. Quality control mechanisms for teaching tools and materials for ESD exist		
Numbers	Sub-indicators	Numbers used in UNECE (2008)
4.2.1	Do you have quality criteria and/or quality guidelines for ESD-related teaching tools and materials that are: a) Supported by public authorities? (yes/no); b) Approved by public authorities? (yes/no); c) Tested and recommended for selection by educational institutions? (yes/no).	4.2.1
4.2.2	Are ESD teaching tools/materials available: a) In national languages? (yes/no); b) For all levels of education according to ISCED? (yes/no).	4.2.2
Indicator 4.3 Teaching tools and materials for ESD are accessible		
Numbers	Sub-indicators	Numbers used in UNECE (2008)
4.3.1	Does a national strategy/mechanism for dissemination of ESD tools and materials exist? (yes/no).	4.3.1
4.3.2	Is public authority money invested in this activity? (yes/no).	4.3.2
4.3.3	Are approved ESD teaching materials available through the Internet? (yes/no).	4.3.3
4.3.4	Is a register or database of ESD teaching tools and materials in the national language(s): a) Accessible through the Internet? (yes/no); b) Provided through other channels? (yes/no).	4.3.4

Table 6 presents a set of ESDI for assessing and monitoring ESD research (under the UNECE general objective 5 of the ESD strategy). The ESDI and sub-indicators defined by the UNECE are considered relevant to assess the role of research in promoting educational practices aligned with ESD.

Table 6: Grid of indicators and sub-indicators, adapted from UNECE list (UNECE 2008), to be applied to evaluate and monitor projects that support promotion, development and divulgation of ESD research (included in the categories "facilitative" and "effect according to Tilbury and collaborators 2008).

Objective 5. Promote Research on and Development of ESD		
Indicator 5.1. Research on ESD is promoted		
Numbers	Sub-indicators	Numbers used in UNECE (2008)
5.1.1.	Is research that addresses content and methods for ESD supported? (yes/no).	5.1.1
5.1.2.	Does any research evaluate the outcome of the implementation of the UNECE Strategy for ESD? (yes/no).	5.1.2
5.1.3	Are there post-graduate programmes available on ESD: a) For Masters level? (yes/no); b) For Doctorate level? (yes/no). Are post-graduate programmes addressing ESD: a) For Masters level? (yes/no); b) For Doctorate level? (yes/no).	5.1.3
5.1.4	Are there any scholarships supported by public authorities for post-graduate research in ESD: a) For Masters level? (yes/no); b) For Doctorate level? (yes/no).	5.1.4
Indicator 5.2. Development of ESD is promoted		
Numbers	Sub-indicators	Numbers used in UNECE (2008)
5.2.1	Is there any support for innovation and capacity building in ESD practice? (yes/no).	5.2.1
Indicator 5.3 Dissemination of research results on ESD is promoted		
Numbers	Sub-indicators	Numbers used in UNECE (2008)
5.3.1	Is there any public authority support for mechanisms to share the results of research and examples of good practices in ESD among authorities and stakeholders? (yes/no).	5.3.1
	5.3.2. Are there any scientific publications: a) Specifically on ESD? (yes/no); b) Addressing ESD? (yes/no).	5.3.2

Table 7 presents a set of ESDI within the assessment and monitoring strategies that are relevant to promote ESD cooperation at all levels (under the UNECE general objective 6 of the ESD strategy). The ESDI and sub-indicators defined by the UNECE are considered relevant to promote ESD international cooperation.

Table 7: Grid of indicators and sub-indicators, adapted from UNECE list (UNECE 2008), to be applied to evaluate and monitor the ESD cooperation at all levels within the UNECE region (included in the categories "facilitative" and "effect" according to Tilbury and collaborators 2008).

Objective 6. Strengthen Cooperation on ESD at All Levels Within the UNECE Region		
Indicator 6.1. International cooperation on ESD is strengthened within the UNECE region and beyond		
Numbers	Sub-indicators	Numbers used in UNECE (2008)
6.1.1	Do your public authorities cooperate in/support international networks on ESD? (yes/no).	6.1.1
6.1.2	Do educational institutions/organizations (formal and non-formal) in your country participate in international networks related to ESD? (yes/no).	6.1.2
6.1.3	Are there any state, bilateral and/or multilateral cooperation mechanisms/agreements that include an explicit ESD component? (yes/no).	6.1.3
6.1.4	Does your Government take any steps to promote ESD in international forums outside the UNECE region? (yes/no).	6.1.4

Conclusions

This chapter presents a set of guidelines concerning ESDI definitions, purposes and data sources that are relevant to the European geopolitical context. This review highlighted the importance of ESDI to evaluate and monitor ESD practices, according to assumptions and criteria that were outlined referring to the DESD context and achievement needs concerning the MDG. By selecting a number of ESDI and organizing them in tables their roles were clarified. They are essential to guide educational policies at different levels, from global to local, including decisions at school and classroom levels, e.g. to clarify which educational approaches are necessary (in particular, regarding formal science education). That is in a systemic view, like the one suggested by Rode and Michelsen (2008) based on the German context, the "macro level" corresponds to the more general component of the system that provides the framework for ESD, comprising the international and national or regional initiatives and regulations. Education programmes offered by formal institutions, e.g. schools and universities, come next defining the "meso level" and the implementation of ESD defines the "micro level" and concerns actual education activities and "individual measures themselves" (Rode and Michelsen 2008: 22), much dependent upon individual teachers without neglecting the other levels.

Taking into account initiatives and regulations at the macro level, notably those concerning the DEDS (UNESCO 2010, GC-UNESCO 2009) and the MDG (UNESCO 2005d), as well as science education reforms that appear interlinked and may be related to them, it is important to explore "the macro-politics of ESD" and "the micro-politics" (Huckle 2009: 1). This is required to reflect upon and refine approaches to ESDI to help convince "teachers and others that the struggle for a more relevant and empowering ESD, linked to a more appropriate indicator, is part of a wider struggle for greater social justice and democracy" (Huckle 2009: 13). Furthermore, exploring the macro and micro-politics of ESD appears essential to bridge the gap between the ESD rhetoric and philosophy and the practices in schools that require specific measures, "for instance by giving the stakeholders ownership in the development of indicators" (Mogensen and Schnack 2010: 71). Concerning geopolitical contexts comprising Portugal and Spain, Vilches and Gil-Pérez (2010) stress that socio-environmental degradation issues have been increasingly addressed in all editions of the Iberian STS Seminars on science teaching, from 2000 to 2010, which have contributed to generate a climate of increased attention and commitment for a sustainable future. In this context, ESDI are valued based on the roles they can play to help develop science education practices guided by ESD perspectives and to assess their quality with regard to their relevance to promote SD.

The analyses of ESDI developed by different international organizations (e.g. UNECE, ARIES) and countries (e.g. Germany, UK), point out both lack of clarity and focus that hamper an effective application to specific geopolitical contexts (e.g. UNECE 2008). They also stress the need to adopt an adequate set of ESDI covering important perspectives such as those highlighted in the objectives (see Introduction) of this chapter (e.g. ARIES, Germany, UK). Moreover, many EU countries (e.g. Portugal) appear lacking in relevant strategies towards the adoption of ESDI to promote educational and cooperative practices aligned with ESD. However, in either case the dynamic nature of SD and ESD implies that as a result of changes in national policies, concepts or methodologies, ESDI should be reviewed based on pertinent research and coherently adjusted to such changes.

Last but not least, when selecting and compiling ESDI within a specific geopolitical context, e.g. in the European Union, or country, e.g. Portugal, policy priorities at formal education level should be met, but above all, articulations with UN objectives established on a global scale must be envisaged, e.g. MDG, EFA and DL. Finally, in order to incorporate ESD in curricular contexts it is crucial that ESDI guide curriculum development and resource materials, for instance for science disciplines, teacher education programmes, should then be used to evaluate ESD practices at the meso level (school) and at the micro level (classroom).

References

Aikenhead, G.S. (2005), Research into STS Science Education. *Educación Química*, 16, pp. 384-397 (http://www.usask.ca/education/people/aikenhead/research _sts_ ed.pdf).

Bossel, H. (2001), Assessing viability and sustainability: a systems-based approach for deriving comprehensive indicator sets. Conservation Ecology 5(2), pp. 12 (http://www.consecol.org/vol5/iss2/art12/).

CEC (Commission of the European Communities) (2004), Progress towards the common objectives in education and training: Indicators and Benchmarks, Commission staff working paper, 114 p. Commission of the European Communities, Brussels, Belgium (http://ec.europa.eu/education/policies/2010/doc/progress_towards_common_objectives_en.pdf).

EC (European Commission) (2000), European report on the quality of school education: Sixteen quality indicators. Report based on the work of the Working Committee on Quality Indicators. 82 p. Luxembourg (http://ec.europa.eu/education/policies/educ/indic/rapinen.pdf).

– (2002), European report on quality indicators of lifelong learning. Fifteen quality indicators. 95 p. European Commission, Brussels, Belgium (http://www.bolo gna-berlin2003.de/pdf/Report.pdf).

EC-Eurostat (2009), Sustainable development in the European Union: 2009 monitoring report of the EU sustainable development strategy. Eurostat Statistical Books. 311 p. European Commission, Luxembourg (http://epp.eurostat.ec.europa.eu/cache/ITY_OFFPUB/KS-78-09-865/EN/KS-78-09-865-EN.PDF).

EU (2004), Progress towards the common objectives in education and training. Indicators and Benchmarks. Commission staff working paper, 113 p. Commission of the European Communities, Belgium, Brussels.

– (2007), Measuring progress towards a more sustainable Europe – 2007. Monitoring report of the EU sustainable development strategy. Eurostat statistical books. 330 p. European Commission, Luxembourg (http://ec.europa.eu/sustain able/docs/estat_2007_sds_en.pdf).

GC-UNESCO (German Commission for UNESCO) (2009), UNESCO World Conference on Education for Sustainable Development. Proceedings. 31 March – 2 April 2009. 124 p. UNESCO/BMBF/German Commission for UNESCO (Org)., Bonn, Germany (http://www.esd-world-conference-2009.org/fileadmin/download/ESD2009ProceedingsEnglishFINAL.pdf).

Hodson, D. (2003), Time for action: science education for an alternative future. International Journal of Science Education, 25(6), pp. 645-670 (doi: 10.1080/09500690320000766 43).

http://ec.europa.eu/education/policies/2010/doc/progress_towards_common_objectives_en.pdf
http://www.bne-portal.de/coremedia/generator/pm/en/Issue__001/Downloads/01_Contributions/Raaij.pdf

Huckle, J. (2006), Indicators for Education for Sustainable Development: Engaging the Debate [stimulus material for breakout groups]. Presented at the Bath Royal literacy and Scientific Institute UK, Indicators for education for sustainable development: engaging the debate. 17 March 2006. 3 p. University of Bath, Bath, UK (http://www.bath.ac.uk/cree/resources/esrcesd/huckle.pdf).

– (2009), Consulting the UK ESD community on an ESD indicator to recommend to Government: an insight into the micro-politics of ESD. Environmental Education Research, vol. 15(1), pp. 1-15 (doi: 10.1080/13504620802578509)

Mogensen, F. and Schnack, K. (2010), The action competence approach and the 'new' discourses of education for sustainable development, competence and quality criteria. Environmental Education Research, vol. 16 (1), pp. 19-33 (doi: 10.1080/13504620903504032)

OECD (Organization for Economic Co-operation and Development) (2006), Assessing Scientific, Reading and Mathematical Literacy. A Framework for PISA 2006. Programme for international student assessment. 188 p. OECD publishing (http://www.oecd.org/dataoecd/63/35/37464175.pdf).

– (2009), PISA 2009: Assessment Framework. Key competences in reading, mathematics and science. Programme for International Student Assessment. 192 p. OECD publishing (http://www.oecd.org/dataoecd/11/40/44455820.pdf).

Ortega-Cerdà, M. (2005), Sustainability indicators as discursive elements. Paper submitted at the 6th International Conference of the European Society for Ecological Economics in Lisbon, 14-17 June 2005. 17 p. Lisbon, Portugal (http://www.w3.ent.cat/articles/indicators.pdf).

Pedrosa, M.A. and Leite, L. (2004), Educação Científica, Exercício de Cidadania e Gestão Sustentável de Resíduos Domésticos – Fundamentos de Um Questionário. In ENCIGA (ed.), XVII Congreso de ENCIGA. BOLETÍN DAS CIENCIAS, pp. 91-93. Coruña (Ribeira). Ano XVII. Número 56, Novembro 2004 (http://www.enciga.org/congreso/2004/congreso17.htm).

Pigozzi, M.J. (2008), Indicators of the quality of education. In Making education for all. A report from the International Working Group on Education (IWGE), Steve Packer (ed.). International Institute for Education Planning – UNESCO, pp. 103-110. Glen Cove, Long Island, New York, USA (http://unesdoc.unesco.org/images/0018/001853/185398e.pdf).

Ramos, T.B. and Caeiro, S. (2010), Meta-performance evaluation of sustainability indicators. Ecological Indicators, 10, pp. 157-166 (doi: 10.1016/j.ecolind.2009.04.008).

Reid, A., Nikel, J. and Scott, W. (2006), Indicators for sustainable development: a report on perspectives, challenges and progress. Centre for Research in Education and the Environment. University of Bath. December 2006. 88 p. Anglo-German Foundation for the study of industrial society, London, UK (http://www.agf.org.uk/cms/upload/pdfs/CR/2006_CR1515_e_education_for _sustainable_development.pdf).

Rode, H. (2006), Different indicators for different contexts? Developing indicators for ESD Germany [stimulus material for break-out groups]. Presented at the Bath Royal Literacy and Scientific Institute, 17 March 2006. 2 p. University of Bath, Bath, UK (http://www.bath.ac.uk/cree/resources/esrcesd/rode.pdf).

Rode, H. and Michelsen, G. (2008), Level of indicator development for education for sustainable development. Environmental Education Research, 14(1), pp. 19-33 (doi: 10.1080/13504620701843327).

Siemer, S., Elmer, M.S. and Rammel, C. (2006), Pilot Study: "Indicators of an Education for Sustainable Development" (English Summary). 27 p. Umweltdachverband (Environmental Protection Umbrella Association Austria) (eds.), Vienna, Austria (http://www.bmukk.gv.at/medienpool/18317/bine_in dicators.pdf).

Sustainable Development Commission (2006), SD indicators for education. SDC proposal to the DfES and Defra, June 2006. 3 p. Sustainable Development Commission, London, UK (http://www.sd-commission.org.uk/publications/ downloads/Education_sd_indicators.pdf).

Tilbury, D. and Janousek, S. (2006), Development of a National Approach to Monitoring, Assessment and Reporting on the Decade of Education for Sustainable Development. Stage 1 Identification of National Indicators. Summarizing documented experiences on the development of ESD indicators and networking with expert groups on ESD indicators. 76 p. Australian Research Institute of Education for sustainability (ARIES) and Australian Government Department of the environment and water resources, Sydney, Australia (http://www.aries.mq.edu.au/projects/esdIndicators/files/ESDIndicators_Feb0 7.pdf).

Tilbury, D., Janousek, S., Elias, D. and Bacha, J. (2007), Asia-Pacific Guidelines for the Development of National ESD Indicators. UNESCO Asia and Pacific Regional Bureau for Education (Publ). 122 p. UNESCO Bangkok. Bangkok, Thailand (http://www2.unescobkk.org/elib/publications/121/Guidelines.pdf).

UN-DESA (United Nations – Department of Economic and Social Affairs) (2006), The Millennium Development Goals Report. United Nations. 32 p. United Nations Department of Economic and Social Affairs DESA – June 2006, New York, USA (http://unstats.un.org/unsd/mdg/Resources/Static/Products/ Progress2006/MDGReport2006.pdf).

– (2007), Indicators of sustainable development: Guidelines and Methodologies. October 2007. Third Edition. 99 p. United Nations Department of Economic and Social Affairs DESA – October 2007, New York, USA (http://www.un.org/esa/sustdev/natlinfo/indicators/guidelines.pdf).

– (2010), The Millennium Development Goals Report 2010. 80 p. United Nations Department of Economic and Social Affairs DESA – June 2010, New York (http://www.un.org/millenniumgoals/pdf/MDG%20Report%202010%20En%20r15%20-low%20res%2020100615%20-.pdf).

UNECE (United Nations Economic Commission for Europe) (2005), Background Paper on Development of Indicators to Measure Implementation of the UNECE Strategy for ESD. First meeting. 26-28 September 2005. 80 p. UNECE Expert Group on Indicators for Education for Sustainable Development, Ede, Netherlands (http://www.unece.org/env/esd/inf.mee ting.docs/Discussion%20paperIndicators.3.doc).

– (2006), EG on ESD Indicators-3/3. Draft outline of indicators (consolidated version). ANNEX III (CEP/AC.13/2005/9. Rev. 12 May 2006). Draft. 15 p. UNECE Expert group on Indicators for Education for Sustainable Development. Hague, Netherlands (http://www.mesdwg.ca/pdfs/indicators_framework.pdf).

– (2008), Guidance for reporting – (EG on ESD Indicators – 8/2). Draft. Eighth meeting, 15-18 September 2008. 78 p. UNECE Expert Group on Indicators for Education for Sustainable Development. Paris, France (http://www.unece.org/env/esd/inf.meeting.docs/EGonInd/8mtg/GRconsolidated_EG-ESD_8_2.pdf).

UNECE-CEP (United Nations Economic Commission for Europe – Committee on Environmental Policy) (2006), Indicators for Education for Sustainable Development: Reporting Format. Second meeting, 4-5 December 2006. 19 p. UNECE Steering Committee on Education for Sustainable Development, Geneva, Switzerland (http://www.unece.org/env/documents/2006/ece/cep/ac.13/ece.cep.ac.13.2006.5.add.1.e.pdf).

UNESCO (United Nations Educational, Scientific and Cultural Organization) (2005a), Guidelines and recommendations for reorienting teacher education to address sustainability. Education for sustainable development in action. Technical paper nº2 – 2005. 71 p. UNESCO, Paris, France (http://unesdoc.unesco.org/images/0014/001433/143370E.pdf).

(2005b), Report by The Director-General on the United Nations Decade of Education for Sustainable Development: International Implementation Scheme and UNESCO's Contribution to the Implementation of the Decade. Draft International Implementation Scheme for the United Nations Decade of Education for Sustainable Development (2005-2014). 9 p. UNESCO, Paris, France (http://unesdoc.unesco.org/images/0014/001403/140381e.pdf).

– (2005c), International Implementation Scheme. United Nations Decade of Education for Sustainable Development (2005-2014): International Implementation Scheme UNESCO Education Sector October 2005. 31 p. UNESCO, Paris, France (http://unesdoc.unesco.org/images/0014/001486/148654e.pdf).

– (2005d), UN Decade of Education for Sustainable Development 2005 – 2014: The DESD at a glance. 31 p. UNESCO, Paris, France (http://unesdoc.unesco.org/images/0014/001416/141629e.pdf).

– (2008), Quality education, equity and sustainable development: A holistic vision through UNESCO's four world education conferences 2008-2009. Education sector. November 2008. 9 p. UNESCO, Paris, France (http://unesdoc.unesco.org/images/0018/001818/181864e.pdf).

Van Raiij, R. (2007), Indicators for Education for Sustainable Development. BNE Journal – Online-Magazine "Education for Sustainable Development", Issue May. 23 p. Leuphana University of Lüneburg, Lüneburg, Germany.

Vilches, A. and Gil-Pérez, D. (2010), Educación para un nuevo orden socioambiental: Reflexiones acerca del futuro del Seminario Ibero-americano CTS. In Universidade de Brasília (ed.). II Seminário Ibero-americano Ciência-Tecnologia-Sociedade no Ensino das Ciências (VI Seminário Ibérico CTS no Ensino das Ciências), pp. 1-8. Universidade de Brasília, Brasília (ISBN: 978-85-62810-01-5).

Wade, R. and Parker, J. (2008), EFA-ESD dialogue: Educating for a sustainable world. Education for Sustainable Development. Policy Dialogue no. 1. 63 p. UNESCO, Paris, France (http://unesdoc.unesco.org/images/0017/001780/178044e.pdf).

Wals, A. (2009), Review of Contexts and Structures for Education for Sustainable Development Learning for a Sustainable World. 81 p. UNESCO, Paris, France (http://unesdoc.unesco.org/images/0018/001849/184944e.pdf).

Chapter 6

Experimental Science Activities and Environmental Education: A pivotal bridge

Sofia Marques, Ruth Pereira, Fernando Gonçalves

Abstract

The problem-solving activity described in this chapter began with a visit to a little stream flowing near the school. This place was already known by students and therefore it was the object of a more careful observation. After the visit, students used their notes to write a report. The students' report revealed a striking fact – all of them believed that the stream was polluted. Their arguments were described, and polluting agents were identified. This preliminary field trip to the river allowed students to raise questions about the role of different freshwater species, and how these groups were affected by different anthropogenic activities (e.g. agriculture; physical disturbance of stream bed, etc.). Toxicity tests with freshwater molluscs were planned and performed by students to test the formulated hypothesis. The comments of students during the visit and their written reports formed the basis upon which a questionnaire was elaborated, focusing on environmental pollution. Students answered the questionnaire before and after the activities had been performed. Our purpose was to evaluate how the inquiry-based activity was successful in the acquisition of scientific knowledge. Students' deep involvement during the different phases of the work indicated how important experimental activities are to science learning and how they can contribute to the development of skills among basic school students, such as critical thinking.

Introduction

The expression "experimental activity" is used to refer to a practical activity in which variables are controlled and manipulated (Leite 2001; Martins 2003). According to Martins (2003) and Rehorek (2004), investigations or inquiry-based learning activities, requiring the engagement of students to find an answer to a problem through the execution of experimental work, are included within the group of practical/experimental science activities.

The main question of the research can be an environmental problem. In agreement with Giordan and Souchon (1997), since these investigations promote the acquisition of environmental concepts, they attract the attention of students for environmental problems and contribute towards changing attitudes and behaviour in line with an environmental and citizenship education. The central objective of environmental education is the development of responsible behaviour towards the environment (Santos et al. 2004), and the role of teachers is to help students to understand the world they live in, the threats posed to the future of the humanity, as well as thinking of strategies to face them (Pedrosa 2004). In summary, inquiry-based science activities, whose starting point is an environmental problem, may contribute to environmental education, as they help students to become aware of environmental problems and to develop essential skills for a better understanding of the environment and for problem-solving.

Moreover, teaching activities, developed under an environmental education perspective, can be a favourable framework to apply theoretical concepts learnt in science classes (Muñoz and Ortega 2004) and to mobilize science procedures to daily life (Cachapuz et al. 2000a, 2000b, 2002), favouring significant learning. Environmental education contributes towards making curriculum subjects more attractive. These could otherwise be boring and useless if not taught using real situations for students.

This kind of activity is an important tool in science teaching and environmental education as it stimulates problem-solving (Zervanos and Maclaughlin 2003; Pato et al. 2004) and allows students to observe and experiment key concepts that are not accurately transmitted in text books (Zervanos and Maclaughlin 2003), thereby making the acquisition of knowledge and skills easier (Pato et al. 2004). Moreover, field trips motivate and stimulate team work and, at the same time, encourage the development of new attitudes towards the environment (Pato et al. 2004).

The main objective of the work described in this chapter was to carry out a problem-based learning activity with basic school students involving them in all the planning steps in order to promote the acquisition of scientific knowledge, to develop skills in different domains, such as critical thinking and communication, and to change attitudes towards the environment. At the same time, we tried to attain the following specific objectives:

- To increase students' motivation for science learning and to promote responsibility and ownership in the development of these types of practical activities;
- To develop the ability of students to criticize and evaluate the impact of some human actions and, in particular, one's own actions on the environment.

Methodology

Sixty-five students, aged between 13 and 16 years old and spread among 8th grade classes of a basic school located in Sernancelhe (Viseu, Central Portugal), participated in this project. The majority of Sernancelhe residents work in agriculture or have emigrated. Hence, many students live with their grandparents during the school term.

The starting point of formulating the problem was a field trip.

A small stream, flowing near the school, called Ribeira do Medreiro (an affluent stream of the Távora River) was found to be an appropriate natural and secure area for this outdoor class that occurred in October. Before the field trip, each group received and discussed a worksheet in which students were asked to take notes about their observations, the most interesting aspects they found in the margins and in the water, detaching the most negative and positive ones, and to write their questions.

Students walked along the stream for 45 minutes (fig. 1). In the class, after this activity, they wrote a report based on their notes. The main objective of the report was to perceive students' main questions and interests and also to help them to think about their observations.

Fig. 1: Students exploring the stream during the field trip.

Based on the reports and on the comments of students, a questionnaire was designed in order to assess the effectiveness of the teaching activity in promoting changes in the cognitive state of students. The questionnaire was filled in before and after the problem-solving learning activity and it comprised four questions about the pollution issue. One was an open question, and the other three were close-ended questions with more than one possible answer (Oppenheim 1966). These questions also included the option "*others*" to account for possible answers/categories not taken into consideration in the questionnaire design. Open questions were qualitatively evaluated, while close-ended questions were encoded for statistical analysis. Contingency tables were constructed and answers to closed-ended questions were analyzed using Chi-square test to measure independence between the answers of groups of students before and after the problem-solving research activity (Calder 1996; Pestana and Gageiro 2000).

The questions asked by students on the reports were so diverse that the majority of them could not be used to start the investigation. However, as most of the questions were related with animals observed in the stream and with water quality, two main questions were selected:

"Which are the sources of pollution of Ribeira de Medeiros?"
"What is the effect of the identified pollutants on stream organisms?"

An investigation like this one was new to these students. Hence, to help them to think about it and to plan it, a worksheet with several questions was given by the teacher in order to stimulate their thinking abilities. This worksheet asked students to make decisions about some aspects that should be considered: i) the species to be used in toxicity tests; ii) the appropriate conditions to maintain the test organisms in laboratory; iii) the material students would need to perform the activity; iv) the appropriate medium to maintain the organisms and to perform the test and, v) the pollutants and concentrations to be tested. A planning chart proposed by Martins (2003) was described in the worksheet and helped students to design the experimental activity as it allowed them to identify dependent and independent variables and which of the later variables should be kept controlled.

After having planned the activity, students discussed when and how to capture the organisms from the selected species to be tested. Students chose freshwater snails from the *genus Lymnaea,* for toxicity tests, because they were abundant in the stream, easy to maintain in laboratory and changes in their behaviour, nutrition rate and creeping, were also easy to observe.

Since the simulation of natural conditions in the laboratory was considered difficult, students decided to collect snails a few hours (maximum 24h) before the beginning of the tests. To collect snails and water from the stream, a group of students went there with the teacher during lunchtime.

After the main sources of pollutants had been identified based on the reports, each class chose a pollutant and, working in groups, each group within the class tested a different concentration. Thus, class A decided to test rust because they had observed rusty nails and cans in the stream, and the floodgates of a little dam were also rusty. The relationship between rust and oxidized iron was made by the teacher and it was suggested to use ferrous sulphate ($FeSO_4$) to replace rusty objects. Glass flasks were filled with 200 ml of water from the stream, and each group of students weighted and dissolved a different amount of ferrous sulphate. Group 1 prepared the control flask, group 2 dissolved 1 g of ferrous sulphate, group 3, 2 g, and group 4, 4 g (Fig. 2). Two round pieces of lettuce, 2 cm approx. in diameter, were previously weighted and placed in the bottom of the flasks using a paper clip to keep them submerged. The test began with the addition of 10 snails to each flask. After 24 hours, live and dead snails were counted and the remaining pieces of lettuce (lettuce disks) were dried with absorbing paper before being weighted.

Fig. 2: Experimental design for the toxicity test with ferrous sulphate performed by class A.

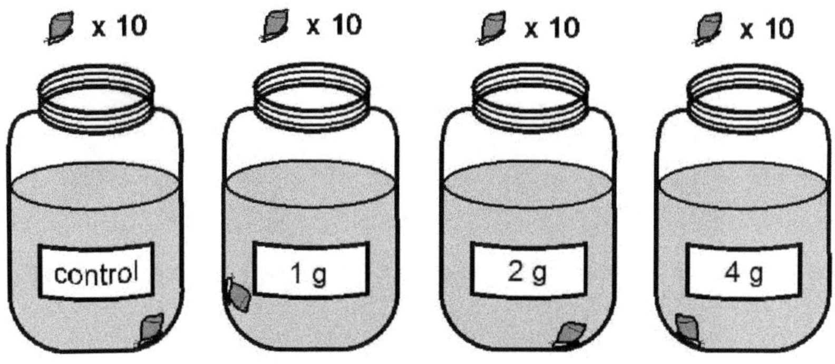

Class B studied the influence of a powder laundry detergent (any kind of powder detergent for laundry washing can be used) on snails because they argued that people still wash their clothes in the stream and also because they had observed a discharging pipe of domestic effluents flowing into the stream. The test design of this class was the same as previously described for class A. The amount of detergent weighted for groups 2, 3 and 4 was 1 g, 2 g and 4 g, while group 1 prepared the control flask (Fig. 3). In class C students decided to test the effects of an apple tree pesticide on snails. Students recorded the existence of agricultural fields in the stream margins. Moreover, they suggested that pesticides applied to those fields were probably dragged by rain to the stream. Therefore, according to students, it was necessary to simulate this process. To attain this purpose they weighted different amounts of the pesticide (group 2, 0.5 g, group 3, 1 g, and group 4, 2 g) and dissolved them in 200 ml of water from the stream. Each group collected 200 g of soil in the fields near the stream, put it in a glass funnel with filter paper, and the river water with different amounts of pesticide was used to irrigate it. The leachate was collected in a graduated cylinder as shown in Figure 2. As group 1 was working with the control flask, the procedure was the same as for the other groups except for the pesticide addition to stream water (fig. 4). After the leachate was collected the test was performed following the design already described for the other classes (fig. 5). According to pesticide caution instructions on the package, this product should not be inhaled or come into contact with the skin, therefore the students used gloves and surgical masks to manipulate the pesticide.

Fig. 3: Experimental design for the toxicity test with detergent performed by class B.

Fig. 4: Simulating the leaching of the pesticide from the soil trough irrigation.

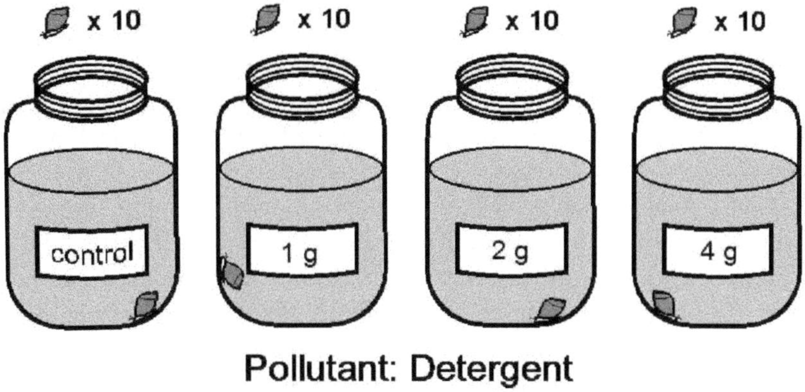

Fig. 5: Experimental design for the toxicity test with a pesticide performed by class C.

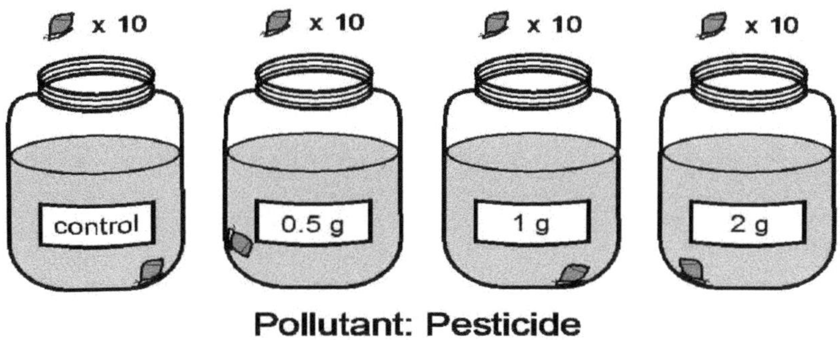

Fig. 6: Experimental design for the toxicity test with car oil performed by class D.

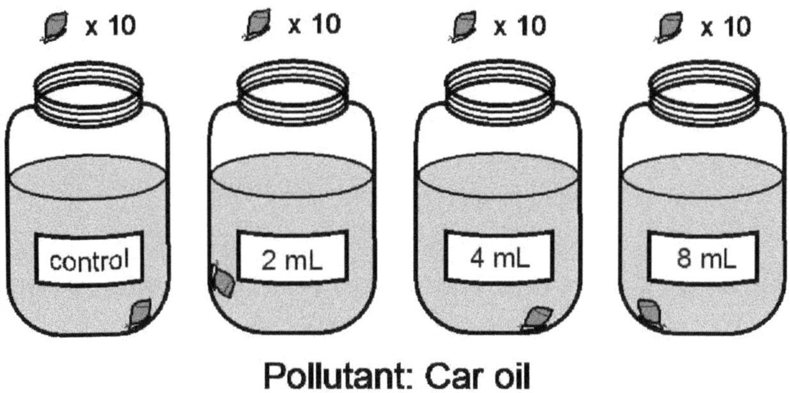

As a car workshop existed near the stream and washing water with car oil was recorded by students flowing to the river, class D decided to test the effects of car oil on freshwater snails. Each group of students prepared a flask with 200 ml of stream water; group 1 did not add any oil, group 2 added 2 ml of oil to the stream water, group 3 added 4 ml and group 4 added 8 ml, as is described in Figure 6. Ten snails and the round pieces of lettuce were added to each flask as was done by the other classes. All students worked with surgical gloves, and class C also used a surgical mask to assure security in the manipulation of pollutant agents. If available, it was suggested to use a laboratory laminar air flow chamber (that did not exist in our school) or a hotte to prevent inhalation of toxic aromatic hydrocarbons.

Twenty-four hours after the beginning of the toxicity tests, students recorded the results. To count the number of dead snails, it was considered that those snails that did not crawl and that were inside their shells were dead. After reading the results, each group communicated them to the rest of the class to be analyzed and discussed. Each class constructed a table with the results of all the groups (tables 1 to 4) and then discussed it.

A report of the activity was elaborated at home by each group allowing students to practice scientific writing.

Results

During the field trip students were very excited, because outdoor activities for exploring the surroundings were not common for them. They took notes, pictures and collected some aquatic plants for identification. Most of their comments were about the water and its pollution, as well as about animals and plants they observed. After taking notes, students wrote a report during the following science class, focusing mainly on freshwater snails, probably because they were unknown to them, on plants and on water pollution. All of the reports showed that students perceived the pollution of the stream water and their suspicions about potential deleterious effects on aquatic organisms. Therefore, most of the questions written in the reports were about water pollution. Two examples of those questions were:

"Why is the river polluted?"
"How do animals survive in such polluted water?"

Another interesting aspect that stood out from the reports was the difficulty shown by the great majority of students in selecting the main aspects to be described in each of its sections.

The worksheet provided by the teacher gave students some orientation in the planning of the experimental activity (Annex 1). Several specific questions were asked in this worksheet regarding identification of pollutants and animals surveyed in the stream. Based on the observations during the field trip and on previous knowledge of students about the stream, ten groups of pollutants were identified: i) rust; ii) pesticides and fertilizers; iii) dead animals; iv) food scraps; v) excrements; vi) organic matter; vii) car oil; viii) detergents; ix) dust from a house under repair, near the stream and x) different kinds of garbage. From these pollutants each class chose one to test in the experimental activity. In the worksheet students were also asked to formulate a hypothesis to the problem – "*What is the effect of pollutants on stream organisms*?" All the groups, except one, agreed that pollutants modify the activity of snails or kill them. One group postulated that "Pollutants do not affect aquatic organisms". Although these students still considered that the stream was polluted, they supported their hypothesis on the existence of snails in the stream.

During the accomplishment of the experimental activity, students were motivated. Although they showed some autonomy, some difficulties were evident when it was necessary to allocate tasks among the members of the group.

Table 1: Results of the toxicity test with ferrous sulphate obtained by the 8th grade, class A

Group	Ferrous sulphate (g)	Lettuce initial weight (g)	Lettuce final weight (g)	Snails initial number	N.º of dead snails
1- Control	0	0.5	0.5	10	2
2	1	0.5	0.3	10	10
3	2	0.5	0.3	10	10
4	4	0.5	0.5	10	10

After the analysis of table 1, students explained unexpected results such as two dead snails in the control and the final weight of lettuce disks in flasks 1, 2 and 3. Despite some disappointment with the results gathered, all the groups suggested repetition and/or modification of the tests in order to eliminate possible errors.

Table 2: Results of the toxicity test with detergent obtained by the 8th grade class B

Group	Detergent (g)	Lettuce initial weight (g)	Lettuce final weight (g)	Snails initial number	N.º of dead snails
1- Control	0	0.3	0.2	10	2
2	1	0.3	0.3	10	10
3	2	0.3	0.2	10	10
4	4	0.3	0.2	10	10

Students from class B inferred that the snails in flasks 2, 3 and 4 (table 2) died because of the high concentration of detergent added. The death of snails in the control flasks was an unexpected result, as was the lettuce final weight in flasks

3 and 4. This class suggested some explanations for these results and mentioned that it was necessary to repeat the test being more cautious with possible sources of error.

Table 3: Results of the toxicity test with the pesticide obtained by the 8th grade class C

Group	Pesticide (g)	Lettuce initial weight (g)	Lettuce final weight (g)	Snails initial number	N.º of dead snails
1 – Control	0	0.4	0.4	10	0
2	0.5	0.3	0.2	10	10
3	1	0.3	0.6	10	9
4	2	0.3	0.3	10	10

When Table 3 was discussed by class C, students argued that all the snails died because the pesticide tested was very toxic, even at low concentrations. According to students, the only organism that survived (flask 3) was probably adapted to pesticides. Results for lettuce final weight in flasks 1, 2 and 3 were unexpected. These results were discussed and explanations were suggested. Despite the results gathered, students argued that the concentration of the pesticide in the stream was lower than those tested because the snails were alive there.

Table 4: Results of the toxicity test with car oil obtained by the 8th grade class D

Group	Car oil (ml)	Lettuce initial weight (g)	Lettuce final weight (g)	Snails initial number	N.º of dead snails
1 – Control	0	0.3	0.4	10	10
2	2	0.3	0.4	10	10
3	4	0.4	0.5	10	10
4	8	0.3	0.3	10	10

After having constructed Table 4 on the blackboard, students felt surprised with the results and immediately said that it was all wrong and it was necessary to repeat the test. They immediately identified unexpected results such as, the death of the snails in the control and the lettuce final weight in flasks 1, 2 and 3. After suggesting explanations for these results, students inferred that they can not make conclusions about the cause of the death of snails in the test. This proved that students perceived the role of the control in a toxicity test.

Regarding questions asked on the worksheet, students inferred that the consequences of the pollution in that stream (question 15) were modifications in the activity of organisms or their death and the impairment of water quality. To prevent pollution (question 16), people should stop releasing garbage, oil and other residues into the stream. On the other hand, to solve pollution problems (question 16), students suggested the cleaning of the stream. In students' opinion (question 17), the experimental activity they performed was important to learn more about nature, to understand why people should not release toxic substances in freshwater resources and to make science classes even more interesting.

A total of 65 and 63 questionnaires were filled in before and after the problem-solving activity. Most of the answers of students to question 1 (*In your opinion, what the term pollution means?*) were related with: i) garbage; ii) the release of garbage in rivers/water; iii) gases emitted by cars and industries. Some students also mentioned that toxic wastes and pollution was something harmful to the environment, human beings and other organisms. They also emphasize, in both questionnaires, that man is responsible for pollution. Regarding question 2 (*How can you identify that a river is polluted?*), most of the students selected options A [97% (before and after)], E [88% (before) and 83% (after)] and J [71% (before) and 67% (after)] in both questionnaires (Fig. 7). Statistical results showed no significant relationships between the students' answers and the period in which the questionnaire was performed (before and after the activity) for all the classes (Class A: $\chi^2 = 6.44$, d.f. = 11, p = 0.842; Class B: $\chi^2 = 5.28$, d.f. = 11, p = 0.917; Class C: $\chi^2 = 6.72$, d.f. = 11, p = 0.821; Class D: $\chi^2 = 3.90$, d.f. = 11, p = 0.973; All of the 8th students: $\chi^2 = 6.20$, d.f. = 11, p = 0.860).

Fig. 7: Frequency of the options chosen by all the students of 8th grade for the question 2, "*How can you identify that a river is polluted?*"

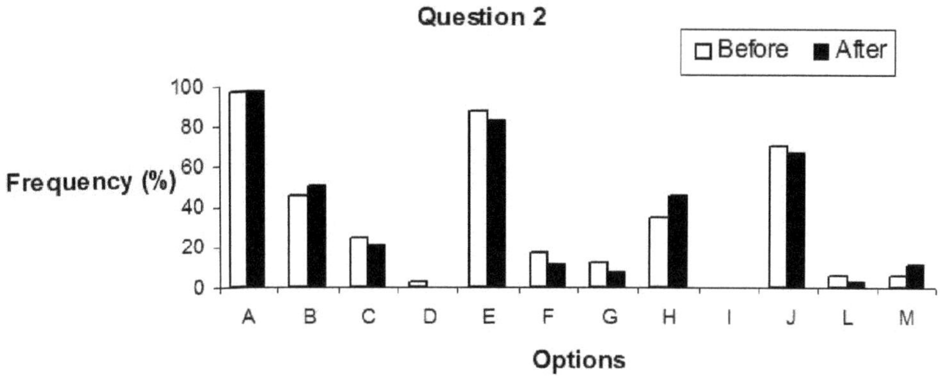

(A- Because there is garbage; B- Because there are excrements; C- Because of the stagnant water; D- Because the water is clear; E- Because the water smells; F- Because there is mud in the river's bed; G- Because there is "verdigris" in the river's rocks; H- Because you can not see the river's bed; I- Because you can see the river's bed; J- Because there is foam in the water's surface; L- Because there are organisms in the water; M- Other)

Figure 8 shows the frequency of the options selected by students to answer question 3, ("*From the listed agents select the ones that, in your opinion, pollute rivers:*"). The options most frequently chosen were practically the same in both questionnaires. Qui-square test showed no significant relationships between the answers given by students on both questionnaires (Class A: $\chi^2 = 6.91$; d.f. = 14, p = 0.938; Class B: $\chi^2 = 5.57$; d.f. = 14, p = 0.976; Class C: $\chi^2 = 3.30$; d.f. = 14, p = 0.998; Class D: $\chi^2 = 5.45$; d.f. = 14, p = 0.979; All of the 8th grade students: $\chi^2 = 4.72$; d.f. = 14, p = 0.989).

Fig. 8: Frequency of the options signed by all the students of 8th grade on question 3, "*From the listed agents select the ones that, in your opinion, pollute rivers:*"

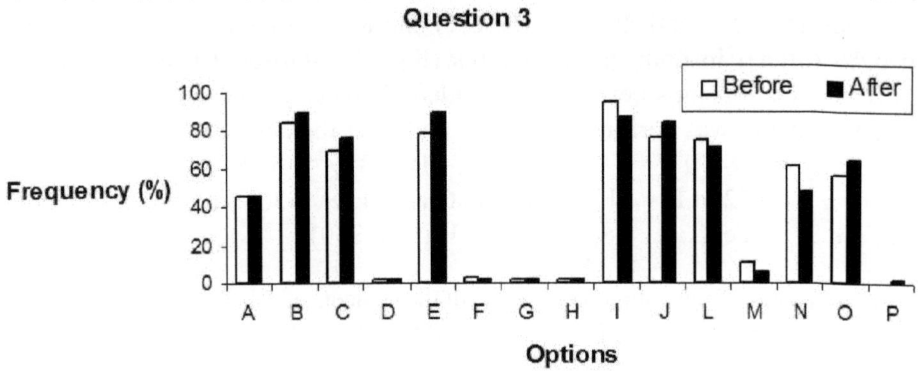

(A- Excrements; B- Pesticides; C- Fertilizers; D- Insects; E- Oil; F- Plants; G- Algae; H- Fish; I- Garbage; J- Detergents; L- Food scraps; M- Verdigris; N- Animal corpses; O- Nails and other iron materials; P- Others)

Question 4 asked students to select the option(s) that would complete more correctly the sentence "*Pollution in a river can affect* ..." (Fig. 9). All the options were signed in both questionnaires. Despite some differences recorded in the graph, no statistically significant relationships existed among students' answers and the period in which the questionnaire was filled in (Class A: $\chi^2 = 3.7$; d.f. = 11, p = 0.978; Class B: $\chi^2 = 3.26$; d.f. = 11, p = 0.987; Class C: $\chi^2 = 2.3$; d.f. = 11, p = 0.997; Class D: $\chi^2 = 1.17$; d.f. = 11, p = 1.000; All of the 8th grade students: $\chi^2 = 4.52$; d.f. = 11, p = 0.952).

Fig. 9: Frequency of the options signed by all the students of 8th grade on question 4, *"Pollution in a river can affect ...:"*

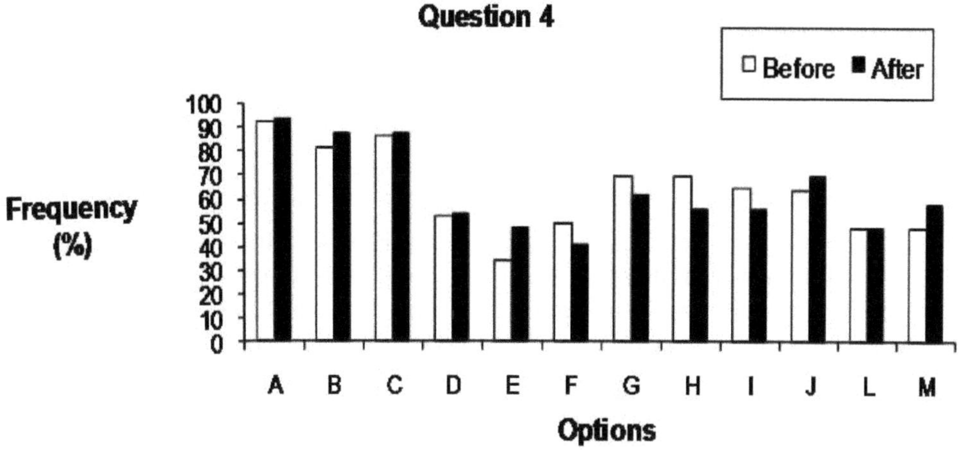

(A- The animals that live in the river; B- The plants in the river; C- The animals that drink river's water; D- The planting that are irrigated with that river's water; E- Soils that are irrigated with that river's water; F- All human beings; G- Human beings that feed from animal and/or plants from that river; H- Human beings that fish on that river; I- Human beings that swim on that river's water; J-Animals that feed on river plants; L- Atmosphere; M- Sea)

Discussion

Liggit et al. (2004) notice that children like water. In fact, in our study the students' interest during the field trip to the stream and their concerns with its polluted conditions showed that it would be easy to promote students' consciousness of environmental problems (Giordan and Souchon 1997), especially those concerning freshwater resources and their degradation. Students' previous knowledge about pollution was significant, which was demonstrated by the list of pollutants elaborated in the worksheet since some of them were not at all evident.

During the accomplishment of experimental activities, students worked hard and autonomously. Luckie et al. (2004) argue that problem-solving activities are important to increase the engagement of students in the different tasks that have to be performed. At the same time, these activities increased the self-esteem of students as they perceive learning as its own responsibility (Ekborg 2003).

Despite unexpected results from the toxicity tests, they evoked discussion, and all the classes were able to suggest explanations for them. Repetition of tests was suggested with the modification of some aspects on their experimental design in order to eliminate possible sources of errors. According to students, concentrations of chemicals to be tested were one of the aspects to be changed, since they were probably not representative of those existing in the stream. This behaviour demonstrated that problem-based investigations are important tools to develop knowledge of procedures, thinking skills and to improve attitudes like creativity, perseverence and self-questioning (Galvão 2001a; 2001b). The results gathered by students did not provide an answer to the problem formulated at the beginning of the experimental activity but led students to perceive that their hypothesis cannot be validated. Hence these science-learning activities may promote the understanding of the scientific method, creating a solid foundation for science literacy (McPherson 2001). As it was argued by other authors (e.g. Proulx 2004) and mentioned by students, who answered question 17 of the worksheet, a better understanding of the pollution subject was also attained. During experimental design planning, students were able to identify and perceive the importance of a control, to validate results gathered in the treatments. However, they did not mention the necessity to replicate treatments and the control.

Although no significant relationships were found between the answers of students to the different questions of the questionnaire and the period in which it was filled in (before or after the experimental activity), it is important to mention that students' perception was reinforced about pollution as something harmful to the environment and to human beings and about the responsibility of man's actions in the development of pollution problems. In fact, one single problem-solving activity focused on the environment is probably not enough *per se* to increase the cognitive acquisition by students about the issue being studied. It would be necessary to start earlier in the students' curriculum, performing these science-teaching activities and doing it regularly in order to take all the advantages they can be offered, even for those who only want to be scientific literate citizens, able to face daily events. As teachers, we must probably change our objectives from knowledge to acquiring abilities such as critical thinking, technical procedures, planning of experimental designs or data collection to test hypotheses and solve problems. Probably the most appropriate approach to evaluate the acquisition of these scientific abilities will be to encourage problem-solving learning activities on a regular basis and the observation of different groups of students throughout their school years. A methodology that has gained prominence in science education research (Méheut 2004).

Conclusions

The problem-based science activity promoted the acquisition of students' procedural and critical thinking abilities. Although teachers helped by giving students a worksheet in which specific questions were posed in order to guide their thinking abilities, students got involved and were responsible for all the steps. The main difficulty for students was in selecting pollutant concentrations to be used in each test, which is justified by the fact that they had never done this kind of test before. In fact, this is usually a difficult decision, requiring range-finding tests in order to obtain the appropriate range of concentrations within which the toxicity parameters (lethal or sub-lethal), will be found. That was not suggested by the teacher because planning should be done, as far as possible, by students without the teacher's intervention. For the same reason, control and treatments were not replicated as students did not suggest it.

Toxicity test results were not conclusive, and all the groups suggested its repetition. Nevertheless, this was not possible because the school year was about to end and a lot of science classes had already been spent with this investigation.

Students' difficulties to interpret oral and written messages and in expressing themselves in writing were expressed in the reports. Nevertheless, they were able to interpret test results and suggest explanations for the unexpected ones. Science learning and scientific writing difficulties shown by the students were probably not due to science itself but to difficulties in using the mother language which could have been an obstacle to the answering of the questionnaire too.

Questionnaires were designed to access concept knowledge acquisitions. However, the results showed that there were no statistically significant relationships among the students' answers and the period in which they filled in the questionnaire (before and after the teaching activity). Teachers are still more concerned with concept acquisition rather than acquiring procedures and skills. This is not because teachers do not realize their importance, but they still do not know how to evaluate them with an objective approach.

It was demonstrated with answers to question 1 of the questionnaire and to question 15 and 16 of the worksheet that this problem-based activity promoted the ability to criticize and evaluate the impact of human actions on the environment. Nevertheless, the effectiveness of these abilities to change the behaviour of students cannot be predicted. Additionally, it will be necessary to survey students' attitudes and behaviours towards the environment in the future in order to be sure that potential behavioural changes persist.

Developing problem-based science activities poses some problems to the students, but also to teachers, namely: i) time problems, they need more classes than other types of activities; ii) evaluation problems, since teachers still do not know how to evaluate abilities acquired and changing behaviours and, iii) teacher training problems, most basic school teachers are not well prepared to develop this kind of practical activity.

References

Cachapuz, A., Praia, J. and Jorge, M. (2000a), Perspectivas de Ensino das Ciências. In Perspectivas de Ensino – Formação de Professores – Ciências – Textos de apoio, n.º 1, pp. 1-79. Organized by A. Cachapuz. Centro de Estudos de Educação em Ciências, Porto.

– (2000b), Reflexão em torno de perspectivas de ensino das ciências: contributos para uma nova orientação curricular – Ensino por Pesquisa. Revista de Educação IX(1), pp. 69-78.

– (2002), Ciência, Educação em Ciência e Ensino das Ciências. Temas de Investigação, p. 353. 26. Ministério da Educação, Lisboa.

Calder, J. (1996), Statistical techniques. In Data Collection and Analysis. Edited by Sapsford, R. and Jupp, V., pp. 225-261. SAGE Publications and The Open University, London.

Ekborg, M. (2003), How student teachers use scientific conceptions to discuss a complex environmental issue – Case study. Journal of Biological Education. 37 (3), pp. 126-132.

Galvão, C., Neves, A., Freire, A.M., Lopes, A.M., Macedo, G., Neves, I., Encarnação, L., Matos, M., Pinho, M., Pereira, M. and Oliveira, M.T. (2001a), Competências específicas das Ciências Fisicas e Naturais. In Currículo Nacional do Ensino Básico – Competências Essenciais. General Coordination Abrantes, P., pp. 124-146. Departamento da Educação Básica – Ministério da Educação, Lisboa.

Galvão, C., Neves, A., Freire, A.M., Lopes, A.M.S., Santos, M.C., Vilela, M.C., Oliveira, M.T., Pereira, M. (2001b), Ciências Físicas e Naturais – Orientações Curriculares para o 3º ciclo do Ensino Básico. 40 pp. Departamento da Educação Básica – Ministério da Educação, Lisboa.

Giordan, A. and Souchon, C. (1997), Uma Educação para o Ambiente. 244 pp. Instituto de Inovação Educacional and Instituto de Promoção Ambiental, Lisboa.

Leite, L. (2001), Contributos para uma utilização mais fundamentada do trabalho laboratorial no ensino das ciências. In Cadernos Didácticos de Ciências, vol. 1. Organized by Caetano, H.V. and Santos, M.G., pp. 79-97. Departamento do Ensino Secundário – Ministério da Educação, Lisboa.

Liggit, P., Moore-Hart, M. and Daisey, P. (2004), Get Wet: Bringing water and wetland education to after-school programs. The American Biology Teacher 66(2), pp. 92-100.

Luckie, D.B., Maleszeweski, J.J., Loznak, S.D. and Krha, M. (2004), Infusion of collaborative inquiry throughout a biology curriculum increases student

learning: a four-year study of "teams and streams". Advances in Physiology Education 287, pp. 199-209.
Martins, I.P. (2003), O trabalho científico no ensino básico. Extracted from the Report of the Basic Level Science Didactics Subject, presented to Aggregation Proofs in Education. 19 pp. University of Aveiro. Aveiro.
McPherson, G.R. (2001), Teaching and Learning the Scientific Method. The American Biology Teaching 63(4), pp. 242-245.
Méheut, M. (2004), Teaching-learning sequences: aims and tools for science education research. International Journal of Science Education 26 (5), pp. 515-535.
Muñoz, M.B. and Ortega, J.A.P. (2004), Educación Ambiental: Praxis científica y vida cotidiana. Descripción de un proyecto. Revista Eureka sobre Enseñanza y Divulgación de las Ciencias 1(3), pp. 233-239.
Oppenheim, A.N. (1966), Questionnaire design and attitude measurement, pp. 40-44. Basic books, Inc., publishers, New York.
Pato, A., Azeiteiro, U.M.M. and Gonçalves, F. (2004), Actividades de campo em educação ambiental. In Tendências actuais em Educação Ambiental. Edited by Azeiteiro, U.M., Pereira, M.J., Leal Filho, W., Caeiro, S., Bacelar-Nicolau, P. and Gonçalves, F., pp. 437-457. Universidade Aberta, Lisboa.
Pedrosa, M.A. (2004), Environmental education and citizenship – Challenges for scientific communities and science teaching. In World Trends in Environmental Education, 14. Edited by Azeiteiro, U.M., Gonçalves, F., Leal Filho, W., Morgado, F. and Pereira, V., pp. 175-194. Peter Lang, Frankfurt.
Pestana, M.H. and Gageiro, J.N. (2000), Análise de dados para Ciências Sociais – A complementaridade do SPSS. 2nd ed. 570 pp. Edições Sílabo, Lda., Lisboa.
Proulx, G. (2004), Integrating scientific method and critical thinking in classroom debates on environmental issues. The American Biology Teacher 66(1), pp. 26-33.
Rehorek, S.J. (2004), Inquiry-based teaching. An example of descriptive science in action. The American Biology Teacher 66(7), pp. 493-499.
Santos, L.S., Caeiro, S. and Azeiteiro, U.M. (2004), Evaluation of an environmental education action about water resources conservation. In International Perspectives in Environmental Education, 16. Edited by Leal Filho, W. and Littledyke, M., pp. 223-239. Peter Lang, Frankfurt.
Zervanos, S.M. and Mclaughlin, J.S. (2003), Teaching biodiversity and evolution through travel course experiences. The American Biology Teacher 65(9), pp. 683-688.

Chapter 7

Eco-Schools: Development of assessment tools

João Gomes, Sandra Caeiro, Filomena Amador

Abstract

Eco-Schools is an international programme for education for sustainability, which was first developed in the 1990s and implemented in Europe, America and Africa.

It aims to implement Local Agenda 21, applying educational and environmental management concepts and ideas in the daily life of schools in the process. The United Nations Decade of Education for Sustainable Development, which began in 2005, is a unique opportunity to put the issue on the national agenda, and the Eco-Schools Programme is an important tool for this purpose.

The study herein reported aims to facilitate understanding of the successful implementation of this Programme's methodology via the development of an assessment tool that reflects the work done and facilitates further corrections and improvements of future activities.

Based on qualitative and quantitative research, this involved a feedback survey given to about 150 Portuguese Eco-Schools.

The research was undertaken between December 2007 and December 2008, and involved the national coordinators of the Eco-Schools Programme and the Regional Education Authority. The proposed tool monitors and evaluates the programme to be applied in Portuguese Eco-Schools. Its use will enable the school community to improve its implementation of the Eco-Schools Programme and to achieve greater pupil involvement, whilst achieving sustainability.

Introduction

The challenges that environmental issues pose today demand profound changes in our vision and the way society functions. Moving in the direction of a human society in harmony with nature implies changing certain structural issues, both on a conceptual and practical level in order to increase the chances of success. The United Nations Decade of Education for Sustainable Development, which began in 2005, is a unique opportunity to include the issue on the national agenda. The Eco-Schools Programme (ES) is an important tool for this purpose, contributing towards the education of children and young people who shape changes in behaviour.

This programme was created in response to the need for developing a School Agenda 21 (SA21). In Portugal, it has been implemented since 1996/97, and 784 schools registered when this study was initiated. It is part of a structure that involves major international involvement and the participation of thousands of schools in 44 countries.

Since 2003, it has been recognized by the UNEP (United Nations Environment Programme) as one of the fundamental tools for making sustainable development and Local Agenda 21 work.

This programme proposes that schools adopt a working methodology that, combining activities that explore different issues, contributes to an overall improvement in the school and the community environment (Gomes 2004a; Mogensen and Mayer 2005; ABAE 2008). This methodology is presented in the form of the "7 steps" (fig. 1).

Fig. 1: The 'seven steps of an Eco-School'

STEP 1 – Eco-Schools Committee; STEP 2 – Environmental Review; STEP 3 – Action Plan; STEP 4 – Monitoring and Evaluation; STEP 5 – Curriculum Work; STEP 6 – Informing and Involving; STEP 7 – Eco-Code. Adapted from ABAE (2008).

We undertook a study that is intended to help understand the implementation of the programme's methodology from the perspective of the development of an assessment tool, while reflecting upon both the work done and eventual corrections/improvements to future activities. The Eco-Schools programme was chosen because its methodology is based on the LA21, since it urges the effective participation of all the members of the school community, and because it is an important tool for education for sustainable development. Moreover, in Portugal, this Environmental Education (EE) programme is popular and contributes to effective citizenship education.

The study was carried out between December 2007 and January 2009 with the cooperation of the national coordinators of the Eco-Schools Programme.

Methodology

On the basis of the Tuckman (2000) principles, we began by defining the issue to be researched: How to create a tool to assess the Eco-Schools Programme?

We applied a mixed methodology (qualitative and quantitative). We worked with a sample of schools, while teams created from within the National Eco-Schools Committee carried out school visits.

An open-ended questionnaire was used because, as Tuckman (2000) mentions, this type of questionnaire is more appropriate when the aim of the investigation is not only to discover attitudes or attributes of those surveyed, but also to learn something about the context in which they formed their opinions.

When devising the interview format, we worked from a theoretical framework (the Eco-Schools seven steps methodology) which was governed by thematic nuclei and their respective topics, and used these to guide interaction during the interviews, while meeting the objectives of the said interviews.

This research tool was subjected to validation by a group of Eco-Schools Programme experts (members of the National Eco-Schools Committee). It is intended that the data collection tool used in this study can later be applied in all schools involved in the programme in Portugal. In other words, apart from the general acceptance of results sought by typical research projects, the issue of transferability needs to be considered.

In the development of the feedback form and the user manual used in the visits to the Eco-Schools, the research was divided into two parts:

- The auditing process of the Eco-Schools, undertaken by the coordinators of the Eco-Schools Programme, in conjunction with the respective National Committee.

- The work carried out directly by the researcher, which included a bibliographical review, discussion with Eco-Schools Programme "experts" (at the headquarters of the National Eco-Schools Committee), school visits and monitoring the auditing process coordinated by the National Eco-Schools Committee.

Within the National Eco-Schools Committee, a feedback form proposal was presented by the national Eco-Schools coordinator during the first meeting to discuss the school assessment process, which took place on 27th November 2007. Those present had the opportunity to comment on the suitability, content and implementation method of an Eco-Schools assessment process. The researcher was also introduced to the Committee as the person who would participate in the process of developing the Eco-Schools feedback and assessment tool. This first version of the feedback form, which constituted the starting point for the development of an Eco-Schools assessment tool, Figure 2, contained:

- An area for the identification of the school
- An area for collecting data on the implementation of the Eco-Schools Programme in terms of:
  - Methodology: the parameters to delineate/evaluate, in accordance with the proposal presented, were the Eco-Schools Committee; environmental audit; action plan; monitoring and assessment; communication/reporting; eco-code; green flag.
  - Results: the parameters to delineate/evaluate, in accordance with the proposal presented, were the school grounds; indoor area of the school; community intervention; school administration; student involvement.

  The scale for this version of the feedback form ranged from 1 to 10.
- An area for ranking the school based on a number of ESD criteria. This field is intended for collecting the opinion of the Eco-Schools coordinator teacher and of the director of the school executive board on six ESD criteria. These criteria were selected from the list presented by Breiting et al. (2005) and translated for the Portuguese context by Gomes (2006). In this part of the feedback form a scale of 1 to 10 was used, as this had already been tested and validated in previous studies done by Gomes (2006).

When the second meeting of the National Committee took place on 14th January 2008, it was possible to collate the ideas and proposals of all of its members, leading to a revision of the first version of the feedback form. On this date, the process of developing the user manual began, as it was considered necessary to define a set of criteria in order to make the assessment and classification of each item included in the Eco-Schools feedback form more objective.

Since that time, the work undertaken by the researcher has occurred in parallel to the tasks of the National Committee. In each administrative region, teams of observers were formed whose function was to contact schools and pay a preparatory visit for the implementation of the Eco-Schools Programme. The main aim was to collect data on the relevance of the parameters included in the feedback form and on the applicability of the tool.

Fig. 2: First version of the Eco-Schools programme evaluation form, presented to the National Eco-Schools Commission on 27/11/2007

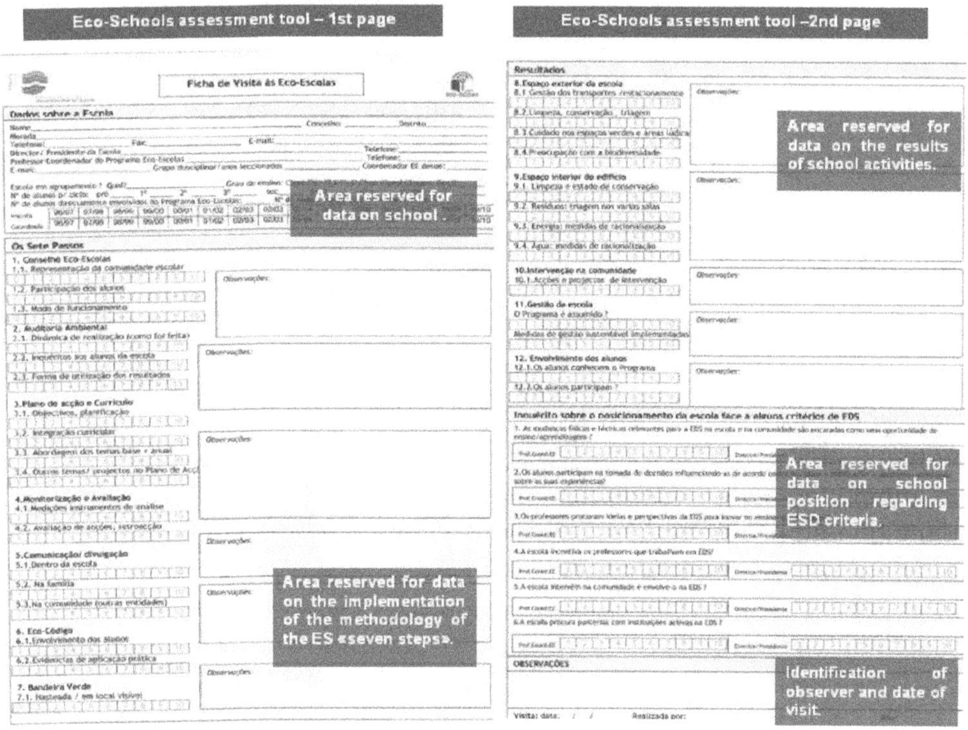

See in Annex A the assessment form in its final version.

Visits to the Eco-Schools took place between February and June 2008. The researcher visited four schools (table I) and, during the same period, the teams from DRA (acronym for the Azores Regional Education Authority) and SRAM (acronym for the Madeira Regional Education Authority) visited schools in the Azores and Madeira respectively. In mainland Portugal, the observations were carried out by teams from the DRE (acronym for the Regional Education Authority). A total of 142 establishments were visited that had been registered for the programme for at least 3 years (table II). This represented approximately 50% of the total number of Eco-Schools.

Interviews were chosen for data collection because of their various attributes: they expand and verify scientific knowledge; they allow reflection on the part of the interviewee, as an object of study, regarding situations in which, until that moment, they have probably never had the opportunity to organize mentally and/or be aware of; and they also allow the systematic collection of data (Carmo and Ferreira 1998).

The interview questions (open-ended questions) were based on the content of the Eco-Schools feedback form and the criteria put forward for each item. Opting for a relatively unstructured type of questionnaire allowed us to avoid excessively conditioning the interviewees, which would have eliminated potentially important data.

The data collected by the researcher during the visits were object of a content analysis based on Carmo and Ferreira (1998) and Bardin (2004).

Table I: Schools involved in the study according to research topics.

Research topics	Period	Stage of the study	Schools involved
How to develop a tool to assess the Eco-Schools programme?	February to June 2008	**Creation and development of the Eco-Schools assessment tool**	– D. Carlos I School / Nursery School – Colmeias School – Nossa Senhora do Rosário College – Francisco Arruda School 2,3
	November 2008 to January 2009	**Application of the final version of the Eco-Schools assessment tool**	– Évora Children's Activity Centre – Campo Flores College – Dr. Joaquim de Barros School – D. Manuel Martins Upper-secondary School

In accordance with the data collected during this stage of the research as "field notes", we continually revised the feedback form and user manual. It is worth highlighting that a scale of 1 to 5 was used, instead of the scale 1 to 10 used in the first versions, as we believed that the latter caused problems of objectivity in ranking schools. We decided to revise the scale used after discussion and analysis with the supervisor. We based this on the Likert scale, considering, both in this case and generally speaking, five response levels: poor, unsatisfactory, satisfactory, good, and very good. We then separated the part of the questionnaire that dealt with the ranking of the schools regarding to a number of ESD criteria from the feedback form. This then became an appendix to the above-mentioned feedback form. Considering the questions in this part are completed individually by the Eco-Schools coordinator and by the director of the executive board, unlike all other entries – which are the responsibility of the observer – we decided that it made no sense for them to be on the same page. As a result, the assessment form fitted on a single page. It included the "seven steps" items and "results" and was to be completed by an outside observer. Using this tool as a basis, the researcher visited the schools involved in the study.

Table II: Schools registered for the Eco-Schools programme for at least 3 years, and schools visited as part of the auditing process undertaken by the ABAE/National Eco-Schools Committee.

	Number of schools		
Region	Registered for the Eco-Schools programme for at least 3 years	Schools visited as part of the auditing process undertaken by the ABAE	**% of representation**
Azores	30	29	97%
Alentejo	6	1	17%
Algarve	7	0	0
Centre	71	23	33%
Lisboa and Vale do Tejo	57	14	25%
Madeira	16	16	100%
North	95	59	62%
Total	**282**	**142**	**50,35%**

After finalizing the auditing process within the National Eco-Schools Committee, we cross-referenced the data collected by the researcher and the observations made on the feedback forms by the DRE/DRA/SRAM observer. This data allowed us to arrive at the final version of the feedback form (see Annex) and user manual. To conclude our research, we visited four schools. Our objective was to test the final version of the feedback form and user manual, allowing us to evaluate their coherence, validity and reliability.

Results and Discussion

146 schools were visited (4 by the researcher and 142 by the DRE, DRA and SRAM teams), and it was possible to involve all the types of schools in the Eco-Schools network. The data collected by the researcher through interviews using open-ended questions were the object of a content analysis. The data was then cross-referenced with the notes taken by the DRE, DRA and SRAM teams, which then served as the basis for the final version of the feedback form, as well as the revision of the user manual, taking into account the type of schools and the criteria regarding the quality of work in the schools. On 16th September 2008, at a meeting of the National Eco-Schools Committee, the creation and development stage of the Eco-Schools feedback and assessment form was signed off.

Fig. 3: Eco-Schools programme evaluation form. See in Annex A the assessment form in its final version

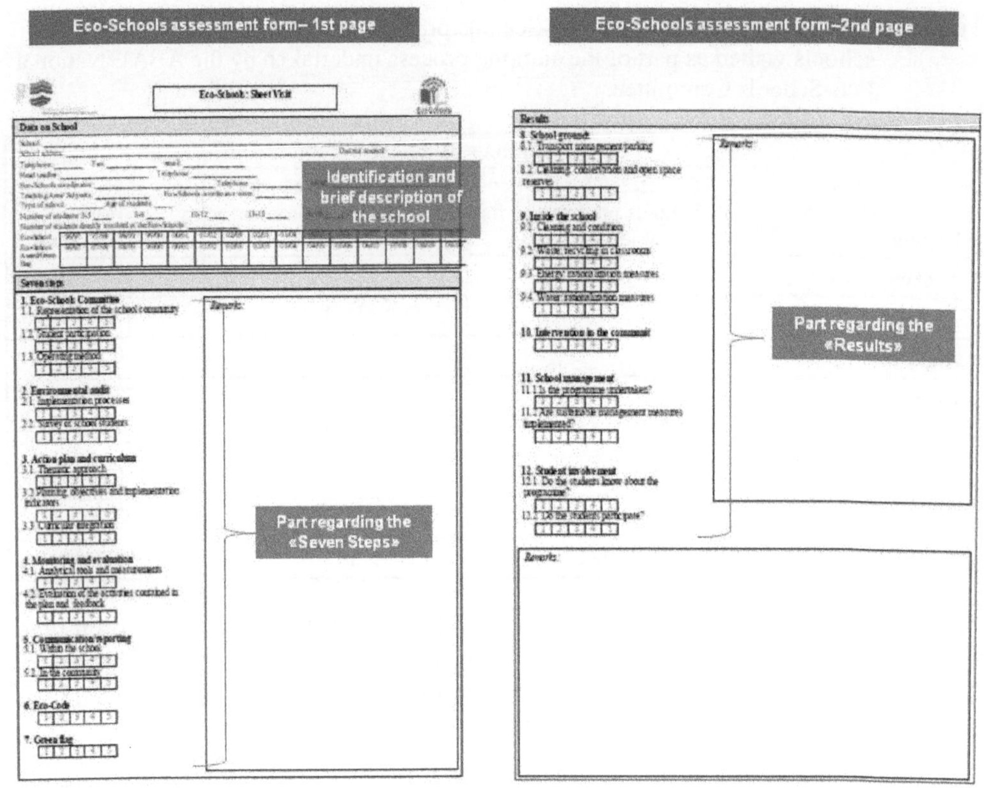

As a result of this process, we present our proposal for an Eco-Schools assessment tool. This consists of a feedback form (fig. 3 and annex A) and a user manual to help complete it. The feedback form is divided into two parts: "the seven steps" and the "results", which are divided into items and sub-items, as described in Table III. These are assessed and classified according to a scale of 1 (minimum) to 5 (maximum). The grade given to a school should be based on the feedback form user manual. In view of the way it was developed, we believe that this manual adheres to the principles of the Eco-Schools Programme via the seven steps, it is suitable for the Portuguese context and is well adapted to the type of schools in the Eco-Schools network.

Table III: Content of the assessment form.

Part	Item	Sub-item
«Seven steps»	**1. Eco-Schools Committee**	– Representation of the school community – Student participation – Operating method
	2. Environmental audit	– Implementation processes – Survey of school students
	3. Action plan and curriculum	– Thematic approach – Planning, objectives and implementation indicators – Curricular integration
	4. Monitoring and evaluation	– Analytical tools and measurements – Evaluation of the activities contained in the plan and feedback
	5. Communication/reporting	– Within the school – In the community
	6. Eco-Code	(no sub-item defined)
	7. Green flag	(no sub-item defined)
«Results»	**8. School grounds**	– Transport management/parking – Cleaning, conservation and open space reserves
	9. Inside the school	– Cleaning and condition – Waste: recycling in classrooms – Energy: rationalization measures – Water: rationalization measures
	10. Intervention in the community	(no sub-item defined)
	11. School management	– Is the programme undertaken? – Are sustainable management measures implemented?
	12. Student involvement	– Do the students know about the programme? – Do the students participate?

The Eco-Schools Assessment Tool in Excel

We recognize that implementing an assessment tool, such as the one presented in this study, can be time-consuming. As such, we have adapted it to an Excel spreadsheet, which we believe facilitates task completion and enables easier handling of data.

The spreadsheet includes formulas that allow the performance of the school to be quantified in each item. This value corresponds to the fraction between the points attributed to the school in each item and the maximum score in that field.

Table IV: Results of the assessment tool, in relation to the «seven steps" (graded from 1 to 5). The shaded boxes correspond to grades considered unsatisfactory.

The seven steps	Described/assessed items	Schools	EBI Dr. Joaquim Barros		Évora Children's Activity Centre		Campo Flores College		D. Manuel Martins Upper-secondary school	
			Grade	School performance	Grade	School performance	Grade	School performance	Grade	School performance
	1. ES Committee	1.1 Representation of the school community	3		5		4		5	
		1.2 Student participation	3	66.7%	5	93.3%	4	80.0%	5	87%
		1.3 Operating method	4		4		4		3	
	2. Environmental Audit	2.1 Implementation processes	4	70.0%	4	90.0%	4	70.0%	5	100%
		2.2 Survey of school students	3		5		3		5	
	3. Action plan and curriculum	3.1 Thematic approach	3		5		5		4	
		3.2 Planning, objectives and implementation indicators	3	73.3%	3	86.7%	4	93.3%	4	73%
		3.3 Integration into the curriculum	5		5		5		3	
	4. Monitoring and evaluation	4.1 Analytical tools and measurements	1	30.0%	3	70.0%	2	60.0%	5	90%
		4.2 Evaluation of the initiatives contained in the plan and feedback	2		4		4		4	
	5. Communication/ dissemination	5.1 Within the school	3	60.0%	4	70.0%	4	90.0%	4	80%
		5.2 In the community	3		3		5		4	
	6. Eco-Code	6. Eco-Code	3	60.0%	3	60.0%	2	40.0%	5	100%
	7. Green Flag	7. Green Flag	a)	–	2		4		5	

Table IV: Results of the assessment tool, in relation to the «seven steps" (graded from 1 to 5). The shaded boxes correspond to grades considered unsatisfactory.

	Described/Assessed items \ Schools		EBI Dr. Joaquim Barros		Évora Children's Activity Centre		Campo Flores College		D. Manuel Martins Upper-secondary school	
			Grade	School performance	Grade	School performance	Grade	School performance	Grade	School performance
Results	8. School grounds	8.1 Transport management/parking	2	50.0%	2	70.0%	3	80.0%	5	90%
		8.2 Cleaning and green spaces	3		5		5		4	
	9. Inside school	9.1 Cleaning and condition	4	55.0%	5	75.0%	5	80.0%	5	95%
		9.2 Waste: recycling in classrooms	3		5		5		5	
		9.3 Energy: rationalization measures	2		2		3		5	
		9.4 Water: rationalization measures	2		3		3		4	
	10. Intervention in the community	10. Intervention in the community	2	40.0%	3	60.0%	2	40.0%	3	60%
	11. School management	11.1 Is the Programme undertaken?	2	40.0%	2	60.0%	2	50.0%	3	80%
		11.2 Sustainable management measures implemented	2		4		3		5	
	12. Student involvement	12.1 Do the students know about the programme?	3	70.0%	5	100.0%	4	80.0%	4	80%
		12.2 Do the students participate?	4		5		4		4	

Implementation of the Final Version of the Eco-Schools Assessment Tool

To validate our research, we visited four schools between November 2008 and January 2009 to pilot the assessment tool. For data collection purposes, the Eco-Schools coordinator was interviewed and then, at a later date, we visited the school in their company in order to verify its performance. The results are presented in Tables IV and V.

At this stage of the research, there were no problems applying the assessment tool in the four schools visited. We also considered that the criteria defined in the manual was helpful in objectively classifying the school's performance for each of the items. The analysis of these results leads us to highlight the following aspects:

- As a tool for SA21, the participation of all sectors of the school community was significant in improving the quality of the physical and social environment of the schools visited. However, there are indicators of very irregular involvement of non-teaching staff and parents in some of the schools visited, not allowing a shared experience among all school stakeholders and which is essential for creating collective responsibility (Bonito et al. 2007).

  With regard to the procedures adopted by the ES Committee in the schools involved, rarely do they involve the participation of the various members of the community. In addition to this, activities such as exhibitions, open days etc. are important ways of promoting the project to the community, but are characterized by the relative unimportance of the participatory processes involving the community. Within this context, advantage should be taken of the fact that the National Commission of UNESCO has defined this area as one for mobilization/participation. For this reason, it is necessary to identify and mobilize all important institutions, target groups and agents to ensure the success of the initiatives undertaken as part of the ES programme, taking into account the considerable lack of awareness, information, training and education for sustainable development in the country (CNU 2006; IA 2006).
- When questioned about the action plan and themes that the school focuses on, references are very often made to participation in other projects, some of which are undertaken in close coordination with the Eco-Schools Programme. Some examples of this include the «Young reporters for the environment[1]», «The Environment and Innovation Project[2]» and «Escola solar Rock

1 *http://www.youngreporters.org/.*
2 *http://www.eco-schools.org/innovation2008/index.php.*

in Rio'08[3]», among others. In many schools, we see that the ES programme functions as a core and unifying factor in the work done in the school in EE, encouraging participation in many projects. Consequently, this makes it possible to introduce other concepts, activity areas and processes related to EE such as education, training, communication and environmental awareness, environmental values and ethics, changes in behaviour, environmental activism and citizenship, resources, and environmental information and data (Giordan 1996; Gaudiano 2006; Leal 2008).

The transversal nature and complexity of environmental and sustainability issues mean that there is no «one way» to achieve these goals and, most likely, one of the forms of achieving it is via the involvement of schools in various projects.

- In terms of the Action Plan, we found that social and economic aspects were largely ignored in terms of diagnosis and the action plan. We believe that this flaw should be analyzed and discussed at various levels, from schools to national/international coordination. Although the ES programme is an SA21 implementation tool, it is not possible to work towards sustainability with a foundation that is almost always sustained on only one of its pillars; in this case, the environmental component. This imbalance can prevent schools from being real centres of information production and dissemination regarding education for sustainable development (Gomes et al. 2007; Mogensen and Mayer 2005).
- In some items, there seems to be poorer performance from schools, with grades as low as level 1 and/or 2. In particular, we identified that:
  - Formal and regular monitoring processes are rare. Of all the schools visited, only D. Manuel Martins Upper Secondary School collects periodic data regarding the consumption of particular resources, such as electricity, water, gas, batteries, etc. ABAE needs to make an effort so that schools improve their performance in this area.
  - The strategies on how the eco-code is implemented need to be improved. In the majority of schools studied, an illustrated poster is designed with a number of sentences that encourage certain types of behaviour. However, there is little in the way of reflection, discussion and communication of these behaviours before making the said posters.
  - In relation to "results", waste is the issue with most obvious impact in daily school life. For example, what is noticeable in the majority of schools is the number of separate waste bins in classrooms and playgrounds, as well as posters raising awareness about this issue. On the other hand, «transport

---

3 *http://rockinrio-lisboa.sapo.pt/.*

management/parking» and «energy rationalization measures» are issues where results are less evident. Regarding the former, schools often justified this by pointing out that they had no authority to deal with problems associated with this issue; in relation to the latter, the reason most often cited was a lack of funds to install certain equipment.

Generally, the data presented in this study are similar to those of the study undertaken by Nave and Guerra (2006), who attempted to describe what was actually done in ES on a national level. In an analysis that involved around 2,300 schools, these authors identified waste as the second most popular choice (the first being conservation), corresponding to 42% of all work done, while the issue of transport/sustainable mobility was not mentioned by any entity.

It is also worth mentioning that in relation to energy (point 9.3) the results obtained do not correspond, in practice, to the recognition of the ES programme in promoting measures to reduce/save energy expressed by the school community.

- Intervention in the community should be a subject for reflection. As the ES programme is an integral part of LA21, schools must be an example in their communities, and coordination between the various members of the educational community should be consolidated. By doing so, there is a greater sense of integration. The data presented herein could imply that work done on the ES programme is very much focussed on the school community. In other words, the school takes an introspective view. We agree with Fidélis' ideas (2006) when she highlights the role of schools as a community that influences not only itself, when teaching, but also the community it is associated with.
- The item «Programme undertaken?» also leads us to reflect upon the importance of the programme in the day-to-day operations of the school. The ES programme is seen by the school executives as important for the efficient management of resources, as well as for the development of citizenship education. However, the time needed to organize the various initiatives and activities throughout the academic year is not made available, and the money required to develop the project is limited. It is also rare to see the «Eco-Schools» logo used on school documents/correspondence. Despite the recognition of the importance of the ES programme, it seems that there is little match between what is espoused and practical application.

According to the data collected, we can say in general that the ES programme is seen as a good tool for promoting citizenship education (Gomes 2004a, 2004b; Santos 1997). On this issue, Mogensen and Mayer (2005) highlight a strong link between integration in the curriculum and the practice of citizenship education, which leads us to believe that the ES methodology contains aspects that encourage education for sustainable development.

The Assessment Tool as a Mechanism for the Self-Assessment of Work in Schools

Assessment is considered to be one of the most important stages in any EE project. It forms part of a process of innovation and training, providing more complete feedback on practice (Benavente 1993; Giordan 1996; Giordan and Souchon 1997).

We have devised an assessment tool for the ES programme and believe this should be primarily geared towards schools as a mechanism for self-assessment, providing data that contribute to continuous improvement of the work undertaken. Secondly, there is the possibility of external assessment.

We believe that the tool proposed in the study reported herein should be made available for all ES via direct download on the ABAE webpage. In this way, whenever the school deems fit, it can use the form and user's manual as a way of identifying possible errors, regulate certain processes and/or foresee certain operations. Generally speaking, the main aims of implementing a process of self-assessment are: to improve schools' self-awareness and the awareness of the activities that take place within them; to improve decision-making processes; to produce systematized information; to promote a culture of dialogue and develop competences in the area of assessment.

Each school should create their own internal assessment mechanisms. However, we believe that they should be transparent, clear, continuous and participative, involving various members of the school community, thus stimulating a reflective culture that strives for permanent improvements.

Final Considerations

Following the study reported herein, it was possible to create an assessment tool that we believe, due to the methodology applied, can contribute to improving the work done in schools. It is important to create a culture of reflection and self-assessment in relation to the work undertaken in schools. Quantifying the level of performance of schools in various areas, such as results, is a necessary process that aims to assess the relevance of certain measures and their efficiency in particular. This assessment can be particularly useful for the bodies that provide funding for the implementation of certain work, such as schools, local authorities, and companies. As a result, the quality of the work improves and conditions are created so that all members of the community feel involved in the project.

Furthermore, the creation of a decision-making process that involves all stakeholders is enhanced, which generates collaborative solutions that are more effective and efficient for the environmental management and sustainability planning of schools (Vasconcelos 2001). In terms of the national Eco-Schools network, the implementation of such a process helps the national operator detect failures and/or successes, which should then be corrected/improved or communicated, as a way of encouraging better school performance. At the same time, this encourages higher standards in the work done by schools, as well as providing information that makes the green flag award process more objective and fairer, allowing the work of each school to be more effectively differentiated.

Within the context of a new relationship with the environment and, consequently, the development of environmental citizenship, schools should be able to encourage an attitude of critical vigilance within schools and the community regarding policies and practices that go against ecological values. Through the ES programme, we believe that schools can become a paradigmatic example to the institutional and legal powers involved in the decision-making process.

Ensuring long-term sustainability requires the involvement of everyone sharing an attitude of active intervention and responsible behaviour. We believe that the ES programme is more than a contribution to Agenda 21 and part of effective education for sustainable development.

References

ABAE (2008), Eco-Schools Program. Associação Bandeira Azul (http://www.abae.pt/programa/EE/descricao.php – accessed 26/11/2007).

Bardin, L. (2004), Análise de Conteúdo. Edições 70. 3ª ed. 223 p.

Benavente, A. (1993), Mudar a escola, mudar as práticas – Um estudo de caso em educação ambiental. Cadernos de Inovação Educacional. Lisboa: Texto Editora.

Bonito, J., Cid, M. and Matos, G. (2007), O significado dos rótulos alegadamente «ecológicos» presentes em produtos comercializáveis em Portugal: um estudo com alunos universitários. in Educação. Temas e Problemas. Nº 3. Edições colibri, pp. 125-155.

Breiting, S., Mayer, M. and Mogensen, F. (2005), Quality Criteria for ESD-Schools: Guidelines to enhance the quality of Education for Sustainable Development. Austrian Federal Ministry of Education, Science and Culture (http://www.ensi.org/Publications/ENSI_related_publications – accessed 21/09/2007).

Carmo, H. and Ferreira, M.M. (1998), Metodologia da investigação. Guia para auto-aprendizagem. Universidade Aberta.

Comissão Nacional da UNESCO Portugal (CNU) (2006), Década das Nações Unidas da Educação para o Desenvolvimento Sustentável Contributos para a sua dinamização em Portugal (http://www.unesco.pt/cgi-bin/home.php – accessed 10/07/2006).

Fidélis, T. (2006), A Agenda 21 Local e a Educação para o Desenvolvimento Sustentável – sinergias e requisitos no contexto português. Comunicação apresentada no encontro nacional de Eco-escolas. Janeiro de 2006. Santarém.

Gaudiano, E. (2006), Educação Ambiental. Horizontes Pedagógicos. Instituto Piaget.

Giordan, A. (1996), A educação ambiental na Europa. 1ª ed. Cadernos de Inovação Educacional. Instituto de Inovação educacional. Instituto de Promoção Ambiental.

Giordan, A. and Souchon, C. (1997), Uma educação para o ambiente. 1ª ed. Instituto de Inovação educacional. Instituto de Promoção Ambiental.

Gomes, M. (2004a), Programa Eco-Escolas: Guia de auditoria ambiental. Fundação para a educação ambiental. Associação Bandeira Azul (http://www.abae.pt/programa/EE/guias.php – accessed 26/11/2007).

– (2004b), Programa Eco-Escolas: Guia do Professor. Fundação para a educação ambiental. Associação Bandeira Azul (http://www.abae.pt/programa/EE/guias.php – accessed 26/11/2007).

Gomes, M.C. (2006), Critérios de qualidade para escolas – EDS: guia para a melhoria da qualidade da educação para o desenvolvimento sustentável. 52 pp. Austrian Federal Ministry of Education, Science and Culture – Dept. V/11c, Environmental Education Affairs.

Gomes, M.C., Almeida, F. and Fernandes, S. (2007), Preparar o mundo para a mudança. Um exemplo de aplicação da Carta da Terra. Comunicação apresentada nas XIV Jornadas Pedagógicas de Educação Ambiente: Ambiente, Saúde e Qualidade de Vida. Lisboa, 26 e 27 de Janeiro de 2007 (www.aspea.org/XIV%20J%20-%20Aplicacao%20Carta%20Terra%20-%20Manuel%20Gomes.pdf – accessed 29/09/2008).

Instituto do Ambiente (IA) (2006) , Estratégia da CEE/ONU para a educação para o desenvolvimento sustentável. Tradução e Edição: Instituto do Ambiente.

Leal, R.M. (2008), Educação e formação ambiental. Futuro sustentável – plano estratégico de ambiente do grande Porto (www.futurosustentavel.org/fotos/ambiente/Educacao_e_Formacao_Ambiental_Rui_Leal.pdf – accessed 2/05/2008).

Mogensen, F. and Mayer, M. (eds.) (2005), Eco-schools: trends and divergences. A comparative study on Eco-school development processes in 13 countries. Austrian Federal Ministry of Education, Science and Culture. SEED.

Nave, J. and Guerra, J. (2006), Educação Ambiental: balanço, oportunidades e perspectivas. OBSERVA, ISCTE-ICS. Seminário anual Eco-Escolas em Santarém. 12 Janeiro de 2006.

Santos, E. (1997), Eco-Escolas. Aprovadas com distinção. in Cadernos de Educação Ambiental. nº 3. Agosto 1997. Instituto de Promoção Ambiental.

Tuckman, B.W. (2000), Manual de Investigação em Educação. Fundação Calouste Gulbenkian. Lisboa.

Vasconcelos, L. (2001a), O Envolvimento do Cidadão na Conservação da Biodiversidade – Rumo à Democracia Deliberativa. 2º Congresso Nacional da Conservação. 2, 2001, Lisboa. Actas. Lisboa: Fundação Calouste Gulbenkian, Outubro de 2001.

Annex A: Assessment form

Eco-Schools: Sheet Visit

Eco-Schools

Data on School

School: __________
School address: __________ District council: __________
Telephone: _____ Fax: _____ email: _____
Head teacher: __________ Telephone: _____
Eco-Schools coordinator: __________ Telephone: _____ email: __________
Teaching Area/ Subjects: _____ Eco-Schools coordinator since: _____
Type of school: _____ Age of students: _____
Number of students: 3-5 _____ 5-9 _____ 10-12 _____ 13-15 _____ 16-18 _____ Total: _____
Number of students directly involved in the Eco-Schools:

Eco-School	96/97	97/98	98/99	99/00	00/01	01/02	02/03	02/03	03/04	04/05	05/06	06/07	07/08	08/09	09/10
Eco-School Award/Green flag	96/97	97/98	98/99	99/00	00/01	01/02	02/03	02/03	03/04	04/05	05/06	06/07	07/08	08/09	09/10

Seven steps

1. Eco-Schools Committee

1.1. Representation of the school community

1	2	3	4	5

1.2. Student participation

1	2	3	4	5

1.3. Operating method

1	2	3	4	5

2. Environmental audit

2.1. Implementation processes

1	2	3	4	5

2.2. Survey of school students

1	2	3	4	5

3. Action plan and curriculum

3.1. Thematic approach

1	2	3	4	5

3.2 Planning, objectives and implementation indicators

1	2	3	4	5

3.3. Curricular integration

1	2	3	4	5

4. Monitoring and evaluation

4.1. Analytical tools and measurements

1	2	3	4	5

4.2. Evaluation of the activities contained in the plan and feedback

1	2	3	4	5

5. Communication/reporting

5.1. Within the school

1	2	3	4	5

5.2. In the community

1	2	3	4	5

6. Eco-Code

1	2	3	4	5

7. Green flag

1	2	3	4	5

Results

8. School grounds

8.1. Transport management/parking

1	2	3	4	5

8.2. Cleaning, conservation and open space reserves

1	2	3	4	5

9. Inside the school

9.1. Cleaning and condition

1	2	3	4	5

9.2. Waste: recycling in classrooms

1	2	3	4	5

9.3. Energy: rationalization measures

1	2	3	4	5

9.4. Water: rationalization measures

1	2	3	4	5

10. Intervention in the community

1	2	3	4	5

11. School management

11.1 Is the programme undertaken?

1	2	3	4	5

11.2 Are sustainable management measures implemented?

1	2	3	4	5

12. Student involvement

12.1. Do the students know about the programme?

1	2	3	4	5

12.2. Do the students participate?

1	2	3	4	5

Remarks:

Remarks:

Chapter 8

Education for Sustainable Development in Primary Schools: Contributions from the Development of a Teacher Training Program

Patricia Sá, Isabel P. Martins

Abstract

At the beginning of this new century, characterized by complexity, unpredictability and uncertainty, human beings face severe and multiple socio-economic and environmental problems. The Decade of Education for Sustainable Development, proclaimed by the United Nations General Assembly for the period 2005-2014, launched the challenge of an education based on a correct perception of the current planet situation and oriented towards solidarity, social equity and environmental sustainability. International Education for Sustainable Development (ESD) guidelines and orientations emphasize the need for: life-long learning; formal, non-formal and informal education; strategies, activities and resources diversity and inter and trans-disciplinary approaches. Teachers, of all educational levels, are key elements to achieve these goals. However, national and international literature (Freitas 2004, 2006; Ko and Lee 2003; Martins 1998; Praia, Edwards, Gil-Pérez and Vilches 2001; Sá and Martins 2005) shows that teachers' perceptions regarding the current planet situation tend to be fragmented and superficial. The importance of teacher training programmes in the field of ESD is accentuated by different authors (Arima, Konaré, Lindberg and Rockefeller 2004; Fien and Tilbury 1998; Freitas 2006; Gil-Pérez and Vilches 2005, 2006; Hesselink, van Kenpam and Wals 2000; Hopkins and McKeown 2001, 2005; Pedrosa and Leite 2005; Pellaud 2001, 2002, 2003; Scoullos 2004), highlighting the importance of teacher education to consistent practice (re)orientation. Being aware that teachers' perceptions are fundamental to practice regulation and that ESD should begin from the very early years of schooling and continue throughout life, research has been conducted based on the development of an ESD training programme for primary school teachers. This chapter focuses on the aforementioned research, presenting the teacher training programme conception, implementation and evaluation and sharing some conclusions.

Introduction

In the world of today with evident and profound imbalances, conflicts that disrespect human rights, humans having to face multiple socio-economic and environmental problems, the proclamation of the Decade of Education for Sustainable Development (DESD) for the period 2005-2014 by the United Nations General Assembly in December 2002, comes as the maximum recognition of the situation of planetary emergency that humanity is facing, emphasizing education as "… *an indispensable element for achieving sustainable development*" (Arima, Konaré, Lindberg and Rockefeller 2004: 7).

At the beginning of this new century, the DESD launched the challenge of an education for solidarity, based on a correct perception of the planet situation and promoting responsible attitudes and commitments which are both socially fair and environmentally sustainable. Although education on its own is not sufficient to achieve these attitudes, it has come to be recognized in several documents under the responsibility of UNESCO (United Nations Decade of Education for Sustainable Development 2005-2014: Draft International Implementation Scheme (2004); Guidelines and Recommendations for Reorienting Teacher Education to Address Sustainability (2005); the United Nations Decade of Education for Sustainable Development (2005-2014) – contributions for its implementation in Portugal (2006); …) and by various authors (Fien and Tilbury 1998; Freitas 2006; Gil-Pérez and Vilches 2005, 2006; Hesselink, van Kenpam and Wals 2000; Hopkins and McKeown 2001, 2005; Pedrosa and Leite 2005; Pellaud 2001, 2002, 2003; Scoullos 2004; Tilbury, Stevenson, Fien and Schreuder 2002; Varga, Kószó, Mayer and Sleurs 2007; Vilches and Gil-Pérez 2003), as an essential prerequisite for the promotion of a change to individual and collective attitudes and, consequently, to the current critical situation.

To achieve the desired goal, it is vital that Education for Sustainable Development (ESD) is implemented from the very early years of schooling, within a framework of life-long learning, in contexts of formal, non-formal and informal education, considering content, contexts, strategies and competences within a multi, inter and trans-disciplinary perspective and highlighting the interactions between the key areas of sustainability – environment, society and economy (Hopkins and McKeown 2001, 2005). In this way, all educators are vital to encourage understanding of the problems we face. The awareness of human beings' responsibility in the current situation facing our planet and the importance of individual actions to reverse the situation, through the exercise of responsible citizenship, are important issues to work from the first years of schooling.

The importance of teacher training programmes in the field of ESD can be seen in all the above-mentioned documents, and UNESCO has emphasized that "*the ongoing training of teachers should take into account professional skills in the field of education for sustainable development, recommending that all training*

(...) from 2010, should include an education module for sustainable development" (UNESCO 2006a:18).

In this context, the perceptions that teachers have are fundamental in their pedagogical practice regulation. However, in terms of the planet situation, these perceptions have been fragmented and are superficial, showing lack of understanding of the problem in a global context (Freitas 2004, 2006; Ko and Lee 2003; Martins 1998; Praia, Edwards, Gil-Pérez and Vilches 2001; Sá and Martins 2005). On the other hand, being aware that ESD should begin from the very early years of schooling and continue throughout life and in the various contexts of learning, research was conducted involving a training programme for primary teachers focusing on ESD.

This chapter presents the aforementioned research, the conception and development of the teacher training programme, as well as its implementation and evaluation, and shares some conclusions.

Research Presentation

The main purposes of this research were: to identify primary school teachers' conceptions concerning science and technology, citizenship and sustainable development; to conceive, develop and evaluate a primary school teacher training programme focusing on ESD; to conceive and produce didactical resources based on ESD guidelines and recommendations, and to identify primary school teachers' training needs concerning ESD practice (re)orientation.

To pursue these aims, the research was organized in four phases, as presented in Table 1.

Table 1: Research phases

Phase I	Phase II	Phase III	Phase IV
Theoretical framework establishment ↓ Content analysis	Identification of primary teachers' conceptions concerning science and technology, citizenship and sustainable development ↙ ↘ Inquiry / Content analysis	Teacher training program: conception and development – Themes definition; – Didactical resources development and validation; – Strategies and activities selection	Teacher training program: implementation and evaluation – Researcher-trainer evaluation; – Participating teachers' evaluation.

During Phase I, national and international ESD guidelines were analyzed and compared. Besides establishing a theoretical framework, this analysis allowed a national primary school curriculum review searching for ESD orientations.

Phase II focused on the identification of primary schools teachers' conceptions concerning science and technology, citizenship and sustainable development. Participating primary school teachers' conceptions were identified by inquiry. For this particular research, a questionnaire was conceived and validated in order to identify primary school teachers' conceptions. The aim was to obtain the same information from all the research participants.

The questionnaire conceived had twenty-three multiple-choice questions, organized in four groups. The first group of questions intended to characterize the participating teachers regarding aspects like, for instance, age or academic degree. The aim of the second, third and fourth groups of questions was to identify primary school teachers' conceptions about science and technology, citizenship and sustainable development respectively.

This questionnaire was administered to one hundred primary school teachers. An interview was conducted among eight of the one hundred primary school teachers who responded to the questionnaire. The interview purpose was to clarify some of the conceptions identified during the analysis of the questionnaire data collected.

Phase III focused on the teacher training programme conception and development. To achieve the stated intention, the theoretical framework (Phase I) and the identified primary school teachers' conceptions (Phase II) were taken into consideration. The themes, strategies, didactical resources and activities that integrated this teacher training programme were defined according to the two previous research phases and will be presented in section 2 of this chapter.

Phase IV was committed to the implementation and evaluation of the teacher training programme.

The training programme was implemented in a private educational school with a group of primary school teachers. This group was composed of eight members, seven female and one male. Their ages were between twenty-four and forty-nine years, and only two had less than five years of service. For each year of schooling in this primary school, there were two classes and, consequently, there were two teachers allocated to each year (one per class): two taught 1st year, two taught 2nd, two taught 3rd, and two taught 4th. Their initial training was of a diverse nature: three out of the eight teachers graduated in elementary education – primary or equivalent; four received their vocational training through a Primary Diploma course and one completed secondary school and obtained a diploma for education in the private sector.

The teacher training programme evaluation involved the researcher-trainer and the participating teachers. This evaluation was done at two distinct moments (during implementation of the teacher training programme and at the end) and supported by different assessment techniques and instruments. The documents used for this analysis were mainly notes from direct observation of the workshop sessions and participating teachers' answers to a questionnaire. Evaluation issues will be developed further in section 3 of this chapter.

Teacher Training Programme: Conception and Development

To conceive, plan and develop this teacher training programme, the recommendations of recognized authors, scientific associations and national and international organizations were analyzed with reference to the role of basic education in the promotion of ESD, as well as the guidelines recommended for teacher training. In addition, educational policy guidelines regarding ESD as applied to intended curricula, primary school programmes and textbooks were also taken into consideration for this level of schooling.

In addition to the educational policy guidelines, primary school teachers' conceptions about science, citizenship and sustainable development were also identified and analyzed.

The results of this analysis gave rise to the definition of both the orientation of the training programme, as well as the content, strategies and resources to focus on.

Recommendations for ESD Implementation

As far as the recommendations put forward by national and international documents for the implementation of ESD are concerned, a large degree of consensus was noted concerning the principal guidelines. In short, a problematic approach within a framework of ESD should:

i) promote the approach of the content within a multi, trans and interdisciplinary perspective (Arima et al. 2004; Morin 1999; Summers et al. 2005; Scoullos and Malotidi 2004; Tilbury et al. 2002; Vargas 2000);
ii) be based and guided by values of respect, solidarity and cooperation (Arima et al. 2004; Ospina 2000; Tilbury et al. 2002);
iii) be supported by diverse methodologies, favouring active methodologies which are diversified and suitable for the contexts, and topics to be worked on (Arima et al. 2004; Ballantyne et al. 2001; Hopkins and McKeown 2005; Huckle 2006);

iv) be guided by democratic principles and processes (Fonseca 2000; Imbernón 2007; Imbernón et al. 2002; Santos 2005a, 2005b);
v) promote scientific and technological problem dimensions understanding (Pedrosa, Gonçalves, Henriques and Mendes 2003; Pedrosa and Leite 2004; Vega-Marcote, Freitas, Álvarez, Suaréz and Fleuri 2007);
vi) encourage questioning and debate (Arima et al. 2004);
vii) make available didactical resources which are relevant and suitable both for the topics to be worked on as well as for the targeted population (Arima et al. 2004; Hesselink et al. 2000).

With regard to teacher training programmes, recommendations are: provision and/or creation of didactic resources to accompany topics of relevance; demonstration of strategies that promote the development of decision-making capacities and encourage the formulation of questions; challenge students to be pro-active and participate in decisions which affect them; include the discussion of issues related to social equality, tolerance and discrimination, and provide students with opportunities to reflect on their own daily values and attitudes, as well as the impact they have at both local and global levels (UNESCO 2006a, 2006b).

In Portugal, since 2001, there has been a concern for curriculum (re)structuring based on guidelines oriented towards ESD.

The document, which lays down the curricular guidelines for primary education – National Curriculum for Primary Education – Basic Competences (ME-DEB 2001), details ideas and principles which underlie ESD (Pedrosa and Leite 2004) recognized in various national and international documents, and which are summarized above. This document, as well as demonstrating that "*an integral part of curriculum is the approach of themes which cut across several disciplinary areas, particularly in the context of education for human rights, environmental education and health education and welfare*" (ME-DEB 2001: 10), defines the topic "Sustainability on Earth" as one of four organizing themes of science teaching in the three cycles of basic education. This clearly growing importance of the theme is reinforced when the document outlines the intention that it is associated with: "*In the third theme – Sustainability on Earth – it is intended that students become aware of the importance of taking action with regard to the system Earth so as not to cause imbalance, and in this way contributing to a careful management of existing resources*".

Total autonomy is also given to the teacher with regard to curriculum management, and teachers from different subject areas are encouraged to do joint lesson plans.

Primary School Teachers' Conceptions

Data collection of primary school teachers' conceptions regarding science and technology, citizenship and sustainable development was done by two instruments – a questionnaire and an interview – designed and validated within the scope of the research that is being presented.

The analysis of the data collected has allowed us to identify the issues that the teacher respondents considered appropriate for working with their students, as well as how to characterize their conceptions. Thus, regarding:

i) *Science and technology*. Most teachers surveyed, 90%, consider working with students on issues related to science and technology. The aspects mostly emphasized by teachers in the survey, both in the questionnaire and in the interview, concern the advantages and limitations of science and technology use. The increase in comfort and quality of life offered by the use of knowledge and scientific-technological tools was the most often cited advantage, and the environmental impact of their use was the most emphasized limitation.

   However, these same teachers expressed a belief in the possibility of unlimited use of science and technology in addressing the major problems of today, which once again, focus on the environment. Its use in the control of environmental contamination and in the search for alternative energy sources were the examples mentioned most often. This duality reveals a paradox in the image of science and technology shared by these teachers: science and technology are at the same time both the cause and the solution to our most serious current problems.

   Teachers in the survey have established a rather vague relationship between science and technology use and the possibility of resolving or minimizing the impact of the main problems we currently face. The sense of responsibility attached to science and technology for the serious and widespread impact of economic growth is well known. This assignment of blame leads to a lack of individual and collective responsibility through our contemporary society's consumer choices and their impacts. There appears to be a lack of awareness that the use of scientific and technological artifacts on a massive scale is carried out by human beings and that we make our own decisions on the basis of a benchmark social paradigm, which underlies the framework of values that guides attitudes and choices. This handing over of responsibility, seen in many other studies, is an easy simplification of the role we humans play in our current situation on the planet. By pointing the finger at science and technology, human beings offload all sense of blame, and thus, all responsibility of both the current situation as well as resolving it. In this way, the belief persists in the possibilities of resolving current problems that science and technology will discover at some stage in the future.

Following the alternatives chosen by the teachers in the questionnaire and the replies given during the interviews it was also possible to perceive the coexistence of naive conceptions about the nature of science and scientific knowledge with conceptions which are closer to the current paradigm. The teachers' group considered science to be: i) a knowledge in construction; ii) a product of human activity which influences and is influenced by contexts; iii) an objective knowledge, constructed on the basis of the application of the scientific method and iv) a knowledge based on evidence and which leads to the truth.

ii) *Citizenship*. Most teachers surveyed, 98.8%, considered using lesson strategies which encourage students' participation skills. The most common engagement strategies referred to by these teachers were the selective separation of waste material, the implementation of measures to save water and the holding of class assemblies, identified as an opportunity for dialogue about solving problems (the "problems" most often mentioned by these teachers refer to student behaviour issues and need for conflict management). These results coincide with those found by Thomaz (2007), where growing importance was given by future primary teacher trainees and their supervisors to teaching strategies and hands-on education for citizenship framed within an experiential approach. The strategies most often reported by teachers participating in the above-mentioned study (Thomaz 2007) are consistent with those that we found, and refer mainly to: i) democratic classroom rules; ii) class assembly; iii) shared management and resolution of conflicts through teacher/student and student/student dialogue; iv) daily plan; v) group work; vi) project work and vii) debate.

In this study it was also possible to identify among the teachers questioned, through a triangle of questionnaires, interviews and results, a clear sense of the growing importance of contributing towards environmental protection. These data support results emerging in the context of other research: on the one hand, the recognition and valuation of environmental issues, reflecting a fragmented view of reality, which in itself represents an obstacle to the necessary holistic and systemic approach to current issues (Freitas 2004; Gil-Pérez et al. 2000a, 2000b, 2003; Gil-Pérez and Vilches 2005, 2006; Martins 1998; Praia et al. 2001). On the other hand, this overvaluation also reflects a concern that teachers have in relation to the environment and the damage to which it is liable. This concern towards the environment amongst teachers has also been found in studies carried out by Borges, Duarte and Silva (2007).

However, the majority of teachers in this survey did not make any reference to teaching strategies planned specifically to encourage: i) awareness of students to some of the problems existing locally and/or globally; ii) consideration of the interaction that has been established at various levels for the problems identified; iii) reflection on their causes and consequences and iv) discussion and implementation of measures appropriate both to the problem in question and to the development and autonomy levels of the students involved.

We, therefore, conclude that Education for Citizenship is not reflected in a cross-curricular way, but is dealt with sporadically and remains on the sidelines of the remaining course contents. On the other hand, the prospect of participation resulting from the discourse of this group of teachers is circumstantial (because it is occasional), locally limited (focuses on school) and is decontextualized.

iii) *Sustainable Development.* The current problems that surveyed teachers acknowledged as being most serious and related to sustainable development concept have essentially to do with environmental issues. Widespread pollution, water shortages, depletion of the ozone layer and the greenhouse effect were some of the most common problems referred to by these teachers, both in the questionnaire as well as during the interviews. However, issues such as population growth and distribution and imbalances in consumption were not mentioned by any of them at all.

These results coincide with those of other research mentioned in the literature. For instance, a low percentage of teachers point out aspects such as overconsumption in developed societies, population explosion on a planet which is limited and finite in terms of space and resources or human rights (Edwards et al. 2001; Gil-Pérez et al. 2000, 2003; Praia et al. 2001). On the other hand, teachers tend to show fragmented and incomplete perspectives of current problems that society faces, in most cases focused almost exclusively on issues of environmental contamination and depletion of natural resources (Edwards et al. 2001; Freitas 2004; Gil-Pérez et al. 2000, 2003; Praia et al. 2001).

After the identification and characterization of educational policy guidelines present in international and national documents which regulate ESD-driven pedagogical implementation, and characterization of primary school teachers' conceptions about science and technology, citizenship and sustainable development, we come to the development of the training programme itself. This presentation will be carried out on the basis of defined stages, selected strategies and resources which have been designed and put into practice.

Teacher Training Programme Stages, Strategies and Resources

The teacher training programme took the form of a workshop and was held over three distinct periods or stages: i) the first stage was essentially one of raising teachers' awareness about the need and importance of ESD; ii) the second stage offered a context which was geared towards the (re)construction of ESD knowledge, and iii) the third stage was for participating teachers to implement EDS resources and methodologies with their students. The first two stages only involved the researcher-trainer and the teachers surveyed. In the 3rd stage these teachers implemented and participated in an activity related to ESD with their primary school students.

The following table sets out the didactic procedures and strategies adopted in the different stages of the teacher training programme.

Table 2: Procedures and learning strategies adopted in different stages of the teacher training programme

	Stage 1	Stage 2	Stage 3
Procedures	– Raising teachers' awareness of the importance of ESD.	– Teacher Training Program Implementation; – Analysis and reflection about consensual aspects related to pedagogical and methodological sessions of the Teacher Training Program.	– Knowledge transposition and mobilization for teaching practice.
Learning Strategies	– Reflection on the relevance of an EDS oriented teacher training program; – Discussion and reflection about the themes relevance – Analysis and discussion of issues to be included in the Training Program; – Reflection on the adequacy and relevance of the themes chosen for primary school.	– Debate activities; – Activities based on using teaching resources; – Documents analysis and discussion; – Clarification Sessions; – Group work.	– Debate activities; – Activities based on using teaching resources; – Documents Analysis and discussion; – Potential multiple resources exploitation demonstration; – Group work.

In accordance with the guidelines considered and following the nature of the training programme, a methodological pluralism was implemented with regard to learning strategies. In all the workshop sessions strategies such as debate, discussion, individual and group reflection and group work were used when carrying out the activities. The choice of these strategies was based on the principle of giving participants the opportunity for conceptual (re)construction based on previous experiences and knowledge.

Analysis, reflections and discussions were carried out by the working groups with a basis on theoretical texts (e.g. Gil-Pérez and Vilches 2006) and teaching resources designed and produced for this programme (e.g. Courseware SeRe. This digital didactical resource was developed by a multidisciplinary team, integrating several software typologies within the educational activities. This courseware allows teachers to work energy-related issues with their students.).

Another strategy used was the systematic exposure of facts and arguments about a few topics included in the programme (in particular the concept of sustainable development and its implications; the current situation on our planet; the ozone layer; the greenhouse effect and ESD), as well as the relationship established between them. This strategy choice is justified by the novelty of the focus of this programme and also by the complexity of a holistic adoption perspective by these teachers in relation to the programme and to its use. Participants felt the need to include in the programme opportunities to share information in a structured and systematic way. These occasions have enabled the exchange of views between the group of teacher trainees, as well as between them and the researcher-trainer; the acceptance or rejection of what was being discussed; the arguments to defend points of view, and the questioning, and fundamental procedures leading to understanding and acquiring information.

The use of Information and Communication Technologies (ICT) was another strategy used. New technologies have been used not only for courseware SeRe development, but also in the search, selection and organization of information in the activities involving research (for example, regarding the ozone layer depletion).

Demonstrations were also used throughout the programme, especially in stage 3, when teachers implemented the programme activities with their students. It was intended to demonstrate how resources could be used with primary school pupils, clarifying what teachers can do and for what purpose(s). We believe that using modelling and exploiting didactical resources in some training situations was not limiting for teachers. It is important that teachers can see how to make use of didactical resources not conceived by them, and which call for a change in their teaching practices. "*In the training program curricular materials play a central role; they are extensively used for the purposes of demonstration and practice. These materials act as a means of support for teachers to view the change that is desired, and not as something that will limit or control them*" (Berg 1997: 4).

Another training strategy used was the provision of books, journals, articles and websites related to ESD and to the topics arising from each of the workshop sessions. It was felt that in this way it would facilitate teachers' access to sources of information while respecting individual interests and training requirements.

Didactical resources development enables a dialogue between theory and the innovative practices that are aimed at (Praia and Cachapuz 1999). It is essential to include the design, production and exploitation of didactical resources during the teacher training process if a change in teaching habits is the goal. This is supported by Martins (2000), who considers that teacher training programmes "*... should include pathways of exploration into the design, development and evaluation of didactic resources and teaching strategies, ...*" (p. 107). Also Ellis (1995 in Vieira 2003) believes that teachers should have the opportunity to explore new materials and teaching strategies. However, he emphasizes the need for individual monitoring in the classroom.

In this study the didactical resources were designed and developed by the researcher-trainer, whose primary aim was a contextualized, appropriate and creative approach of the issues defined for the teacher training programme. Flexibility of use with regard to the context and the exploitation were also factors taken into account in the design phase.

The design of these resources, as well as their use, were founded on a holistic and dynamic perspective of the sustainable development concept, whereby this is understood as a process of change driven by a set of values and/or principles in which the interdependence between human and non-human aspects is recognized.

As each didactical resource was developed for a specific theme of the teacher training programe, these will be presented with direct reference to the corresponding session. The following table identifies the didactical resources used:

Table 3: Identification of the didactical resource according to session topic

Theme/topic for the session	**Didactical resource produced**
Sustainable Development	The Bean Game
World Human Population	Us and the Planet
Human Rights	Human Rights Cards
Ecological Footprint	What is the size of your footprint?
Life Cycles Analysis	Let's have a picnic?
Human Impact on Natural Resources	Courseware SeRe
Ozone Layer Depletion	Ozone Layer Kit

Teacher Training Programme Organization

The teacher training programme in question was based on teacher training guidelines, on reflection purposes definition for each workshop session and on didactical resources exploitation as a starting point for the work to be done with the teachers. For each session, activities were defined, training strategies were selected, resources were developed, and support documents for each topic were selected and/or created. All these decisions were based on the analysis of teaching practice guideline documents and on the identification of the conceptions and needs of the participating teachers mentioned above.

Each session followed a similar structure: first the topic to be worked on was put into context, and then the activity planned and prepared by the researcher-trainer was put into practice, and finally, there was a synthesis/reflection stage which takes place in a large group and is mediated by the researcher-trainer.

In fact, to promote participation, the contextualization provided the opportunity to raise teachers' conceptions about the ongoing topic, as well as an opening for working sessions in an environment of sharing, dialogue, openness and trust. After the contextualization of each topic an activity was proposed, aimed at stimulating the involvement of every teacher, whether it was through personal involvement (individual work) or through interpersonal relations necessary for the development of a working group (in a small or a large group).

Activities were organized based on the didactical resources designed for each session and/or the analysis of available documents. This should be done individually or in a small group at an early stage of the session, and then, later, they should discuss in a large group not only the procedures adopted by each person, but also the conclusions reached. All sessions ended with a period of synthesis/reflection involving the whole group and were moderated by the researcher-trainer. The primary objectives of this moment of reflection were: to emphasize the meanings of those ideas that had come from the sessions; to reflect on the conceptions manifested by teachers before and after the end of the activity; to establish relationships between the subject worked on during the session in question and those covered in previous sessions; to discuss the suitability of the themes and teaching resources used amongst primary school pupils and discuss alternative ways of using these resources.

The teacher training programme was devised in such a way as to bring out permanent interactions between human beings and the planet by choosing consumption as the interface/mediator of this interaction. Our aim was to highlight the impact of growth and unequal distribution of the human species on a planet which is limited and finite in terms of space and resources, using consumption and its impact as evidence of the growing pressure to which human beings are subjecting our planet.

The teacher training programme started with a session dedicated to the characterization of the current planetary situation which also made teachers aware of the severity of the situation, for them to think about cause-effect relations of the problems identified and the general nature of their impact. In addition, the analysis of documents, discussion and organization of information gave a suitable context to the teacher training programme, and in this way it emphasized the dimensions of the problems in question and the extreme situation that we face.

In previous sessions we had identified the human population explosion as one of the most serious problems we face today, due to the ensuing environmental, social and economic impacts. Thus the theme of population growth and inequalities in the distribution of the population and in the levels of consumption became essential for a reflection on the "quality of life" and the possibilities for "development". The change in our levels and patterns of consumption as well as the guarantee of the respect for human rights thus emerged as inseparable from the promotion of more sustainable ways for the future.

To understand the relationship between human beings and the planet, consumption plays a very important role, both in respect of the need to exploit the planet's natural resources, as well as the consequent impact such as poisonous gas emissions, contamination and desertification of land or depletion of natural resources. The discussion of the ecological footprint concept showed that each living being needs a minimum amount of natural, productive soil and water in order to live. The survival of human beings depends on the existence of food, an energy source, the capacity of various waste products to be absorbed, as well as availability of raw materials for the production processes. The didactic use of the ecological footprint concept led to reflections about imbalances in the possibilities of consumption, addressing the relationship between the various categories of consumption and the fertile soil available, different lifestyles and their environmental impact.

The products' life cycle analysis facilitated the understanding of the environmental, social and economic impacts on products we consume, starting with the exploitation of necessary raw materials and ending with waste disposal. Polluting gas emissions during the various stages of the product life cycle provided the necessary context for the sessions dealing with the ozone layer depletion, global warming and acid rain. The relationship between human activity and the consequences raised awareness of our individual responsibility in the present situation. Education thus emerged as an essential tool for more conscious and responsible actions towards promoting more sustainable ways for the future.

Teacher Training Programme Activities

The organization of this programme was based on a set of activities using didactical resources designed and produced for this purpose. Although didactical resources have been designed to be used at the primary school, it was considered important to offer the participating teachers the same pathway of reflection that they would eventually give to their pupils. For this reason, resources, strategies, discussions and exercises used in the training programme were the same ones that the teachers were expected to use later with their pupils in the classroom context.

The following table lists the activities carried out in each workshop session, as well as the resources used (in the case of sessions focusing on the use of a resource produced for this purpose) and the expected work sequence.

Table 4: Activities suggested for each training session

Activities proposed for each training session
1st Session: characterization of the current situation on the planet a) Thematic contextualization; b) Survey of teachers' conceptions about the current major problems; c) Discussion about the main problems identified and the interactions between each other; d) Discussion of ideas emerging from the previous discussion; e) Reading and analysis of documents relating to the characterization of the current situation on the planet; f) Reflection in group and synthesis of results.
2nd Session: The Concept of Sustainable Development a) Exploitation, in small groups, of the educational resource "The bean game"; b) Reflection in extended group, about the meaning of the various stages of resource exploitation and the results that each group produced; c) Presentation, reading and discussion of documents relating to the concept of SD and the context in which it arises; d) Summary of the analysis of those documents and conclusions resulting from the activity; e) Critical analysis of the didactical resource used.
3rd Session: The Growth of World Population a) Contextualization of the activity to be carried out, relating it to the previous one; b) Exploitation of the didactical resource "Us and the Planet"; c) Discussion of teachers' ideas concerning the distribution of world population in conjunction with the actual current distribution; d) Analysis and reflection on the imbalances in the distribution of the population and the possibility of access to basic essentials; e) Discussion, in extended group, about the global impact of imbalances in exploitation; f) Synthesis of the results of exploitation and whole group discussion; g) Critical analysis of teaching resource used.
4th Session: Human Rights a) Contextualization of the activity to be carried out based on the previous activity; b) Exploitation of the didactical resource "Human Rights Cards"; c) Discussion about the various generations of Human Rights and the impact of the lack of access; d) Reading, analysis and discussion of documents concerning human rights; e) Reflection on possible relationships between human rights and more sustainable ways in the future; f) Critical analysis of the teaching resource used

Activities proposed for each training session
5th Session: Ecological footprint a) Contextualization of the activity to be undertaken, establishing relationships between levels and patterns of consumption, imbalances, and fertile soil and drinking water; b) Exploitation of the didactical resource "What is the size of your footprint?"; c) Analysis and whole group discussion of what it means to use this resource; d) Reflection on the relationship between imbalances in access to consumption, fertile soil and drinking water needed to maintain current levels and patterns of consumption and fertile soil available; e) Analysis and discussion of documents relating to the ecological footprint; f) Critical analysis of teaching resource used.
6th Session: Life Cycle Analysis a) Contextualization of product life cycles; b) Using the activity "from raw material to waste"; c) Analysis and reflection on the inputs and outputs of the life cycles of products; d) Discussion of the relationship between life cycles, imbalances, levels and patterns of consumption and human rights; e) Summary of the analysis and reflection carried out; f) Critical analysis of the teaching resource used.
7th Session: Ozone Layer a) Contextualization of the activity by means of the relationship between some outputs of life cycles under analysis and their environmental impact; b) Survey of teachers' conceptions about the depletion of the ozone layer and its impact (use of cartoons); c) Preparation and carrying out of experiments in small groups (using a planning chart); d) Analysis, comparison and discussion of results obtained by each group; e) Synthesis of the key features of this session.
8th Session: Greenhouse Effect a) Identification of teachers' conceptions about the greenhouse effect; b) Reading, analysis and discussion of various documents about the greenhouse effect; c) Reflection and synthesis of the main aspects of the session.
9th Session: Acid Rain a) Identification of teachers' conceptions about acid rain; b) Reading, analysis and discussion of various documents about acid rain; c) Reflection and synthesis of the main aspects of the session.
10th Session: Human Impact on Natural Resources a) Courseware SeRe exploitation; b) Reflection about the results and conclusions emerging at each stage of using this courseware; c) Critical analysis of the resource used.
11th Session: Education for Sustainable Development a) Discussion about the importance of education in promoting more sustainable ways in the future; b) Analysis and discussion of the implications of the sessions of the training programme in teaching practice; c) Reading, discussion and analysis of documents relating to education for sustainable development.

Teacher Training Programme: Implementation and Evaluation

Teacher Training Programme Implementation

Work sessions were held at two different times, defined by the stages of the teacher training programme and availability of those involved: the 1st stage took place during the school year 2005/2006, between the months of February and July, and the second stage took place in the next school year, between October 2006 and May 2007.

Sessions had an average duration of 90 minutes each (from a little over an hour to more than two hours). For issues that needed more time, two more work sessions were used. Sessions devoted to the greenhouse effect and the ozone layer depletion amounted to a total of six hours, twice what had been initially assigned for each session.

During stage 3 preparation meetings were held between the participating teachers who chose the same activity and the researcher-trainer. Thus four preparation meetings were held, the same number as the activities chosen, to review teachers' doubts concerning using the resource exploitation and/or topic in question.

Some of the work sessions took more time than anticipated and one of the sessions contained in the initial schedule did not take place. This is justified by the difficulties that arose during the teacher training programme implementation and the consequent restructuring, which were part of the ongoing discussions between the researcher-trainer and involved teachers.

Teacher Training Programme Evaluation

It is important to point out that the data, interpretations and inferences presented in this section are a result of content analysis and triangulation between the data collected during the direct observation of the work sessions (in the case of the assessment carried out by the researcher-trainer), data collected through administration of questionnaires at the end of the second stage of the teacher training programme (in the evaluation of the programme done by participating teachers) and the data from the instrument for analysis and characterization of the teaching practices of these participants, designed and validated as part of this research. For this analysis, the objectives set for each one of the training programme's set topic sessions and the aims on which the programme is based were also considered as a reference.

To this end, the teacher training programme evaluation was carried out in two distinct moments, at different levels, and using various assessment techniques and instruments, which are normal in the qualitative research context. The aim of this procedure was to ensure the validity and accuracy of the assessment of the training programme.

Regarding these moments, the assessment was done throughout the implementation of the programme (continuous evaluation) and at the end (final evaluation). The process of continuous evaluation in which all those involved in the training programme participated allowed:

1. Altering the sequence of work sessions after analysis and discussion of the training plan with the trainee teachers;
2. Matching the time allocated to each of the themes included in the programme to the real needs of the teachers' group. Initially, it was planned that the time devoted to the topics ozone layer and greenhouse effect would have the duration of one session (approximately 90 minutes). However, awareness on the part of participating teachers and researcher-trainer that the group needed more time on the content involved in these issues led to a readjustment of the established plan, meaning that each of these topics was attributed twice the time originally allocated. Thus, there were four sessions dedicated to the topics ozone layer and greenhouse effect, which took up approximately 360 minutes of classroom attendance;
3. Adjusting the use of topics which had been worked on with teachers to the needs of primary school pupils. This cross-matching was done in two phases: i) during the thematic sessions, where the methodologies adopted by the researcher-trainer and teaching resources designed and used were discussed with the teachers' group. These discussions were based on the suitability of the topics, teaching resources and methodologies for those participating in this programme: primary school teachers and students and ii) at classroom intervention preparation meetings;
4. Adjusting content and methodologies in each session according to the needs and difficulties experienced by the teachers.

To carry out this evaluation, we used multiple techniques and instruments: participative observation by the researcher-trainer; a record of interpretations/reflections carried out by the researcher-trainer on the basis of observation; the opinions voiced by the participating teacher in relation to the educational and didactical strategies which had been developed and the topics underlying the programme; the video recording of the sessions and later transcription; the teachers' answers to the questionnaire as well as teacher monitoring, both on an individual and collective basis.

In the two moments of evaluation mentioned above different levels were taken into account: the researcher-trainer evaluation, the programme impact on the participating teachers and the evaluation of these teachers with regard to the relevance and suitability of the training programme.

Evaluation by the Researcher-Trainer

The training programme implementation always included the identification of conceptions, ideas and opinions of participating teachers in respect of various issues that would be addressed throughout the programme. To this effect:

1. Regarding the current situation on the planet, teachers identified the greenhouse effect, global warming, the ozone layer depletion and widespread pollution as the main problems. Even though some referred to hunger and war as serious problems, the main emphasis was placed on our current environmental issues.

   Population growth, current consumption levels and patterns and imbalances in the population were not recognized by these teachers as serious problems.

   The image of our current world situation emerging in this study is fragmented and characterized by a lack of awareness of its true severity which hinders both the construction of a holistic and systemic vision of the planet and its systems as well as the recognition of the urgency for action. These difficulties were also reported in several studies previously conducted by other authors on teachers' conceptions about the current situation on our planet (Edwards et al. 2001; Gil-Pérez et al. 2000, 2003; Praia et al. 2001), all of which emphasize the need for the implementation of training courses for teachers to encourage reflection on the current situation on Earth, the responsibility of human beings in this situation and our role in finding a solution.
2. The teachers who participated in this teacher training programme showed vague ideas about sustainable development. Advanced definitions are based on key ideas such as "balance" and "maintenance" to define the relationship between consumption and the use of natural resources. In other words, the need for environmental protection and management of natural resources was acknowledged, but always with reference to the possibility of maintaining levels and patterns of consumption which are close to them. Thus, the idea of sustainable development shared by this group was based on a conception of nature as a resource; they took into account only the consumption levels and patterns close to their own reality, which are not widespread or generally applicable to the world human population. Consequently, this conception is limited in time and space because it is based on personal, present day and local interests.

   These teachers also showed a strong belief in the future and the possibility of future generations solving the problems we currently face through the use of knowledge and scientific and technological artifacts that they will have at their disposal.

This imprecise idea of the concept of sustainable development and the over-stressing of the environmental domain corroborates the results achieved by Freitas (2004) in a study conducted in the context of initial teacher training concerning future teachers' conceptions of sustainable development.

3. Regarding world human population, participating teachers showed a lack of knowledge about its absolute value and current distribution on the planet. Although they identified China and India as the most populated regions on the planet, these teachers did not show any awareness of the real percentage that these countries represent in terms of the world's population. They were also unaware of how the rest of the world's population was distributed, with the continents of Africa and Oceania generating the greatest controversy – the African continent by having more population in absolute terms than expected and Oceania due to its relatively small number.

   With respect to the ratio between the current human population distribution and the possibility of access to basic goods, teachers were able to confirm some of their ideas and compare others with real figures: a) although the group considered North America as the planet's richest region compared to the others, this region was in fact richer than their initial expectations; b) Europe's population, compared to North America, has fewer resources available than the group imagined; c) Latin America proved to be poorer than initial expectations and d) in Oceania there was more money than they first thought, but less water and less food.

4. Products' life cycle analysis allowed us to conclude that these teachers did not consider the environmental, economic and social impacts related to each stage, from products' production to disposal.

   The members of the group only recognized the environmental impact of products after the moment of purchase, and even this impact was limited to the packaging of products. All the raw materials necessary for the products we need and the impact associated with their extraction; the place of origin of the products and greenhouse gas emissions, the working conditions of all those involved in the production chain, the raw materials necessary until final waste disposal, or the impact of individual choices in the economy (local and global) were aspects that these teachers did not consider in their consumption choices.

   It is essential to promote reflection on the social and environmental impacts of each stage in the life cycle of the products we eat so that the act of consuming becomes an ally in the promotion of more sustainable ways of development.

5. The concept of the ecological footprint was unknown to all members of this teachers' group. The relationship between consumption and the need for fertile soil to sustain it was not a concern for this group. Consumption impacts

were interpreted, above all, at the economic and environmental level, and even these only after the product is purchased.

There was, among the participating teachers, an awareness of the relationship between consumption and utilization of fertile soil and water, on the one hand, and between the impact of imbalances in consumption levels and the access to land use, on the other. It was not until the end of the session that there was some reflection on the fertile space currently existing on the planet and its *per capita* availability, and the consumption of a few which is sustained by a disproportionate use of available soil and water. Only after exploiting and debating this activity was it possible for these teachers to establish the relationship between consumption (products and services) and the use of fertile soil and drinking water, the imbalances in both consumption and in soil use, and between over-consumption and increased emissions of greenhouse gases.

Evaluation by the Participating Teachers

Participating teachers considered that the methodology adopted by the researcher-trainer was "very good", and was one of the most valued aspects throughout the programme implementation. Other aspects emphasized by the group were: the importance of starting the sessions with their own ideas, the availability of a variety of resources; the diversity of the methodologies adopted, and the constant encouragement for participation in different moments of the sessions.

The didactical resources were identified as an asset and the themes worked on were considered "very relevant".

All participating teachers considered that the frequency of the programme contributed towards increasing the knowledge of the topics that had been studied in the course content. The sessions focusing on the distribution of world human population and the ecological footprint were identified by the group as excellent opportunities for raising awareness. The issues most valued by the group of teachers were the greenhouse effect and ozone layer depletion.

The most negative aspect referred to by these teachers was the lack of time to work on the themes in more depth. This may be considered an indicator of the growing awareness and motivation of participating teachers of the importance of education in this area.

Conclusion

The orientation towards ESD in lifelong learning implies that this should be (re)evaluated in terms of a new direction for teaching. The future involves a new educational culture, which goes beyond the bounds of the educational framework and promotes involvement in the community. This is the essence of ESD orientation. To educate in order to promote more sustainable ways of development is to educate for action on a planet which is clearly in crisis and characterized by uncertainty.

Teachers have a very important role in citizens' education for action in a world of complexity and unpredictability. To this end, we believe it is crucial that education with an ESD orientation includes three essential aspects: education about sustainable development; education in sustainable development and education for sustainable development.

For these three variants, training will have to become a special time for reflection, not only about working on the contents and the type of interrelations between them but, essentially, on how to put ESD approaches into practice.

Education about sustainable development. Although we have taken ESD as an organizing principle of the curriculum and teaching practice, we consider it essential that teachers' education includes the identification and consideration of their conceptions about the planet's current situation and sustainable development concept. It is essential to introduce the idea of the complexity of the planet and its systems (interaction, unpredictability), otherwise fragmented conceptions about the real situation and how one's own knowledge might be reinforced. Based on the results achieved in this study, we consider that all the contents worked throughout the teacher training programme are fundamental. However, we would like to stress two due to their organizing nature:

i) *Teachers' conceptions about the current planet situation.* At the beginning of the 21st century it is completely unacceptable to have teachers with scant, naive and fragmented notions about our current situation. This perception of the state of the world, which Gil-Pérez and Vilches (2006) classify as spontaneous, is a barrier to their own understanding of the actual scale of the situation and, consequently, to the promotion of learning situations that might offer this awareness to others. One cannot share with others an awareness of aspects of which he or she is not aware. It is necessary that teachers understand current issues from a systemic and holistic perspective so that they can consciously think about their actions both as teachers and citizens but, above all, as agents of change.

ii) *The sustainable development concept and the controversy that this implies.* To face up to the controversy surrounding the sustainable development concept and discuss its meanings and implications for education seems to us an essential dimension when the aim is to instil ESD sensitivity in teacher trainees. We agree with Freitas (2006), who highlights the importance of maximizing the polysemous and ambiguous character of the concept for teaching purposes. It should be the training institutions (at all levels, albeit with varying degrees of appropriate depth of analysis) to create the conditions necessary to ensure that proper discussion and reflection is encouraged about the concept(s) of sustainable development, its strengths and limitations. We believe that, as far as the in-service teacher training is concerned, it is important to analyze the genesis and evolution of this concept, as well as to identify the main dimensions, how these interact and the generalization of the impacts of these interactions. This exercise is important not only to encourage a holistic and systemic understanding of the world – it is not possible to understand the sustainable development concept in all its dimensions by fragmenting the world and knowledge – but also to develop strong convictions among the teachers involved, thus permitting the (re) orientation of behaviour and attitudes towards more sustainable means of development.

Education in sustainable development. The right conditions will have to be created so that educational practice towards ESD can be experienced, discussed and reflected on by teachers during their education. In this context of practice as knowledge production, it makes sense to refer training as a way of sharing among equal partners, of joint construction between teachers and educational institutions, of collaboration for the resolution of problems, and of monitoring during the transformational phase of teaching practice. An ESD approach involves: looking carefully at documents that underlie educational practices in order to find ways to articulate ESD principles with the content of these documents; looking at textbooks, their contents and their guidelines; working carefully with materials from diverse areas of knowledge as this is a transversal dimension to multiple areas of our daily lives; starting from problematic contexts that make explicit the relationships between different dimensions of the problem under consideration (e.g. environmental, social, economic, ethical, …); resorting to methodological diversity in dealing with contents to be worked on; developing, validating, implementing and evaluating didactical resources during training; involving the immediate community in the training and in future action; rethinking the profile of trainers in training institutions. Trainers should accompany teachers at all stages of training, encouraging them to reflect on their practices in the sense of re-evaluating their practices.

Education for sustainable development. ESD is the ultimate goal of the dimensions mentioned above. Together, the various dimensions of ESD aim to promote the exercise of good citizenship which is geared towards the transformation of socio-economic and environmental realities and their processes so that they become more sustainable.

ESD is related to citizens' attitudes and behaviour, and the promotion of competences is the best way of ensuring the future dimension that comes with the concept of sustainability. Individual decisions must be made in an informed way, namely with knowledge about the environmental, social and economic implications; based on values such as solidarity, equity, justice as well as both inter- and intra-generational cooperation; promoting cultural and biological diversity and with respect for our planet production capacity.

References

Arima, A., Konaré, A., Lindberg, C. and Rockefeller, S. (2004), United Nations Decade of Education for Sustainable Development 2005-2014. Draft International Implementation Scheme. UNESCO, Paris, France.

Ballantyne, R., Fien, J. and Packer, J. (2001), Program effectiveness in facilitating intergenerational influence in environmental education: lessons from the field. Journal of Environmental Education, 32(4), pp. 8-15.

Borges, F., Duarte, M.C. and Silva, J.P. (2007), Atitudes de professores portugueses sobre o ambiente e a problemática ambiental. Revista Electrónica de Enseñanza de las Ciências, 6 (1), pp. 176-190.

Edwards, M., Gil-Pérez, D., Vilches, A., Praia, J., Valdés, P., Vital, M. L., Cañal, P., Del Carmén, L., Rueda, C. and Tricárico, H. (2001), Una propuesta para la transformación de las percepciones docentes acerca de la situación del mundo. Primeros resultados. Didáctia de las Ciencias Experimentales y Sociales, 15, pp. 37-76.

Fien, J. and Tilbury, D. (1998), Education for Sustainability: Some Questions for Reflection. Roots, 1998, 17 (available at: http://www.bgci.org).

Fonseca, A.M. (2000), Educar para a cidadania: motivações, princípios e metodologias. Porto Editora, Porto, Portugal.

Freitas, M. (2004), Concepções de Desenvolvimento Sustentável em estudantes de uma licenciatura em educação em Portugal. Implicações para a reorientação curricular no âmbito da Década das NU. Journal of Sustainability in Higher Education, 4(2), pp. 100-105 (available at: http://www.fchst.unlpam.edu.as/iciels/255.pdf).

– (2006), Educação Ambiental e/ou Educação para o Desenvolvimento Sustentável? Uma Análise Centrada na Realidade Portuguesa. Revista Iberoamericana de Educación, 41, pp. 133-147.

Gil-Pérez, D. and Vilches, A. (2005): Capítulo 14. Qué desafíos tiene planteados hoy la humanidad?. In Gil-Pérez, D., Macedo, B., Martínez, J., Sifredo, C., Valdés, P. and Vilches, A., Como promover el interés por la cultura cien-

tífica?. Una propuesta didáctica fundamentada para la educación científica de jóvenes de 15 a 18 anos. OREALC/UNESCO, Santiago.
– (2006), Algunos Obstáculos e Incomprensiones en Torno a la Sostenibilidad. EUREKA, 3(3), pp. 507-516 (available at: http://www.apac-eureka.org/revista).
Gil-Pérez, D., Vilches, A., Astaburuaga, R. and Edwards, M. (2000), La atención a la situación del mundo en la educación de los futuros ciudadanos y ciudadanas. Investigación en la Escuela, 40, pp. 39-56.
Gil-Pérez, D., Vilches, A., Edwards, M., Praia, J., Marques, L. and Oliveira, T. (2003), A proposal to enrich teachers' perception of the state of the World: first results. Environmental Education Research, 9(1), pp. 67-90.
Gil-Pérez, D., Vilches, A., Edwards, M. and Vital dos Santos, M.L. (2000), Las concepciones de los profesores brasileños sobre la situación del mundo. Investigações em Ensino das Ciencias, 5(3) (available at: http://www.if. ufrgs.br/public/ensino/vol5/n3/13incide.htm).
Hesselink, F., van Kempen, P.P. and Wals, A. (eds.) (2000), ESDebate International Debate on Education for Sustainable Development. IUCN, Cambridge, Switzerland.
Hopkins, C. and McKeown, R. (2001), Education for Sustainable Development: past experience, present action and future prospects. Educational Philosophy and Theory, 33(2), pp. 231-244.
– (2005), Guidelines and Recommendations for Reorienting Teacher Education to Address Sustainability. UNESCO/UNDESD, Paris, France.
Huckle, J. (2006), Education for Sustainable Development. A briefing paper for the Teacher Training Agency. Teacher Training Agency, United Kingdom (available at: http://john.huckle.org.uk/publications_downloads.jsp.).
Imbernón, F. (2007), La formación permanente del profesorado. Nuevas ideas para formar en la innovación y el cambio. GRAÓ, Barcelona, Spain.
– (Coord.), Majó, J., Mayer, M., Mayor, F., Menchú, R. and Tedesco, J. (2002), Cinco ciudadanías para una nueva educación. GRAÓ, Barcelona, Spain.
Ko, A.C.C. and Lee, J.C.K. (2003), Teachers' perception of teaching environmental issues within science curriculum: a Hong Kong perspective. Journal of Science Technology, 12(3), pp. 187-204.
Martins, I.P. (1998), Teachers' Conception about their understanding of societal science issues. ERIC Document Reproduction Service NºED 418949; SE 061 302.
Ministério da Educação – Departamento de Educação Básica (2001), Currículo Nacional de Ensino Básico: Competências Essenciais. ME, Lisboa, Portugal.
Morin, E. (1999), Os sete saberes para a educação do futuro. Instituto Piaget, Lisboa, Portugal.

Ospina, G.L. (2000), Education for Sustainable Development: a local and international challenge. Prospects, XXX(1), pp. 1-12.

Pedrosa, M.A., Gonçalves, F., Henriques, M.H. and Mendes, P. (2003), Educação CTS e Formação de Professores de Ciências via Problemática do Lixo. ENCIGA (ed.). XVI Congreso de ENCIGA, pp. 145-153.

Pedrosa, M.A. and Leite, L. (2005), Educação em Ciências e Sustentabilidade na Terra: Uma análise das Abordagens Propostas em Documentos Oficiais e Manuais Escolares (available at: http://www.enciga.org/boletin/58/Pedrosa_Leite_Educao_en_ciencias_Ponencia.pdf).

Pellaud, F. (2001), Approche didactique du development durable: un concept entre utopie et réalité. Education permanente, 148(3) (available at: http://www.ldes.unige.ch/publi/rech/2003HawaiFPGB.pdf).

– (2002), Le Developpment Durable: D'un Concept Complexe a la Vie Quotidienne. Les sciences de l'éducation pour l'ère nouvelle, 35(1) (available at: http://www.ldes.unige.ch/publi/rech/2003HawaiFPGB.pdf).

– (2003), Conceptions, paradigms, values and sustainable development. Actes du Hawaii International Conference on Social Sciences, Honolulu, Hawaii (available at: http://www.ldes.unige.ch/publi/rech/2003HawaiFPGB.pdf).

Praia, J., Edwards, M., Gil-Pérez, D. and Vilches, A. (2001), As percepções dos professores portugueses e espanhóis sobre a situação do mundo. Revista de Educação, X (2), pp. 29-55.

Sá, P. and Martins, I.P. (2005), Ciência, Cidadania e Desenvolvimento Sustentável: Concepções dos Professores do 1° Ciclo. Enseñanza de las Ciências, Número Extra. VII Congreso (CD-ROM).

Santos, M.E. (2005a), Que Cidadania? Tomo II. Santos-Edu, Lisbon, Portugal.

– (2005b), Que Educação? Tomo I. Santos-Edu, Lisbon, Portugal.

Scoullos, M. and Malotidi, V. (2004), Handbook on Methods used in Environmental Education and Education for Sustainable Development. MIO-ECSDE, Athens, Greece.

Summers, M., Childs, A. and Corney, G. (2005), Education for sustainable development in initial teacher training: issues for interdisciplinary collaboration. Environmental Education Research, 11(5), pp. 623-647.

Tilbury, D., Stevenson, R.B. and Fien, J., Schreuder, D. (eds.) (2002), Education and Sustainability: Responding to the Global Challenge. IUCN, Gland, Switzerland and Cambridge.

Thomaz, A.C. (2007), Supervisão Curricular e Cidadania: Novos desafios à formação de professores. Tese de Doutoramento. Universidade de Aveiro. Aveiro, Portugal.

UNESCO (2006a), Década das Nações Unidas da Educação para o Desenvolvimento Sustentável. Contributos para a sua Dinamização em Portugal.

Comissão Nacional da UNESCO, Lisbon. Portugal (available at: http://www.unesco.pt/pdfs/docs/contributosdeds.doc).

– (2006b), Estratégia da CEE/ONU para a Educação para o Desenvolvimento Sustentável. Instituto do Ambiente, Lisbon, Portugal (available at: http://www.iambiente.pt/portal/page).

Varga, A., Kószó, M., Mayer, M. and Sleurs, W. (2007), Developing teacher competences for education for sustainable development through reflection: the environment and school initiatives approach. Journal of Educational for Teaching, 2007, 33(2), pp. 241-256.

Vargas, C. (2000), Sustainable Development education: Averting or mitigating cultural collision. International Journal of Educational Development, 20, pp. 377-396.

Vega-Marcote, P., Freitas, M., Álvarez Suaréz, P. and Fleuri, R. (2007), Marco Teórico y Metodológico de Educación Ambiental e Intercultural para um Desarrollo Sostenible. Revista EUREKA, 4(3), pp. 539-554 (available at: http://www.apac-eureka.org/revista).

Vieira, R.M. (2003), Formação Continuada de Professores do 1° e 2° Ciclos do Ensino Básico. Para uma Educação em Ciências com Orientação CTS/ PC. Tese de Doutoramento. Universidade de Aveiro. Aveiro, Portugal.

Vilches, A. and Gil-Pérez, D. (2003), Contruyamos un futuro sostenible. Diálogos de supervivencia. Cambridge, Madrid, Spain.

Chapter 9

The Relevance of Learning Strategies Relying on Investigative and Experimental Premises for Science Teaching: A Case Study with Junior School Students

Fernanda Luísa Lopes, Joana Luísa Pereira, Fernando Gonçalves, António Veríssimo, Rui Ribeiro

Abstract

Natural Sciences are a spreading field for the development of innovative and diverse teaching strategies envisaging improvements in learning processes. Moreover, most of the formal contents of the natural sciences' curricula can be closely linked to environmental issues, and thus used as environmental education promoters. The achievement of such a cooperative platform requires the teacher to invest in establishing the links between formal contents to be taught and processes that students can observe and feel in their everyday life, and requires a shift towards the use of more practical/experimental teaching strategies that can enhance the actual meaning of acquiring concepts. In line with these principles, this study aimed to evaluate whether an innovative teaching strategy (based on experimental procedures) would promote better learning than traditionally conducted practical lessons. Departing from the same problem (related to microalgae growth), a 7th grade class (26 students) was divided into two groups, which were driven to solving the problem according to distinct teaching strategies: (i) the "participative group", where an investigative-like learning was promoted by the teacher; (ii) the "expositive group", where practical lessons were conducted by the teacher. Levels of procedural and conceptual knowledge were evaluated quantitatively in both groups before and after the intervention (inquiry and report-based evaluation), and the students' performances were compared between and within groups. The results showed that the participative group often achieved higher levels of both conceptual and procedural knowledge when compared with the "expositive" group. In addition, the former group was more efficient in identifying and solving problems concerning the activity, and registered better overall quality in laboratorial procedures. Although some attitudinal changes were qualitatively observed in students from the "participative" group, the sample size and the evaluation design constrained conclusions on this item.

Introduction

Environmental Education (EE) is a sound issue worldwide given its social role in the postmodern world. Wide environmental changes meet everyday life, and society is frequently asked to participate in pro-environment actions as a result of broad and diverse ecological concerns. In this way, EE becomes a powerful tool to provide people with the necessary basis to understand and take action on environmental issues. Nowadays, knowledge acquisition is a keystone in the context of the global patterns of worldwide social changes. Environmental literacy and education for citizenship as processes involving knowledge acquisition are two major challenging areas for sustainable development, and constitute baseline accomplishments for EE (Stables and Bishop 2001). However, the role of the educators (e.g. formal teachers) in these processes and the way they drive students through learning are often biased by methodological shortcomings. One of the most significant examples of this problem is grounded in the trend to require absolute conceptualization, and strictly compartmented knowledge is frequently observed (*fact-based tradition*; e.g. Sund and Wickman 2008). Teaching strategies more focused in relating concepts and observations to solve questions following investigative courses should probably be better considered, particularly if close links with EE is intended. These strategies are supposed to be actively engaged in experimental work or at least with its driving assumptions, and should never be restricted to ordinary practical activities where precise guidance and suggestive linear cues for results are provided to students.

Practical activities have been criticized since a mechanical approach is always followed and there is no valorization of knowledge integration. Consequently, comprehension and integrated perception of processes are minimal (Miguens and Garrett 1991; García and Joaquín 2000). Moreover, it is often hard to encourage students forward into *a posteriori* discussion on the rationale behind a given practical activity, and hence laboratorial work modelled within these premises is of little use as a pedagogical strategy (Hodson 1990, 1992, 1994). It should be noted that the use of science practices as pedagogical strategies presupposes stimulation for structured thinking and reflection, where the acquired knowledge and skills condition the directions of the investigation as well as the intervention of the student in the process (García and Joaquín 2000; Hodson 1992; NRC 1995). In this way, theoretical background, more than ungrounded laboratory work, plays a pivotal role when using experimental practices in education. Experimental activities surpass practical work and direct observation activities in the way that theoretical speculation, debate and arguing on general ideas is required, through the construction of a theoretical frame that allows the establishment of an effective experimental design and testing (Solomon 1991). The use of experimental approaches in science teaching is recommended in the formal scholar science study subjects in Portugal (Dec.-Lei nº 6/2001, January 18th; Abrantes 2001). There is a clear recommendation for: (i) the stimulation of the student towards comprehension, contextualization and worth perception of the learned concepts; (ii) the enhancement of a more autonomous behaviour of students within their learning processes; (iii) a growing use of exploratory and experimental activities as teaching strategies. In this way, studies focusing and/or evaluating the implementation of teaching-learning strategies based on these premises should be given attention.

The definitions of practical work (PW), laboratory work (LW) and experimental work (EW) are not consensual. Hence here we will follow the approach by Hodson (1992). According to Hodson (1992), every activity requiring active involvement by the student can be generally considered PW. Considering that this engagement can be settled at the cognitive, affective or psychomotor level, e.g. laboratorial, field, problem-solving and internet searching activities can be assumed as PW. LW would be any activity requiring the use of laboratorial materials and carried out either in a specialized room or in an ordinary classroom. Finally, any activity involving control and/or manipulation of one or several variables, contextualized in the lab, in the field or in another physical space, can be understood as EW (Hodson 1992). The EW additionally supposes confronting and comparing between what was expected prior to the experiment (which allowed the statement of the testing hypotheses) and actual observations made during and/or after the activities.

This work describes the development and application of an investigative approach as a teaching strategy within formal biology study subjects. The activities were planned and then applied to groups of children aged between 12 and 15, and despite limitations in sample size (a single class of 26 students was studied), some directions for future use of these new approaches were depicted. A number of specific aims were taken into account when planning and carrying out the experiment, and these should be noted: (i) To develop an investigative course (with experimental components) as an attempt to solve a specific problem raised within the study matter, and in such a way that it became suitable for use by other schools or other classes in the same school; (ii) To test whether an investigative-based teaching strategy promotes a better achievement of conceptual and procedural knowledge by the students, when compared to a traditional-like teaching strategy; (iii) To compare changes in the attitudinal skills of the students, and consequently in the reinforcement of positive and pro-active citizenship behaviour, promoted by each of these teaching-learning strategies (e.g. motivation for team work); (iv) To develop and evaluate the efficiency of objective evaluation grids that allow the actual quantification of the referred variables (conceptual and procedural knowledge acquisition, changes in values and attitudinal skills, and changes in teamwork skills).

Background Information: Microalgae

Water resources sustain complex living communities and are of crucial social and economical value for humankind. A great part of the Earth's surface is covered by water (71%), distributed among different aquatic environments such as ice caps and oceans (99% of the water in the biosphere), rivers, lakes and groundwater bodies (the 1% portion remaining – which is the available fraction of water for human consumption) (Molles 1999). Given its relevance in sustaining biodiversity and ultimately humankind, and considering that freshwater is notoriously scarce, efforts should be made towards its conservation.

Freshwater Trophic Interactions

The key actors of the freshwater food web (pelagic zone i.e. water column), and particularly that of lakes and lagoons, can be summarized considering four general groups/trophic levels (e.g. Wetzel 1983; Molles 1999): (i) The phytoplankton (e.g. green algae, cyanobacteria, diatoms) constitute the basis of the food web and are the aquatic (water column) equivalents to plants in terrestrial ecosystems (producers); (ii) The animal fraction of plankton constitutes the zooplankton, which is composed of three general groups with herbivoral activity over phytoplankton (rotifers, copepods and cladocerans); the following levels of the food web include (iii) plankton consumers (planktivorous) and (iv) fish predators (piscivorous). The trophic structure of freshwater ecosystems is regulated by top-down interactions – where the predators assume a control role over lower levels of the trophic web (e.g. Carpenter et al. 1985), and bottom-up interactions, which refer to the competition within the lower trophic levels for nutrients (phytoplankton) or food (e.g. zooplankton) (e.g. McQueen et al. 1986).

Eutrophication – a Relevant Conservation Issue Worldwide

Freshwater ecosystems are threatened worldwide by human activities and a common derived problem affecting water quality is eutrophication (Harper 1995; Smith 2003). Eutrophication is a general term that describes the overall consequences of the natural or artificial (man-induced) overenrichment of water bodies with nutrients, mainly inorganic nitrogen and phosphorus (Harper 1995; Carpenter

2005). Ultimately, the eutrophication process can also be a natural consequence of the evolution of lakes from an initial oligotrophic state (characteristic of a young lake), where productivity is low and the water column is poorly enriched with nutrients (and poorly concentrated in suspended organic matter) (e.g. Wetzel 1983). With ageing, water bodies generally evolve slowly into a mesotrophic state, and then into eutrophic and hypereutrophic states, characterized by growing productivities and increased nutrient enrichment (Wetzel 1983).

Eutrophication has nowadays become a major environmental issue given its dramatic acceleration driven by human activities. The replacement of forests by crop fields and the increase in agricultural production worldwide led to an intensive use of pesticides and fertilizers; its residues and greater amounts of organic matter then reach water bodies standing in the vicinities, therefore contributing to raise nutrient inputs in these systems (Tilman 1999; Carpenter 2005). In addition, point-source pollution events such as industrial, domestic and urban sewage effluent discharges may constitute important sources of nutrients in water bodies, hence contributing to accelerate eutrophication events (Carpenter 2005).

Excessive nutrient inputs induce deleterious effects in the aquatic ecosystem, which include uncontrolled high primary production, e.g. increase in plant growth and harmful phytoplanktonic algal blooms (Carpenter 2005). In spring and summer, algal blooms are more likely to occur: higher temperatures and longer periods of light, together with nutrient availability, dramatically favour algal growth (Wetzel 1983; Harper 1995). High primary productivity leads to great increases in microbial biomass (microbial mediated decomposition processes are favoured by the increase of decaying organic matter), thus promoting oxygen depletion below the water surface at unsustainable rates; anoxic events may actually occur and large species kills (e.g. non-tolerant fish species) can be observed as a direct consequence (Harper 1995; Carpenter 2005). High levels of suspended organic matter, promoted by the increase in productivity, lead to the increase in turbidity, hence reducing light penetration rates in the water column.

These light-limiting conditions will naturally favour rooted macrophytes standing at the surface of the water body rather than phytoplanktonic species suspended in the water column, and macrophytes shadowing over the water column will reduce even more light penetration. At this point, green algae growth is highly disfavoured but cyanobacterial blooms are likely to occur. Cyanobacteria (or blue-green algae) are photosynthetic prokaryotes; despite not having chloroplasts, these organisms produce photosynthetic pigments such as chlorophyll *a*, carotenoids and phycobilins hold in tylacoids that accumulate freely in cytoplasm (Graham and Wilcox 2000). These algae can passively migrate along the water column by controlling cell fluctuation rates, which enable them to surpass light limitations – they migrate into epilimnetic areas where light is available for photosynthesis

(Oliver and Ganf 2000). Several cyanobacterial species (e.g. *Microcystis* sp., *Anabaena* sp., *Aphanizomenon* sp.) produce toxins (cyanotoxins – e.g. neurotoxins, gastrointestinal toxins, irritant toxins, hepatotoxins) that can induce sub-lethal deleterious effects or even lethal effects in the aquatic communities, including mammals, fish and amphibians through contact or ingestion (see the reviews by Hitzfeld et al. 2000; Funari and Testai 2008). The recreational use of a given lake suffering a cyanobacterial bloom may also expose humans to hazardous cyanotoxins that can seriously compromise human health; drinking water and contaminated food (e.g. fish, shellfish and molluscs) can also be sources of human poisoning with cyanotoxins (Hitzfeld et al. 2000; Funari and Testai 2008).

Methodology

Framing of the Case Study

This study was developed with a standard class of 26 students belonging to the 7th regular grade of the Portuguese formal education (12-15 years). The school where the activities were undertaken (Escola Básica 2, 3 Eugénio de Castro) is located within the urban perimeter of Coimbra in the Centre-Littoral of Portugal. This sample of students was characterized considering some social and academic indicators: age, gender, distance to the school from home, formal education and age of parents, and final marks obtained in classes for natural sciences, mother language (Portuguese), and mathematics (these classes were selected given that the activities allow transversal skills training within all these three disciplinary curricular areas). A student t-test was used to compare students' age, parents' age and final marks per disciplinary area (natural sciences, Portuguese and mathematics); students gender and distance to the school from home were compared using a Fisher test; and formal education level of parents was compared between the groups through the Mann-Whitney U-Test (Zar 1996). No statistically significant differences between these two groups were found in respect of social/ academic indicators.

Two groups of students were then established at random prior to any action, and both groups were assigned the same task of solving a study problem. However, distinct teaching strategies were used within each group: (i) the "participative" group (PG), which would carry out an investigative course [i.e. following the model by Hodson (1992) for problem-solving in sciences] to solve the problem, provided that an active role of the student in constructing its own knowledge grounds is assumed; (ii) the "expositive group" group (EG) would undertake the solving of the same problem but via traditional methodologies, where the student assumes a passive role as receptor of transmitted knowledge. In both PG and EG, teamwork (small teams established within a group at random) was used for the study. The main methodological differences between the two groups relate to the strategies for presentation and identification of the problem, hypothesis selection, search for information, planning of practices, and discussion of results (see below for details). Please refer to Appendix I for a group-specific detailed description of specific aims that students are intended to achieve, strategies used in each group, concepts approached within strategy, and for a global overview of the classes' schedule.

Working Problem and Remarks on the Practical Activities

The working problem was chosen in strict accordance with the demands of the Portuguese compulsory *curriculum* for natural sciences (12-15 years) by being integrated as a subject within the theme "Sustainability on Earth – Living Beings-Environment Interactions" and the subject "The influence of environmental factors in populations" (Abrantes 2001): *Sometimes, during summer, massive fish death is noticed in lakes; this is related with a sudden marked growth of algae. Why do algae have such a marked growth?* Then, a series of activities was carried out in EG and PG following distinct strategies within each group towards problem-solving (Appendix I). Appendix I also provides detail on the specific aims and concepts intended to be achieved following the application of both strategies. These aims and concepts relate directly to the indications of the Portuguese curricular recommendations for natural sciences.

The experimental activities consisted generally in analyzing the main differences in algal growth between unconditioned (control treatment) and stressor-conditioned (temperature, light or nutrient availability) cultures. *Chlorella vulgaris* and *Selenastrum capricornutum* were used as test species (concentrated alga were supplied by the Department of Zoology of the University of Coimbra – Ecotoxicology laboratory). Tests were undertaken using transparent bottles filled with still bottled water, and the same standardized quantity of concentrate of nutrients was added to each bottle (this concentrate was also supplied by the Department of Zoology of the University of Coimbra – Ecotoxicology laboratory). Each bottle was then inoculated with a standardized quantity (volume) of algae concentrate.

Each test consisted of three experimental treatments, i.e. 1 bottle containing the control treatment, and 2 bottles containing conditioned treatments: (i) when testing the influence of light in algal growth, one of the bottles was half-covered while the other was completely covered with foil; (ii) when analyzing the effect of temperature, common aquarium thermostats were used in order to keep one of the bottles at 24°C and the other at 30°C (control treatment was kept at 20°C); the two treatments testing the effect of nutrient availability contained increasing concentrations of nutrients relative to the initial standardized concentration used in all tests and all bottles within each test. All bottles were aerated and conveniently illuminated with lamps. Students were scheduled in teams to ensure diary stirring of test bottles and daily records of temperature fluctuations in the room. Tests were kept under these conditions for one week. After this period, records on the experiment were taken by comparing the actual colour of each treatment (and respective control) with a known green-gradient, previously established through successive dilutions of the initial algae concentrate.

The records obtained in each experiment were compared with the expected results and the study problem was then solved based on the knowledge acquired during the experimental trial. Finally, and based on the actual experimental setting followed by the students, the teacher revised and synthesized a series of related scientific terms: study problem, hypothesis, testing, replicate, variable, control treatment (support by school textbooks e.g. Anunciação and Ribeiro 1994; Leal et al. 1996). Moreover, the teacher exposed the main rules guiding the creation of a scientific report, and students were required to report the experiment accordingly.

Evaluation

The achievement of the general aims by the students (i.e. aims (ii)-(iv); see 'Introduction' section) was quantitatively evaluated. For these purposes, a set of variables was established, so that quantitative measurements relying on the student's knowledge acquisition, behaviour and commitment could be taken. The quantitative changes of the variables were ultimately used as a measure of the efficiency of the teaching strategies, which differed considering EG or PG. Three different evaluation methods were used to address the set of variables:

(i) diagnosis evaluation – changes in students' responses to a given inquiry before and after the activities; please refer to Appendix II for the full inquiry;
(ii) continuous evaluation – observations registered by the teacher when following the groups during experimental testing, results registering, discussion and report writing; these records were taken using evaluation grids (per student per class) using simplified qualitative scales for each variable (table I – column 'Evaluation');
(iii) final evaluation – experiments report written by the students and practical exam. Table I depicts the set of variables used and its correspondence either to the main aims set by the students (see 'Introduction' section) or to evaluation methods used. Appendix I should be addressed as a complement to this table for aims and terms/concepts to be achieved by the students from both groups that guided the establishment of some evaluation parameters.

Table I: Correspondence of variables used to address students' evaluation to aims established within this case study (addressed aim) to variables used to address and related components of evaluation methods.

Addressed aim	Variables	Evaluation
Efficiency in acquisition of conceptual knowledge: differences resulting from experiencing a traditional teaching strategy or another based in an investigative course with experimental components.	Acquisition of the following concepts:	
	– scientific method	D
	– problem	Int
	– hypothesis	Int
	– variable	MM
	– experiment	MM
	– control	MM
Efficiency in acquisition of procedural knowledge: differences resulting from experiencing either teaching strategies.	– Acknowledgement the responsibility of the humankind in freshwater bodies' dynamics.	D; Int
	– Identification of environmental factors able to promote algal growth.	D; Int
	– Formulation of hypotheses considering these factors.	D; Int
	– Isolation of experimental variables regarding these factors.	D; MM
	– Planning of experiments regarding the problem and factors.	D; MM
	– Preview of results.	D; MM
	– Record of results.	Res
	– Comparison between previewed and obtained results.	Dis
	– Rejection/Acceptance of hypotheses.	Dis
	– Proposal of solutions for the study problem.	Dis
	– Application of knowledge acquired in solving problems.	D
	– Presentation of creative ideas.	C (**)
	– Manipulation of basic lab material and equipment.	Ex (***)
Efficiency in acquisition of values and attitudinal skills, and in reinforcement of citizenship (respect and participation): differences resulting from experiencing either teaching strategies.	– Trust in colleagues.	C (*)
	– Listening of the opinion of the colleagues.	C (*)
	– Consideration of colleagues' opinions.	C (*)
	– Participation in discussion of ideas.	C (*)
	– Participation in data registering.	C (*)
	– Participation in experimental practices.	C (*)
	– Interest of natural sciences classes.	D
	– Usefulness of natural sciences classes.	D
Efficiency in promoting students motivation for and during group work: differences resulting from experiencing either teaching strategies.	– Balance between time devoted discussion and experiments.	C (**)
	– Discipline during work.	C (**)
	– Independent work of teams.	C (**)
	– Solidarity between teams.	C (**)
	– Competition between teams.	C (**)
	– Balanced task distribution within team.	C (**)
	– Interest of team work.	D
	– Usefulness of team work	D

(*) The qualification scale used for this evaluation variable comprised three categories: 'no'; 'not always'; 'yes'. (**) The qualification scale used for this evaluation variable comprised two categories: 'yes' and 'no'. (***) This variable was evaluated taking into account the time spent by the students in performing a temporary microscopic preparation and in acquiring a focused microscopic image of the microscopic preparation. D – diagnosis evaluation; C – continuous evaluation; Ex – practical exam (final evaluation); Int – introduction in the final report; MM – materials and methods in the final report; Res – results in the final report; Dis – discussion in the final report.

Data Analysis

The outcome from both the first diagnosis inquiry, and particularly those devoted to the concept of scientific method, problem identification and approaches for problem-solving, allowed testing the general similarity between the different groups prior to any activity (EG and PG) through a Fisher test. This same statistical approach was used (i) to compare the answers to the inquiry done before and after the experimental course by the students which could be classified as 'correct' or 'incorrect'; (ii) to address the particular case of the relevance of natural science classes for the acquisition of skills to solve study problems, and hence compare the total number of answers denoting the qualifications 'very useful' and 'essential' between groups (EG and PG); (iii) to analyze the changes in the pattern of answers verified before and after the experimental course (from 'not useful' and 'useful' to 'very useful' and 'essential') within groups; (iv) to compare the records of the parameters addressed in the continuous evaluation between groups (EG and PG); (v) to compare groups regarding the qualitative evaluation of their general practical skills and their ability to accurately measure volumes. The Mann-Whitney U-test was applied to detect statistically significant differences between groups on the time spent in elaborating the microscopic preparation, and a Student t-test was used to detect differences on the time spent by the students in acquiring a suitable microscopic image of the object. A Mann-Whitney U-test was also used to analyze the differences between groups on the accurate accomplishment in the written experimental report. The statistical design adopted to address all the evaluation items and the choice of statistical tests was in accordance with Zar (1996), and the statistical significance level (α) was always set to 0.05.

Results

Inquiry – Diagnosis Evaluation (Related Frequency Table in Appendix II)

The diagnosis inquiry comprised two questions (dual or multiple-choice) relying on a student's motivation for science classes (questions 1 and 5); both groups (EG and PG) tended to find science classes very useful. As to the first question, all students qualified sciences classes within the options 'useful/interesting' and 'very useful/very interesting', regardless of the group or evaluation moment. All students recognized that science classes can be useful in solving problems (question 5) before and after the intervention; when qualifying this usefulness (question 5.1), no significant differences were found between EG and PG in motivation either before intervention ('little useful' and 'useful' vs. 'very useful' and 'essential' – two-tailed Fisher test, P=0.22) or after intervention (increase in frequency of the category 'essential' after intervention in EG vs. PG – one-tailed Fisher test, P=0.5). A third question related to motivation was considered in the inquiry, where students were asked for their opinion on the usefulness of group work (question 3). Results show that students were highly motivated for this working strategy since, regardless of the evaluation moment, the answer categories 'useful/interesting' and 'very useful/very interesting' registered a frequency of 100%.

When the inquiry was first applied, the knowledge level of students on the definition of scientific method was low (neither group achieved percentages of correct answers higher than 15%) and identical between groups (two-tailed Fisher test, P=0.5). After the intervention, the increase of correct answers was significantly higher in PG than in EG (one-tailed Fisher test, P=0.02). As to the identification of the presented problem (question 4.1.), before the intervention neither group answered the question correctly; and after the intervention PG succeeded in answering correctly (>80%), whereas EG kept the record of 100% incorrect answers (comparison of the increase in correct answers between groups – one-tailed Fisher test, $P=10^{-5}$). This pattern was maintained when regarding the problem resolution (question 4.2.); results show that PG evolved significantly more towards answering correctly than EG who were still answering incorrectly after intervention (one-tailed Fisher, $P<10^{-5}$).

Continuous Evaluation (Please Refer to Table I for the Variables Analyzed)

Regardless of the group, students generally showed high trust in colleagues, either before or after intervention, and a single register of 'not always' was recorded in each of the three first classes of PG. There was no record denoting that a given student always refused to listen to a colleague's opinion, but some students in EG did not always listen to colleagues (8, 15, and 31% in each of the 3 classes). The same trend was observed for the parameter 'consideration of colleagues' opinions', where very few negative records ('not always') were taken; the positive records ('yes') from EG and PG were compared, and no significant differences were depicted before (two-tailed Fisher test, P=1) or after (one-tailed Fisher test, P=0.24) intervention. There was a pronounced increase in PG students' 'participation in discussion of ideas' during the intervention, while no great changes were observed in this parameter in EG. Nevertheless, EG and PG were statistically similar regarding the records obtained after the intervention (testing frequency of 'yes'; one-tailed Fisher test, P=0.118).

All students from both groups participated in the experimental practices, but PG was significantly better than EG in data registering (comparison of frequencies for 'yes' achieved in post-intervention classes; one-tailed Fisher test, P=0.048). On the contrary, and despite showing a better relative performance, PG was not significantly superior to EG in 'presentation of creative ideas', either before (comparison of frequencies for 'yes' observed in pre-intervention classes; two-tailed Fisher test, P>0.5) or after (comparison of frequencies for 'no' registered in post-intervention classes; one-tailed Fisher test, P=0.19). Regarding the balance between time devoted to experiments and discussion, no pattern could be depicted within either group since there were teams in both groups that had not respected this balance in different classes. On the other hand, teams in EG were clearly more undisciplined during work than in PG. Teams in both groups generally worked independently of each other, and there were no records of any team showing non-solidarity or being competitive towards each other. As for the parameter 'task distribution within team', all teams in both groups showed a balanced distribution of tasks to all elements before the intervention; in the post-intervention evaluation records, an increase in unbalanced task distribution was observed in both groups – although this increase was higher in EG, no significant differences were detected between groups (comparison between groups in the final two classes; one-tailed fisher test, P=0.243 and P=0.4).

Final Evaluation

Practical Exam

Figure 1 shows the performance of both groups concerning the quantitative parameters defined for evaluation in the practical exam. No significant differences were found between groups on the time spent in performing a microscopic preparation ($U_{13,\ 13(2)}$ = 85.5; P>0.2). PG has taken similar time (average: 2.538 min; coefficient of variation=66%) in acquiring a focused microscopic image of the microalgae than EG (average: 2.538; coefficient of variation=52%), which was confirmed statistically ($t_{0.05(1)}$ =0; P=0.5).

Fig. 1: Time spent by students in (a) performing a temporary microscopic preparation; (b) acquiring a focused microscopic image of the temporary preparation.

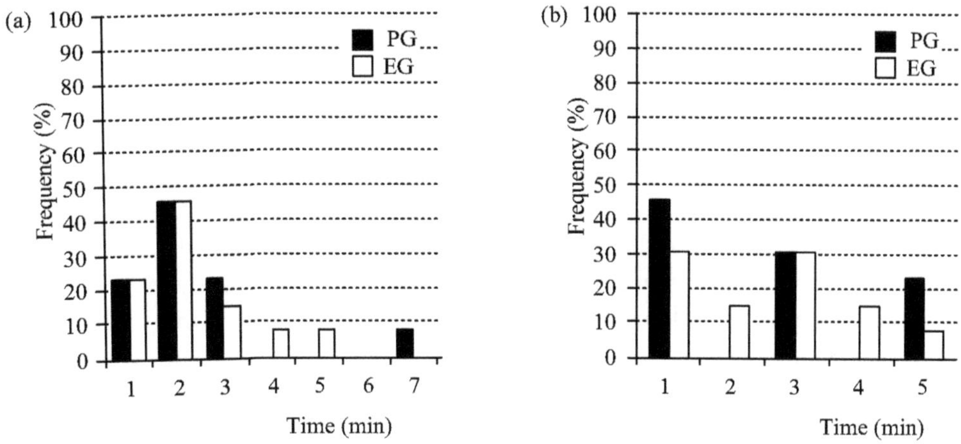

Regarding qualitative records, it should be noted that only 16% of EG students acquired the microscopic image satisfactorily; PG was significantly superior in accomplishing this task since all students succeeded well (one-tailed Fisher test, $P=10^{-5}$). Moreover, students in PG were significantly more accurate in pipetting 2 mL with a 10 mL pipette (77% accurate measurements) than in EG (24% accurate measurements) (one-tailed Fisher test, $P=0.008$).

Final Report (Please Refer to Table I for the Variables Analyzed per Section)

In the 'Introduction' section of the report, PG obtained significantly better records than EG ($U_{4,\ 4(2)}=14$; $P>0.1$), since three teams approached 4 of the 5 related variables correctly. The same trend was verified in 'materials and methods', where three teams from PG approached all related variables correctly ($U_{4,\ 4(2)}=14$; $P>0.1$). There were no significant differences between groups in describing test results ($U_{4,\ 4(2)}=8$; $P>0.2$), while PG was significantly better in the 'discussion' section than EG ($U_{4,\ 4(2)}=10$; $P>0.2$), since teams from EG were only able to approach correctly one out of the three variables defined for evaluation within this section. When comparing globally the mean number of variables approached correctly in the final report by EG and PG teams, statistics depicted a better performance by the latter ($U_{4,\ 4(1)}=7$; $P>0.1$).

Discussion and Conclusions

An investigative course with experimental components, based on problem-solving, was developed and actually implemented with students from a Portuguese school, thus demonstrating its viability and electability as a teaching-learning strategy for formal education. The present case study was not directed towards a traditional view of problem-solving as a learning strategy where a unique solution/answer is required for a common problem. Conversely, on the whole students experienced and understood a practical course where several solutions could be accepted for a given problem, as long as they complied with the observations made during the experimental components of the course (following considerations on scientific problems and its solving by Jaén 2000). Time available per class was found insufficient since work had to be interrupted and restored in a new class later on subsequent days, which is not desirable in most activities; with 90 min classes (foreseen in Portuguese mandatory regulation – Decreto-Lei nº6/01, 18.01) these problems will probably be overcome, since safe programming of activities and previewing of breaks will be facilitated. The students were very demanding and the evaluation grids required permanent attention across the board. We found that reducing the number of students per class should be considered when implementing new teaching-learning strategies centered on students, particularly when these include experimental components.

The investigative course as a teaching-learning strategy was tested against a traditional teaching strategy centered on the teacher by using two groups of students where either was applied. Generally, both conceptual and procedural knowledge were acquired better by the participative group (PG); the evaluation over expositive group (EG) showed that the applied strategy was less efficient for the students to learn concepts and processes. These differences between groups, although graphically visible, were not always depicted as statistically significant, which may indicate that variables chosen for evaluation may need refinement or scaling adjustments. In particular cases (see below), the statistical similarity between groups should be more carefully addressed.

No differences were found between groups when evaluating data registering; students are usually required for data registering in other activities held in different classes regardless of the teaching strategy or the nature of the activity, which may explain this result. For example, in related variables such as 'comparison between previewed and obtained results', 'rejection/acceptance of hypotheses' or 'proposal of solutions for the study problem', PG obtained significantly better records than EG. Moreover, and considering diagnosis evaluation, PG was again significantly superior in applying learned concepts to problem-solving, in identifying problems and in exposing possible solutions for new problems. A correct identification of a given problem will certainly facilitate its solving, and this natural link between the processes may have promoted higher EG failure. No differences between groups were found in creativity. Considering the specificity of this variable, a more extended and directed study comprising more evaluation moments would probably be more adequate to accurately assess a student's progress and to better understand whether investigative courses with experimental components influence their creative thinking. Although creativity and originality cannot be taught, training can be provided in related skills, thereby promoting creative thinking (Jaén 2000).

There was a great variation in the records obtained by the teacher when evaluating the manipulation of lab materials and equipment, which was one of the variables used to address efficiency in the acquisition of procedural knowledge. Repetition is a fundamental conditioner in acquisition of practical and technical skills (Hodson 1990); thus, it is very likely that the time devoted to manipulation of material and equipment was not enough for practice memorizing through repetition in either group. Indeed, the time spent by students from both groups in technical manipulations was identical. Conversely, PG was significantly better if the quality of work done is addressed. The acquisition of technical/laboratorial skills requires repetition of the manipulation process and its clear understanding by the student (Hodson 1990); if not understood, the process will be simplified to a mechanical follow-up of steps lacking judgement (Miguens and Garret 1991). This may explain the lower efficiency observed in EG, particularly when considering the quality of the work done.

Evaluation variables that accounted for the respect and trust of students in each other, and to their participation in the activities, were used to address the efficacy of the proposed strategy in the acquisition of attitudinal skills and values compliant with improvements in citizenship principles. No remarkable differences between groups were depicted for these variables and its unresponsiveness should be noted – mainly due to scaling inadequacy (e.g. in some cases groups were levelled at the maximum records even before the intervention, thus no improvements could be registered) as well as to difficulties inherent to evaluation processes regarding values and attitudes. PG was significantly better in 'participation in data registering', but no feasible conclusions can be taken as to attitudinal skills and values, since this variable is linked to the efficiency of the proposed strategy in acquisition of procedural knowledge. The high records of both groups regarding participation in experimental practices were consistent with records taken when evaluating motivation (diagnosis evaluation), and the low discriminative ability of this variable to detect changes in attitudes or values should hence be recognized.

Group work was adopted as the main working strategy in both EG and PG. In fact, this working strategy has been shown to promote collaboration and knowledge-sharing, esprit de corps, strategic negotiation, reinforcement of oral expression skills, cooperation, solidarity, humility and responsibility (Barron 2000; Galvão et al. 2000; Reigosa and Jiménez 2000; Solomon 1991). Regarding the time devoted to experiments and discussion, the positive evolution of PG should be noted; this group overcame an initial situation of instability, which is common in students facing a new teaching-learning strategy (Lock 1998). Teams in EG did not always succeed well in the strategy processes, and this may have contributed to the observed trend for a decrease in the ability to balance time devoted to experimental tasks and discussion. Results from evaluation of this variable are generally consistent with those obtained for 'discipline during work', which reflects the straight relationship between the variables. Lack of attention and concentration/focusing, as well as the related feeling of inadequate accomplishment of the work, may compromise the student's posture within the team along procedures and ultimately learning processes (Lock 1998). An adequate time scheduling for each task (established taking into account the specificities of the working groups/teams) or the definition of supplementary tasks for groups that eventually finish ordinary tasks earlier are programming items that may prevent such undisciplined events. Discipline records were consistent with those relating to distribution of tasks within teams. It seems that, if disturbed, a group of students will not be able to efficiently distribute responsibilities to its own members for the accomplishment of a common goal. Teams in EG were more dependent on each other than in PG, and records of independence were not found to reflect competitive behaviour i.e. regardless of the degree of its independence, teams always showed solidarity with each other. These results, along with those related with motivation addressed in the diagnosis evaluation, reinforce the benefits of group work as a learning strategy relative to individual work, particularly within the context of problem-solving (see the discussion on the benefits and shortcomings of group work in Barron 2000, 2003).

Particular care was taken when analyzing results of the present study, since the study sample was limited in size and in diversity. The tested teaching-learning strategy does not intend to be the definite solution for efficiency in knowledge acquiring by students, but is rather a contribution to the development of instruments and methods for use in science teaching, and may serve as a basis to promote discussion of ordinary pedagogical practices. A large array of evaluation techniques was employed given the diversity of skills that can be addressed when an investigative course with experimental components is used. Based on this comprehensive evaluation dataset and considering this particular case study, it should be noted that this strategy seems more efficient than traditional teaching-learning practices centered on the teacher. However, more studies are needed in order to conclude accurately the benefits and efficiency of such strategies in learning science subjects.

References

Abrantes, P. (2001), Currículo Nacional do Ensino Básico-Competências Essenciais. DEB – Departamento de Educação Básica, Ministério da Educação, Lisboa.

Anunciação, C. and Ribeiro, E. (1994), Técnicas Laboratoriais de Biologia. Bloco I. Raiz Editora.

Barron, B. (2000), Achieving coordination in collaborative problem-solving groups. The Journal of the Learning Sciences 9(4), pp. 403-436.

– (2003), When smart groups fail. The Journal of the Learning Sciences 12 (3), pp. 307-359.

Carpenter, S.R. (2005), Eutrophication of aquatic ecosystems: bistability and soil phosphorus. Proceedings of the National Academy of Sciences 102(29), pp. 10002-10005.

Carpenter, S.R., Kitchell, J.F. and Hodgson, J.R. (1985), Cascading trophic interactions and lake productivity: fish predation and herbivory can regulate lake ecosystems. Bioscience 35, pp. 634-639.

Galvão, C., Freire, A., Neves, I. and Pereira, M. (2000), Ciências Físicas e Naturais – Competências essenciais. Ministério da Educação – Departamento de Educação Básica, Lisboa.

García, G. and Joaquín, J. (2000), La solución de situaciones problemáticas: una estrategia didáctica para la enseñanza de la Química. Enseñanza de las Ciencias 18(1), pp. 113-129.

Graham, L.E. and Wilcox, L.W. (2000), Algae. Prentice-Hall, Upper Saddle River.

Harper, D. (1995), Eutrophication of Freshwaters – Principles, Problems and Prospects. Chapman & Hall, London.

Hitzfeld, B.C., Höger, S.J. and Dietrich, D.R. (2000), Cyanobacterial toxins: removal during drinking water treatment, and human risk assessment. Environmental Health Perspectives Supplement 108 (S1), pp. 113-122.

Hodson, D. (1990), A critical look at practical work in school science. School Science Review 70 (256), pp. 65-78.

– (1992), Redefining and reorienting practical work in school science. School Science Review 73(264), pp. 65-78.

– (1994), Hacia un enfoque más crítico del trabajo de laboratorio. Enseñanza de las Ciencias 12(3), pp. 299-313.

Jaén, M. (2000), Cómo podemos utilizar en Geología el planteamiento y resolución de problemas? Enseñanza de las Ciencias de la Terra 8.1, pp. 69-74.

Leal, J., Serapicos, M. and Pinto, M. (1996), Técnicas Laboratoriais de Biologia. Bloco I. Asa.

Lock, R. (1998), Fieldwork in life sciences. International Journal of Science Education 20 (6): 633-642.

McQueen, D.J., Post, J.R. and Mills, E.L. (1986), Trophic relationships in trophic pelagic ecosystems. Canadian Journal of Fisheries and Aquatic Sciences 43, pp. 1571-1581.

Miguens, M. and Garret, R.M. (1991), Práticas en la enseñanza de las ciencias – Problemas y posibilidades. Enseñanza de las Ciencias 9(3), pp. 229-236.

Molles, Jr., M.C. (1999), Ecology – Concepts and Applications. Mc Graw-Hill, New York.

NRC, National Research Council. 1995. National Science Education Standards. National Academy Press, Washington.

Oliver, R.L. and Ganf, G.G. (2000), Freshwater blooms. In The ecology of Cyanobacteria – their diversity in time and space. Edited by Whitton, B.A. and Potts, pp. 149-194. M. Kluwer Academic Publishers, Dordrecht.

Reigosa, C. and Jiménez, M.P. (2000), La cultura cientifica en la resolución de problemas en el laboratorio. Enseñanza de las Ciencias 18(2), pp. 275-284.

Solomon, J. (1991), Group discussions in the classroom. School Science Review 72(261), pp. 29-34.

– (2001), Group discussion in the classroom. School Science Review 72(261), pp. 29-34.

Smith, V.H. (2003), Eutrophication of freshwater and coastal marine ecosystems: a global problem. Environmental Science and Pollution Research 10(2), pp. 126-139.

Stables, A.W.G. and Bishop, K.N. (2001), Weak and strong conceptions of environmental literacy: implications for environmental education. Environmental Education Research 7(1), pp. 89-97.

Sund, P. and Wickman, P.O. (2008), Teachers' objects of responsibility: something to care in education for sustainable development. Environmental Education Research 14(2), pp. 145-163.

Testai, E. and Funari, E. (2008), Human health risk assessment related to cyanotoxins exposure. Critical Reviews in Toxicology 38, pp. 97-125.

Tilman, D. (1999), Global environmental impacts of agricultural expansion: the need for sustainable and efficient practices. Proc. Natl. Acad. Sci. USA 96, pp. 5995-6000.

Wetzel, R.G. (1983), Limnology. Saunders, Philadelphia.

Zar, J.H. (1996), Biostatistical Analysis. Prentice-Hall, Upper Saddle River.

Appendix I

The table shows a detailed plan of the experimental course held in the present case study, for both groups of students. 'Specific aims' refers to goals that students are intended to achieve within each teaching strategy; in this column, the schedule for activities is additionally provided in brackets (a class usually corresponds to 90 min devoted to the planned activities).

Specific Aims	Strategies: Control group (CG)	Strategies: Experimental Group (EG)	Terms/ Concepts
(Class 1) Realizes the human influence on the dynamics of freshwater systems. Identifies the environmental factors that can promote algal growth.	Students are questioned on their actual knowledge about pollution issues in freshwater. A set of pictures showing eutrophised freshwater systems is presented by the teacher. An experimental model comprising two aquaria is observed and explored: the first aquarium is overloaded with microalgae and a great part of them is dead and being decomposed (bad odour, dark green coloration); the second has a low microalgae concentration (inodorous, green coloration).		Algal growth Producer
Acquires the concept of study problem.	The study problem is identified by the teacher	The study problem is identified by the students	Problem
	The students fulfil the diagnosis inquiry provided by the teacher (before-intervention diagnosis evaluation). Discussion within group, driven by the teacher, where the two aquaria are compared: extrapolation of the relationship between cause and observed effects – water coloration/algae concentration and odour/presence of decomposers. Fish survival in both conditions should be approached as an eutrophication related issue. The teacher demonstrates procedures for elaborating microscopic preparations and microscope handling.		Decomposer Light Nutrients Temperature
Manipulates lab equipment: microscope, pipette, microscope slides and slips, dissection needle, drop-dispenser.	Students observe unicellular algae under the microscope.	Students elaborate a microscopic preparation and observe unicellular algae under the microscope.	
(Class 2)	Assignment of the students to small groups (teams) of 3-4 students, by chance		
Acquires concepts behind the strategy (please refer to the column 'Terms/Concepts'). Formulates suitable hypothesis to test regarding the study problem. Isolates variables in accordance to the testing hypothesis. Plans an experiment with an adequate experimental design to correctly address the study problem and the formulated hypothesis.	The teacher supplies the students with a written practical protocol, which is read and explained to the group. The teacher questions the students on possible causes leading to the study problem, and exposes the goals intended to achieve with the experiment. The teacher explains that a range of dilutions within test should be established, so that colour gradients can be compared and reliable records can be taken over the results.	Literature search on the problem (websites and magazine/book sources supplied by the teacher).* Within team discussion on the formulation of experimental hypotheses to address the problem. Team brainstorming driven by the teacher, regarding the needs for variables isolation. *Group discussion and establishment of testing* hypotheses and variables. Within team planning of experiments i.e. experimental protocol development (teacher drives team work using a questioning strategy, so that the meaning of control treatments and replicates is addressed and its use referred in the protocol).	Hypothesis Variable Control Replicates Experiments

Specific Aims	Strategies: Control group (CG)	Strategies: Experimental Group (EG)	Terms/ Concepts
(Class 3)	The teacher exposes (using image projection resources) laboratorial safety rules and simulates an emergency exit. As well, the identification of lab equipment and materials is required from the students.		
(Class 4) Correctly manipulates lab equipment and materials need for the experiment.	Students follow the supplied protocol to settle the experiment.	The teacher supplies the teams with materials and equipments they required. Students settle the experiment and leave it running.	
	Students record a colour (green intensity) gradient obtained through dilutions of the algae concentrate; the teacher provides the protocol for this activity and assumes close guidance over the activity.		
(Class 5)	Students observe and record the results obtained with the testing procedure.		
Records the experiment results. Compares expected and obtained results.	The teacher exposes the results achieved by all teams and identifies the solution of the problem.	Within team discussion on the records, considering the expected results.	
Rejects/accepts testing hypotheses properly. Proposes solutions for the study problem.	The teacher exposes the scientific terms that relate to the procedures previously followed.	Group discussion on the results obtained by all teams and on possible solutions for the study problem.	
Understands the concept of scientific method.		Group systematising of procedures; the teacher exposes the related appropriate scientific terms.	
	Report writing (team work – 1 report per team) containing the following sections: title, introduction, materials and methods, results, discussion/ conclusions, and bibliography.		
(Class 6)	Students finalise the report and are then required to fulfil the diagnosis inquiry (after-intervention diagnosis).		
(Class 7)	Practical exam.		

* *Resources provided to the students:*

Louro V (1969). Poluição das águas interiores. Separata da "Agros" 52 (1-3), Jan./Junho

Saraiva I (1997). Água e ambiente – o constatar de uma realidade inquietante. AIP-Associação Industrial Portuense nº 12/13, Dezembro/Janeiro.

http://www.eps-cunha-rivara.rcts.pt/visita-expo/agua/residuais.htm; http://www.eps.ufsc.br/nlgqa/agua.html; http://www.eb23-jacinto-correia.rcts.pt/new-page-5.htm

Appendix II

Diagnosis inquiry provided to the students before and after the intervention allowing comparisons between and within groups. Dual- and multiple-choice questions relate to motivation for Sciences classes, while open questions relate to knowledge levels, hence its answers were classified as 'correct' or 'incorrect' for analysis. The summary table shows the results obtained for this inquiry in the present case study.

1. **What is your opinion about the Natural Sciences classes? (Choose one option within each row)**
   □ Little interesting; □ Interesting; □ Very interesting; □ Other (_______________).
   □ Little useful; □ Useful; □ Very useful.

2. **Expose, briefly and using key-words, the meaning of the expression 'Scientific method'.**

3. **What do you think about group work? (Choose one option within each row)**
   □ Little interesting; □ Interesting; □ Very interesting; □ Other (_______________).
   □ Little useful; □ Useful; □ Very useful.

4. **Consider a given stream, located in the vicinities of a garbage deposit, where no fish exist.**

4.1 **Identify the existent problem.**

4.2 **Do you think that one can blame the garbage deposit for the problem? Explain how would you solve this.**

5 **Do you think that Natural Sciences classes can be useful in solving problems? (Choose one of the options)**
□ Yes; □ No.

5.1 **If you have chosen 'Yes', please specify by choosing one of the options provided:**
□ Little useful; □ Useful; □ Very useful; □ Essential.

SUMMARY TABLE (FREQUENCIES SHOWN IN %)

		Control Group		Experimentation Group	
Question #	**Answer category**	**BEFORE**	**AFTER**	**BEFORE**	**AFTER**
1	100% students classified within 'useful/interesting' and 'very useful/very interesting'				
2	CORRECT	9	9	15	55
	INCORRECT (*)	91	91	85	45
3	100% students classified within 'useful/interesting' and 'very useful/very interesting'				
4.1	CORRECT	0	0	0	84
	INCORRECT (*)	100	100	100	16
4.2	CORRECT	0	0	0	92
	INCORRECT (*)	100	100	100	8
5.1	All students answered affirmatively				
5.2	LITTLE USEFUL	8	0	0	0
	USEFUL	15	23	0	0
	VERY USEFUL	62	62	85	70
	ESSENTIAL	15	15	15	30

(*) inadequate or subjective answer that does not considers essential concepts

Chapter 10

Effectiveness of Outdoor Practices on Environmental Sustainability

Ana Sofia Almeida, Ana Marta Gonçalves, Ulisses Miranda Azeiteiro, Fernando Gonçalves

Abstract

Earth sustainability is a central theme in natural sciences which concerns issues such as the environment and living organisms, energy flows and material cycles, ecosystems equilibrium and disturbances, natural resources and nature protection and conservation. One of the new proposals of the curricular reorganization of the basic school is the implementation of a non-disciplinary curricular area which aspires to develop projects through the articulation of several curricular areas, and moves around research and intervention in problems or themes that coincide with the students' needs or interests. In the context of the theme "Earth Sustainability", students of the basic school (12-15 years) were asked to develop an environmental project, in small groups, concerning an infested aquatic system (*e.g.* a lake) near the school or their homes. The activities proposed in this chapter are related with Pateira de Fermentelos, a lake situated in the Vouga hydrographical basin in the central zone of Portugal, localized in the same area as the students' school and offering great places for family outings. Nowadays, this lake is affected by several problems, notably pollution and eutrophication, especially due to industrialization and agricultural practices which contribute to the enrichment of the water's nutrients. This process culminates in a huge amount of *Eichhornia crassipes* (Mart.) Solms, an infesting plant in Portugal that, on some occasions, covers the entire water surface, depletes the water quality and affects all endemic species. Confronted with this problem, students may research the study area in order to acquire a better knowledge associated with the shallow lake. Then, outdoor activities were performed in order to become more acquainted with the problem and to apply methodologies to evaluate the contamination of the aquatic system. In the classroom, students had the chance to discuss the results obtained and formulate strategies in an attempt to solve or minimize the problems that affect the aquatic system. The project is also involved in environmental public campaigns for the school community.

Introduction

Sustainability has been the central theme of many international conferences, where documents have been signed in order to implement action plans and national policies to achieve an environmental education and a sustainable community development (Tilbury et al. 2002). These concerns were first mentioned in the World Conservation Strategy (IUCN, UNEP, WWF 1980), reinforced in 1987 by the Brundtland Report, in the World Commission on Environment and Development, that catalyzed attention to the environmental problems caused on a major scale by social and economic factors. In 1992, during the Earth Summit in Rio de Janeiro, citizens became aware of the urgency of sustainable development. With the "*Agenda 21*", signature countries compromise to promote sustainability through a set of means, education being one of the main areas (Tilbury et al. 2002). Nevertheless, education is the greatest resource for achieving a sustainable society, where teachers play a crucial role in worldwide changes, increasing people's knowledge about environmental and development issues.

The curricula of natural sciences are organized in themes, one of which is Earth sustainability. This includes aspects related to the environment and living organisms, energy and material cycles, ecosystems equilibrium and disturbances, natural resources and nature protection and conservation. Studying these themes will enable students to become aware of the importance and consequences of their actions on ecosystems, thus contributing to the reasonable use of the resources and also to avoid disturbances. There are some aspects – physical, biological, social, economic and ethical – that should be considered in a sustainable development perspective (DEB 2001). The teaching of natural sciences should contribute to an environmental education that may focus on concrete local daily problems in order to link scientific concepts with real-world experiences (Giordan 1996; Fernandes 2003; Rutledge 2005). During this process, teachers should cooperate with the students, guiding them through the learning process, instead of being mere knowledge transmitters (Almeida 2006). As Rutledge (2005) refers, curricular reforms have focused in a much more investigative and inquiry method with laboratory activities and laboratory experiences, where students may utilize their critical thinking skills. Another novelty introduced by the curricular reorganization is the creation of non-disciplinary curricular areas (NDCA). One of them (Área de Projecto – AP) intends to develop actions through the articulation of several curricular subjects and contents, and moves around research and intervention in problems or themes that coincide with students' needs or interests. It is based on a methodology that gives emphasis on investigation work based on the resolution of problems. The students have an active role during the development of tasks, in which the teacher is the supervisor (Felizardo 2004). They interact in the group, with teachers and also with people outside school, exploring and dealing with an issue of their interest that, at the same time, should be related to the social reality in which such issues have emerged and is inserted (Fernandes 2003; Nova 1994). In this way, the aims of AP is to apply the knowledge from different disciplinary areas and to promote its integration in an interdisciplinary way, thus breaking the limits between the different curricular areas and, moreover, allowing the problem to be treated by several angles (Felizardo 2004). It involves interactively different domains (*e.g.* affective, creative, communicative, intellectual), providing knowledge in a social-cultural context and creating conditions for a dynamic construction of values. The project allows changing school in a set of different experiences that request all students' participation, but, at the same time, leaves them the possibility to choose and give them guidelines to define problems, to gather information, to present possible solutions and to evaluate themselves (Fernandes 2003). On the other hand, it requires students' efforts, in groups or individually, to study the problem and search for a solution to it according to their abilities. The development of a project promotes general skills and attitudes such as: mobilization of cultural, scientific and technological acquaintance in order to understand actual problems; research, select and organize information to change it into mutable information; adopt adequate strategies to solve problems and take decisions; cooperate with others in the group. In this way, the promotion and development of scientific knowledge are proposed in order to form aware citizens about environmental problems, as well as about preservation of natural and cultural patrimony, achieving a responsible, critical, solidary self-identity (DEB 2001; Fernandes 2003). The activities where the inquiry method is used promote the scientific knowledge and provide an accurate understanding of nature, being in agreement with the goal of science education (Rutledge 2005)

Nowadays, one of the most daunting environmental problems is the pollution that affects aquatic resources, which can be chemical (e.g. water acidification), physical (e.g. temperature) and/or biological (e.g. invasive species). These kinds of pollution are also related with the influence of anthropogenic activities (e.g. agriculture activities, domestic effluents, transport, fossil fuel), namely eutrophication, that affects surface waters and compromises the environmental quality of those resources (Salvia-Castellví et al. 2005). Thus, the management and control of water resources by implementing programmes of monitorization and information are important to ensure the effectiveness of measures taken and to minimize the adverse effects on the environment (Parr et al. 2002; Velasco et al. 2006).

In Portugal, the Decreto-lei nº 236/98 establishes standard criteria for water quality according to its use in order to improve it and to protect aquatic environment (Decreto-Lei no 236/98). Such criteria can be physical (density, temperature, solar radiation, turbidity, conductivity and dissolved oxygen), chemical (pH, hardness, nutrients' availability and salinity), microbiological (total coliforms, faecal coliforms and faecal *Streptococcus*) and biological.

Students of the eighth grade of a basic school in the north of Portugal, aware of an issue of concern that affects Pateira de Fermentelos, a shallow lake in their region, defined a project about water quality and its management in AP in order to improve the sustainability of Pateira de Fermentelos. During the development of this activity students had to articulate concepts for several curricular areas, namely natural sciences.

The main source of water of the lake Pateira de Fermentelos is the Cértima River, a sub-tributary of the Vouga River, a watercourse that drains into the Atlantic Ocean (fig. 1). Pateira de Fermentelos is a wetland classified as a sensitive area by the Decreto-Lei nº 152/97, which brings to the Portuguese law the European Directive 91/271/CEE concerning urban wastewater treatment. There is an extraordinary biodiversity and aesthetic value. Surrounding areas are characterized by forests of *Pinus pinaster* and *Eucalyptus globulus* in the upper basin and agricultural fields and pastures in the lower basin. The main industries are from ceramics, wine production and metal-processing sectors (Cerqueira et al. 2005).

Fig. 1: Pateira de Fermentelos topographic map [adapted from sheets 185, 186, 197 and 198 of Portuguese Military charts (1:25 000)] (adapted from Gameiro 1999).

Nowadays, Pateira de Fermentelos is being affected by several problems of pollution and eutrophication especially due to the industrialization and agricultural practices, which contribute to the enrichment of water nutrients (Almeida et al. 2006). This process culminates in the huge amount of *Eichhornia crassipes* (Mart.)

Solms, commonly known as water hyacinth (fig. 2), an infesting plant in Portugal brought from Brazil. This macrophyte frequently reproduces asexually. By this process, a single plant can give origin to a new plant in only two weeks (Unknown 1980), showing high reproduction rates, therefore being remarkably aggressive. *E. crassipes* can accumulate relatively high amounts of metals and organic contaminants, and can easily tolerate environmental variations (Rommens et al. 2003). These biological characteristics explain water hyacinth rapid expansion and density (Unknown 1980) that, on some occasions, covers the entire water surface, depletes water quality and affects all endemic species.

Fig. 2: *Eichhornia crassipes* (Mart.) Solms, commonly known as water hyacinth (Almeida 2006).

Methodology

The activities proposed in this chapter should be performed in heterogeneous groups of 4-5 elements. The tasks are grouped in several sessions (table 1), some of them should be developed in the classroom and others outside (in the field or at a university to acquire a closer knowledge about the study area and about methodologies and techniques respectively). Table 1 also proposes a time for each session. The class where this activity was developed was formed with 18 students (6 girls and 12 boys) of differing ages (12-14 years). Four groups were established, each with 4-5 elements. The teacher of natural sciences and Área de Projecto (NDCA) was the same, which allowed for an easier development of this activity.

According to the methodologies of NDCA, the teacher should propose to students the development of a project that focuses on scientific issues which affect their lives and/or may be related with their region and the sustainability of ecosystems and water resources. After the choice of the theme, students' first task is to research and gather information about the roles of a mayor, an industrial worker, a farmer, a hotel manager and even a tourist, in order to defend and manifest the point of view and interests of the "person" learned. In class, students may encourage a debate where they have to carry out the interests of the person personified by them. During the discussion, the teacher may conduct it to the problem in question – Pateira de Fermentelos, as was defined above – focusing students' attention on the abundance of water hyacinths and consequent problems. Then, a laboratory lesson should be performed in order for students to practise some techniques and know some rules that may be taken into account in order to maintain her/his security and the colleagues' security in the laboratory. An initial questionnaire (Annex I) should be applied due to its importance in order to provide information to the teacher about the basic knowledge and concepts of students. We propose a set of questions with fast responses (multiple choice) that focuses on concepts and ideas about "Water quality evaluation and *Pateira de Fermentelos*". At this stage we do not expect students to know all the concepts referred to in the questionnaire; they may give incomplete or incorrect answers and even "I don't know". Some students retain misconceptions that can be detected in this kind of exercise, helping the teacher to create strategies in a way that allows the use of learners' knowledge and the development of correct concepts (Tunnicliffe 2006). It is important to know and understand how students acquire and structure the knowledge and the new information in order to avoid a poor/defective learning. Thus, to teach major concepts in biology it is necessary to understand students' former knowledge and identify what they are confused about and then create and give them opportunities and strategies for teaching conceptual changes (Tanner and Allen 2005). After this task, the groups may investigate and gather information about the study area – main activities, history of the region, groups of plants and animals that could be observed – and also information about water quality and aquatic resources: its importance to human beings. The search for information about the problem in question is an important stage of this project because everything that is discovered by each group may be classified, and accepted or rejected, to attach to the structure of their work. Students must have the ability to select information, otherwise the development of the following activities and the organization of the theme could become threatened due to a confused and inaccurate selection of information and ideas. Observation and the question-asking process are other skills that should be developed by students and will be helpful in further exercises. Thus, we suggest the promotion of a lesson outside room classes, in the school's gardens for example, where students are asked to formulate problem questions about the aspects that attract their attention in order to assess their perceptions. We suggest that pencil and paper should be given to students to write down their questions, to be easier to do a qualitative evaluation and to examine the kind of questions posed. The type of questions can reveal important information about the theoretical background of each student and also how he/she structures his/her thinking. According to Brill and Yarden (2003), asking questions is an important educational role and a basic skill in the development of scientific research that contributes to the build-up of knowledge by a dynamic process. Rutledge (2005) reports that decisions on issues which affect individuals' lives may be made according to biological and scientific information that should be analyzed critically. In that way, it is important for students to be sensitized to the area where they live by observing and making questions, to understand the real-world problems and base their decisions on evidence (Rutledge 2005). A level of knowledge may be achieved after these activities and we think it to be appropriate to visit the study area (the shallow lake of Pateira de Fermentelos) after these tasks. Nevertheless, before the field activity takes place, we suggest, in class, a lecture and interpretation of water quality legislation (Decreto-lei nº 236/98) and the introduction of some concepts about physical, chemical, microbiological, and biological parameters related with water analyses. If the theme concerns water quality of the shallow lake of Pateira de Fermentelos, it is important that students know how to evaluate the water quality of the lake. It also becomes necessary to teach students how to use military charts in order to have a guide to the field activity. An effective teaching of this subject is to give each student a copy of the military chart of the study area where they can practise and explore the main geological concepts (Annex II). Before the trip, the teacher should inform the class about the details of the activity such as: to wear comfortable and informal clothes, to follow the itinerary and the tasks in group, not to crop plants and living animals, and other advice which the teacher considers important. A report (guide) should be handed out with all information necessary to make the field activity a success (Annex III). One of the goals of the outside activity is to perform a *Herbarium* with the most abundant species of the study area. Thus, students may observe the biodiversity of the field trip and practise techniques of collecting plants (fig. 3) with the assistance of some teachers (from school and/or from the *Herbarium* of a University, as was the case with our students) and with the information acquired from previous tasks mentioned above.

Fig. 3: Students collecting vegetal samples

Each group may also collect water from the lake to perform microbiological analyses in the school laboratory or at a university. Our students visited a university where they carried out water analyses (fig. 4A and 4B) and had the chance to visit the *Herbarium* to observe how people work and do scientific research. Then, at school, the class may analyze the results of the microbiological analyses and answer a set of situations about "Water Quality Evaluation in Pateira de Fermentelos" (Annex IV).

Figs. 4A: Microbiological analyses in Aveiro University

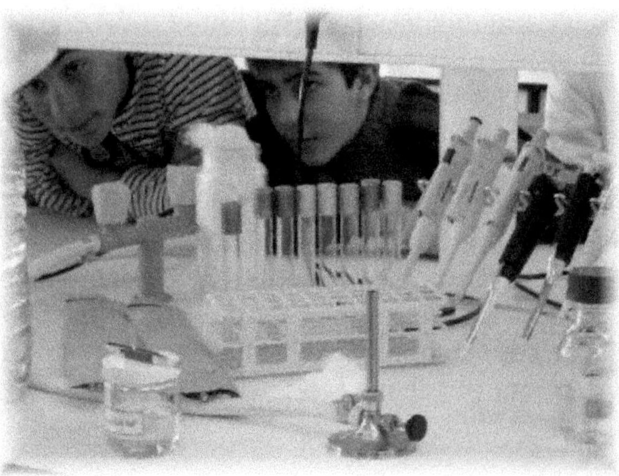

A

Figs. 4B: Microbiological analyses in Aveiro University

B

A lot of information has been acquired during the activities performed in and outside school. What has been learned becomes an important structure and a list of actions plans is made to minimize or solve the problem concerning Pateira de Fermentelos. Thus, the teacher may pose the following questions to the students:

1. **WHAT ACTION PLANS AND WORKING PROGRAMS SHOULD BE IMPLEMENTED TO SOLVE THE CONCERNING PROBLEM OF THE SHALLOW LAKE?**

2. **HOW TO AWARENESS CITIZENS FOR ENVIRONMENTAL PROBLEMS?**

It is important that students evaluate and take into account the interests of the farmers, industrial workers and the residential population when answering these questions. Thus, it seems important that the implementation of activities and working programmes give some "benefits". In this way, a relationship may be developed with these people and talks with them in order to offer some advantages with the implementation of action plans. The farmers, for example, may have to reduce the high amounts of nutrients, like nitrogen and phosphorous, to the soils that drain into the lake causing the eutrophication process. If they reduce these quantities, the water analyses may have no costs to farmers. The tasks and activities may be performed with the knowledge and acceptance of people who are directly related with the environmental problem. During the process a good relationship and discussion must be maintained with farmers, industrial workers and population by holding periodical meetings. Thus, it is a way to improve awareness of environmental problems, but more must be done. Firstly, it is important to make the school community aware and then the general public by means of campaigns, activities in the field and experiences that demonstrate the impact of such problems on the environment and in living organisms. In this class, students proposed developing an activity to celebrate 'Water Day', which consists of carrying out physical and chemical analyses to evaluate the quality of water samples collected by them and school workers, with an explanation to the "visitors" of the meaning of each parameter. The class made a poster "inviting" the school community to analyze a water sample collected by them, which had been well received due to the curiosity of the task.

During the debate, the influence of water hyacinths (*Eichhornia crassipes)* in water quality should also be named as environmental factors that affect the growth of this macrophyte, such as salinity and nutrients available. We suggest a research lesson about the impact of *E. crassipes* in water (physical and chemical parameters) and the influence of water properties in the development of this macrophyte in order to promote an informative debate, where students have the chance to share their ideas by explaining and applying their knowledge. A relevant aspect of this invasive plant is the formation of extensive stands that reduce or extinguish other species of plants and cause changes in water quality (Rommens et al. 2003). The exudates released by *E. crassipes* also affect water quality and populational parameters of some species (van Donk and van de Bund 2002), provoking changes in diverse functional groups which can be related to unexpected indirect effects through the food web.

The next stage of this project consists of an oral presentation in class where each group presents their work and conclusions about the problem of the shallow lake. Then, an oral presentation may be done to school and local communities in order to alert them to the urgency of becoming aware and active citizens about environment and natural resources concerns, recognizing the critical importance of sustainable livelihood. Finally, public campaigns in field activities may be done to sensitize the community to a problem that must be solved to restore the quality of water of the shallow lake. The knowledge acquired during this project should be evaluated by delivering, once again, the worksheets of Annex I and IV to students and comparing the answers with the prior ones.

Table 1: Group activities proposed in the performance of a theme concerning environment sustainability, in a NDCA

Session	Task	Time/duration
1	✓ Choose/propose the theme ✓ Research of information about the interests of a group of people directly related with the study area ✓ Debate	3h
2	✓ Laboratory lesson	1h
3	✓ Questionnaire (Annex I)	30'
4	✓ Research about the study area: – main activities; – history of the region; – groups of plants and animals that lives in the study area; – the importance of water quality and aquatic resources	2h
5	✓ School's garden activity "Question-asking process".	1h
6	✓ Preparation of field activity: – interpretation of water quality legislation; – interpretation of military charts (Annex II); – information about the details of the field activity (Annex III).	2h
7	✓ Field activity (Annex III)	[1]
8	✓ Visit to a University: – water analyses; – visit the Herbarium.	[2]
9	✓ Interpretation of the results of microbiological analyses ✓ Worksheet activity (Annex IV)	1h30'
10	✓ Debate to define strategies to: – implement working programmes to minimize the polluted effects in the study area; – awareness citizens to the environmental problems.	1h30'
11	✓ Oral presentation: – in class; – to school and local communities.	[3]
12	✓ Public campaigns	[4]
13	✓ Worksheets activities (Annex I and IV)	1h

Throughout the development of this project students are evaluated in different activities, promoting qualitative and quantitative evaluations according to the aims of each task. A quantitative assessment takes place by applying worksheets (Annex I-IV) in order to estimate students' knowledge and students' attitudes

1 The time to perform session number 7 depends the distance between school and the study area.

2 The time to perform session number 8 depends the distance between school and the University.

3 The teacher may decide if is done one or several presentation to school and local communities and thus how much time it is expended to this session.

4 The time expended in this session depend the source of public campaigns the students have organised.

towards water quality and Pateira de Fermentelos, in addition to the oral presentations where the class may prove that it has achieved the main goals of this project. A qualitative assessment also takes place every time the teacher registers the most important aspects of each session in the so-called "Teacher's Diary".

Results and Discussion

The students, where this project had been proposed, acquired several skills during the activities. A change of attitudes resulting in the acquisition of knowledge was shown during the course of the tasks. They showed more awareness and were concerned with the problems affecting their region. At the beginning of the activities water hyacinths were only seen as an ornamental plant to be taken and displayed at home. After the development of the activities, students were aware of the harmful effects of *E. crassipes* and knew why they should not touch it. Students' motivation during the activities was expressed by their actions and the way they carried out the work. They became more secure about their abilities and more responsible about the roles that they should assume as citizens. Morgado (2001) came up with similar results when applying field work in geosciences. According to Hodson (1994, 2000), group activities and themes related with students' lives develop a partnership between colleagues and motivate them by having the opportunity to carry out something new and interesting respectively. Studies about the students' region give them a real knowledge and the chance to become more active in the resolution of environmental concerns (Gandra 2003; Silva 2003).

This class, when compared to other classes with a different project, proved to be more proficient with ecological concepts as well as with the environmental problems. Fernandes (2003) also demonstrated a change of attitudes in a group of students due to the knowledge acquired during the activities conducted through an investigative course. Other authors (e.g. Pato 2001; Silva 2003) did not show significant differences in the knowledge of a control group and the experimental one.

The activities proposed in this chapter give the chance to work in groups, promote the development of scientific knowledge by the performance of activities and to share opinions and ideas.

Students' main difficulties concerned formulating questions relating to the problem and methodologies to solve or minimize the effects of pollution in the study area. At this point, the teacher may have a more active intervention.

Conclusions

Projects about school/environment relationships are often subexplored (Esteves et al. 2000), which may be due to the difficulties in undertaking outdoor activities. These activities require more time to prepare than theoretical lessons, where the same concepts are taught but with innovation in learning. Outdoor activities allow students to come into closer contact with the problem; they may take the school to outdoor locations and the environment into the school. These projects are a source of motivation in the development of thinking skills, helping students in the construction of scientific knowledge. It promotes students' interest in scientific research with a more cooperative and active participation where the teacher's role is to guide the performance of how students gain their knowledge correctly. One of the best ways to teach biological concepts is by connecting the real world with the aims of natural sciences and students' interests.

References

Almeida, A.S.R.L. (2006), A Avaliação da Qualidade da Água na Pateira de Fermentelos – caso de estudo na Área de Projecto. 107 p. Master thesis, Universidade de Aveiro, Portugal.

Almeida, A.S., Gonçalves, A.M.M., Pereira, J.L. and Gonçalves F. (2006), The impact of *Eichhornia crassipes* on green algae and cladocerans. Fresenius Environmental Bulletin 15 (12a), pp. 1531-1538.

Begon, M., Harper, J.L. and Townsend, C.R. (1996), Ecology. 1068 p. Blackwell Science. Oxford, UK.

Brill, G. and Yarden, A. (2003), Learning Biology through Research Papers: A Stimulus for Question-Asking by High-School Students. Cell Biology Education 2, pp. 266-274.

Cerqueira, M.A., Vieira, F.N., Ferreira, R.V. and Silva, J.F. (2005), The water quality of the Cértima river basin (Central Portugal). Environmental Monitoring and Assessment 111, pp. 297-306.

DEB (2001) Currículo Nacional do Ensino Básico – Competências Essenciais. 240 p. Ministério da Educação, Departamento da Educação Básica. Lisbon, Portugal.

Decreto-lei no 152/97 de 19 de Junho. In Diário da República no 139 Série A, 19-6-1997, Impressa Nacional Casa da Moeda, Lisbon, 1997 (in Portuguese).

Decreto-lei no 236/98 de 1 de Agosto. In Diário da República no 176 Série A, 1-8-1998, Impressa Nacional Casa da Moeda, Lisbon, 1998 (in Portuguese).

Donk, E. van and Bund, W.J. van de (2002), Impact of submerged macrophytes including charophytes on phyto- and zooplancton communities: allelopathy *versus* other mechanisms. Aquat. Bot. 72, pp. 261-274.

Esteves, M.A., Esteves, P. and Carvalho, R. (2000), Um poço provoca interrogações. In Trabalho Prático e Experimental em Ciências, pp. 651-658, Departamento de Metodologias da Educação e do Instituto de Educação e Psicologia da Universidade do Minho. Braga, Portugal.

European Union (1991) The Urban Waste Water Treatment Directive (91/271/CEE).

Felizardo, D. (2004), Área de Projecto. Porto Editora, Porto, Portugal.

Fernandes, A. (2003), A Influência do Trabalho de Projecto na Aquisição de Conhecimentos Científicos na Área Ambiental. Master thesis, Universidade Aberta, Portugal.

Gameiro, P.J.A.S. (1999), *A* Avifauna da Pateira de Fermentelos.114 p. Master thesis, Universidade de Aveiro, Portugal.

Gandra, J.P.O. (2003), Clube de Ar Livre – impactos de uma actividade de complemento curricular na formação dos alunos. 133 p. Master thesis, Universidade de Aveiro, Portugal.

Giordan, A. (1996), *A Educação Ambiental na Europa*. Cadernos de Educação Ambiental, nº1. Instituto de Promoção Ambiental, Lisbon Portugal.

Hodson, D. (1994), Hacia un enfoque más critico del trabajo de laboratorio. Enseñanza de las Ciencias, 12 (3), pp. 299-313.

– (2000), The place of practical work in Science Education. In Trabalho Prático e Experimental em Ciências, pp. 29-42. Departamento de Metodologias da Educação e do Instituto de Educação e Psicologia da Universidade do Minho, Braga, Portugal.

IUCN, UNEP and WWF (1980), World Conservation Strategy: Living Resources Conservation for Sustainable Development, IUCN, Gland, Switzerland.

Jeffries, M. and Mills, D. (1998), Freshwater Ecology – principles and applications. John Wiley and Sons Ltd. West Sussex, UK.

Moiseenko, T.I. (2005), Ecotoxicological approach to water quality assessment. Water Resources. 32, pp. 163-174.

Morgado, M.M.M. (2001), O trabalho de campo em Geociências: um percurso de investigação com materiais didácticos de orientação construtivista. 290 p. Master thesis, Universidade de Aveiro, Portugal.

Nova, E.V. (1994), Educar para o Ambiente – Projectos para a Área-Escola. 96 p. Texto Editora, Lisbon, Portugal.

Parr, T.W., Ferreti, M., Simpson, I.C., Forsius, M. and Kovács-Láng, E. (2002), Toward a long-term integrated monitoring programme in Europe: network design in theory and practice. Environmental Monitoring and Assessment, 78, pp. 253-290.

Pato, A.R. (2001), A influência das actividades de campo na aquisição de conhecimentos científicos. Master thesis, Universidade de Aveiro, Portugal.

Rommens, W., Maes, J., Dekeza, N., Inghelbrecht, P., Nhiwatiwa, T., Holsters, E., Ollevier, F., Marshall, B. and Brendonck, L. (2003), The impact of water hyacinth (Eichhornia crassipes) in a eutrophic subtropical impoundment (Lake Chivero, Zimbabwe). I. Water quality. Arch. Hidrobiologia. 158 (3), pp. 373-388.

Rutledge, M.L. (2005), Making the Nature of Science Relevant: Effectiveness of an Activity that Stresses Critical Thinking Skills. The American Biology Teacher 67 (6), pp. 329-342.

Salvia-Castellví, M., Iffly, J.F., Borght, P.V. and Hoffmann, L. (2005), Dissolved and particulate nutrient export from rural catchments: A case study from Luxembourg. Science of the Total Environment, 344, pp. 51-65.

Silva, M.F.V.B. (2003), A eficácia do desenvolvimento de trabalho prático de carácter investigativo na construção de conhecimentos. Master thesis, Universidade de Aveiro, Portugal.

Tanner, K. and Allen, D. (2005), Approaches to Biology Teaching and Learning: Understanding the Wrong Answers – Teaching toward Conceptual Change. Cell Biology Education 4, pp. 112-117.

Tilbury, D., Stevenson, R.B., Fien, J. and Schreuder, D. (eds.) (2002), Education and Sustainability: Responding to the Global Challenge, Commission on Education and Communication, 206 p. IUCN, Gland, Switzerland and Cambridge, UK.

Trudgill, T., Walling, D.E. and Webb, B.W. (1999), Water Quality – Processes and Policy. Wiley. West Sussex, UK.

Tunnicliffe, S.D. (2006), The importance of research to biological education. Journal of Biological Education, 40(3), pp. 99-100.

Unknown (1980), Conservação da Natureza, colectânea de textos de publicações da Liga para a Protecção da Natureza. Gabinete de Estudos e Planeamento do Ministério da Educação.

Velasco, J., Lloret, J., Millan, A., Marin, A., Barahona, J., Abellan, P. and Sanhez-Fernandez, D. (2006), Nutrient and particulate inputs into the Mar Menor Lagoon (SE Spain) from an intensive agricultural watershed. Water, air and soil pollution 176, pp. 37-56.

Annex I

This activity is anonymous and confidential and it aims to check your knowledge on the theme: *Water quality evaluation in "Pateira de Fermentelos"*.

Answer to these questions with frankness and try to do it by yourself – it is important to your teacher know how the main concepts are structured in your mind to help you in the construction of your knowledge.

1- Have you ever heard about water quality?
☐ Never ☐ Rarely ☐ Sometimes ☐ Often

2- In your opinion, water quality is important?
☐ No, it is not important ☐ No much important
☐ A bit important ☐ Very important

3- In your opinion, acceptable water quality is:
☐ Water without colour, smell or flavour. ☐ Water where is possible to swim in.
☐ Water where is possible to have fishes in. ☐ Water that obey laws.
☐ I don't know.

4- Have you ever heard about water chemical analyses?
☐ Never ☐ Rarely ☐ Sometimes ☐ Often

5- Do you know what water chemical analyses are used for?
☐ Yes, I do. They are used to ______________________________
☐ No. I don't. ☐ Vaguely

6- In your opinion, water chemical analyses are used to determine the concentration of: (Choose a maximum of eight options)

☐ Pesticides ☐ Nitrate ☐ Total coliforms ☐ Carbonates
☐ Cyanide ☐ Salmonellas ☐ Heavy metals ☐ I don't know
☐ Phenols ☐ Phosphates ☐ Faecal coliforms
☐ Mineral oils ☐ Nitrogen ☐ Faecal streptococcus

7- Have you ever heard about water microbiological analyses?
☐ Never ☐ Rarely ☐ Sometimes ☐ Often

8- Do you know what water microbiological analyses are for?
☐ Yes, I do. They are used to ______________________________
☐ No. I don't. ☐ Vaguely

9- In your opinion, water chemical analyses are used to determine the concentration of: (Choose a maximum of eight options)

☐ Pesticides ☐ Nitrate ☐ Total coliforms ☐ Carbonates
☐ Cyanide ☐ Salmonellas ☐ Heavy metals ☐ I don't know
☐ Phenols ☐ Phosphates ☐ Faecal coliforms
☐ Mineral oils ☐ Nitrogen ☐ Faecal streptococcus

10- Do know an animal that lives in polluted waters?
☐ Yes, I do. Which? ______________________________
☐ No, I don't.

11- Do know a plant that lives in polluted waters?
☐ Yes, I do. Which? ______________________________
☐ No, I don't.

12- Do you know Pateira de Fermentelos?
☐ Yes, I do. ☐ No, I don't. ☐ Vaguely.

13- Do you believe that Pateira de Fermentelos might be polluted?
☐ Yes, I do. Why? ______________________________
☐ No, I don't.
☐ Maybe.
☐ I don't know.

14- Choose the activities that are related with Pateira de Fermentelos.
☐ Agriculture ☐ Fishing
☐ Hunting ☐ Swimming
☐ Canoeing ☐ Diving
☐ Other ______________________________
☐ I don't know

Thank you for your collaboration!!

Annex II

School
"Área de Projecto" – 8th Form
Name:_________________________ **N º**_____**Group:** __________ ***Date:*** __/__/__

The following figure represents Pateira de Fermentelos.

Downstream

Upstream

Pateira de Fermentelos Military chart 1:25 000 of the study area (adapted from Gameiro 1999)

Answer the following questions about the study area. You can discuss your ideas with your group.

1. Identify the scale of the chart. ______________________________

2. How many meters in the field correspond ______________________________

   2.2. 1 mm in the chart? ______________________________

   2.3. 1 cm in the chart? ______________________________

3. Identify the distance between the level curves. ______________________________

4. Copy to a transparency the level curves and the watercourse of the study area.

5. Identify the maximum and minimum levels of the study area.

   ____

6. Identify the maximum and minimum descending slope of the study area.

   ____

   ____

7. Calculate the linear distance, between "Requeixo" and "Óis da Ribeira".

Annex III

School
"Área de Projecto" – 8th Form

Name: ______________________ **Nº** ______ **Group:** ______ **Date:** __/__/__

FIELD ACTIVITY IN PATEIRA DE FERMENTELOS

Itinerary:

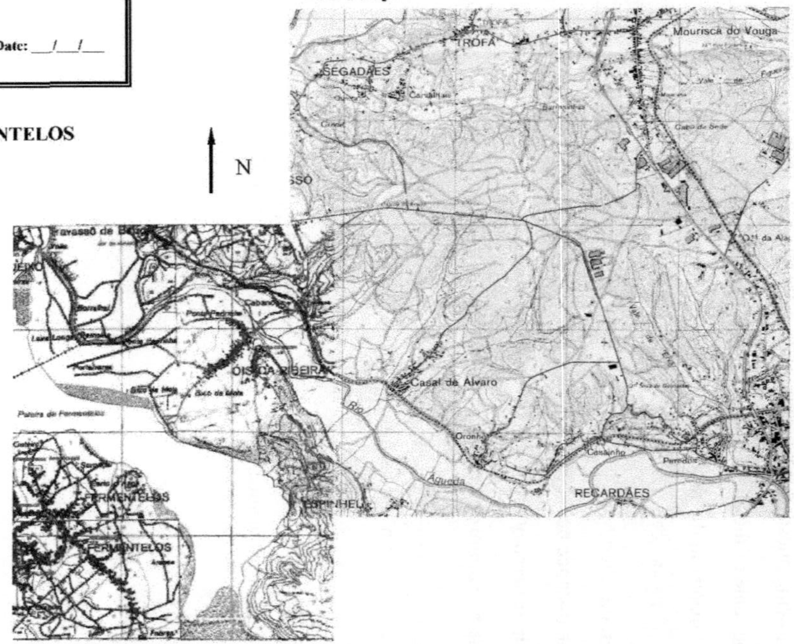

Recommendations:

- Don't collect living animals and/or plants in the field
- Don't go alone and be aware of possible dangers.
- Try to obey the itinerary and the established tasks.
- Follow the existing trails.
- Wear informal and suitable clothes to field work.
- Don't litter.

Programme:

8:45 – Get together in classroom.
8:50 – Set off to the park in Espinhel of Pateira de Fermentelos.
9:10 – Arrival at "Espinhel".
9:55 – Get together and going back to school.
10:15 – Arrival at school.

1. Mark in the map the route you have crossed until you arrived to the study area.

2. Write down the main characteristics that call your attention.

Annex IV

This activity is anonymous and confidential and it aims to check your knowledge about *Water quality evaluation in "Pateira de Fermentelos"*. Answer to these questions with frankness and try to do it by yourself – it is important to your teacher know how the main concepts have been structured during the activities.

No.	SITUATION	I totally agree	I agree	I totally disagree	I don't have an opinion
1	Drinking water from a fountain is a correct procedure.				
2	Drinking water from a community water supply system is a correct procedure.				
3	Drinking water from a fountain is a correct procedure because it is healthy.				
4	Drinking water from a community water supply system is a correct procedure because it is healthy.				
5	Using water from a fountain to wash food supplies and cook is a correct procedure.				
6	Using water from a community water supply system to wash food supplies and cook is a correct procedure.				
7	Using water from a well for bathing is a correct procedure.				
8	Using water from a fountain for bathing is a correct procedure.				
9	Drinking water from a lake is a correct procedure because animals also drink it.				
10	Drinking water from a community water supply is a correct procedure because animals also drink it.				
11	Crop water hyacinths to bring home is a correct procedure.				
12	The growth of water hyacinths in Pateira de Fermentelos is a correct procedure.				
13	Removing/eliminating water hyacinths from Pateira de Fermentelos is a correct procedure.				

Thank you for your collaboration!!

Chapter 11

"Questions in Biology" Designing an Online Discussion Forum for Promoting Active Learning about Evolution

Betina da Silva Lopes, Aurora Coelho Moreira, Maria Helena Pedrosa de Jesus

Abstract

The present context of Portuguese Higher Education as a consequence of the so-called 'Bologna Process' asks for pedagogical innovations in order to promote the acquisition and development of competences such as critical and creative thinking and problem-solving, where questioning plays a central role. The design and implementation of student-centered approaches, in order to enhance deep and active learning, requires an alignment between teaching strategies, learning outcomes and particularly assessment methods, which strongly influence students' approaches to learning.

This chapter describes the process of designing and implementing a teaching-learning assessment strategy called "Questions in Biology" in the context of a curricular unit taught by one Biology professor to first year undergraduates at the University of Aveiro (Portugal) during two academic years (2007/2008 and 2008/2009). "Questions in Biology" started as an online discussion forum (ODF), hosted at the e-learning platform of the university, where students had the opportunity to develop questioning, discussion and argumentation about topics on evolution. During the second year, some ODF improvements were introduced as well as the implementation of presential group discussions during lectures.

Students' performance was assessed during the study. Data collected by interviewing the professor and some of the students provided evidence of the positive impact on student quality learning. Also, as a result of this strategy, the number of teacher-student and student-student interactions increased, enhancing both oral and written questioning and argumentation by students. Results show the positive influence of "Questions in Biology" in attaining the intended learning outcomes and the development of competences by students, with the assessment playing a central role in the teaching and learning processes.

Introduction

Communication is the basis of all teaching-learning processes where questioning is an important element for the continuous exchange of ideas, knowledge and feelings between the teacher and students, and also between the students themselves. Indeed, several studies show that the promotion of a true spirit of inquiry can improve the quality of teaching and learning in several ways (Chin and Osborne 2008).

Students' questions can be very useful for the teacher, providing important evidence on learning gains (Chin 2001; Watts, Gould and Alsop 1997; White and Gunstone 1992), generating opportunities for the identification of incorrect conceptions as well as the main obstacles for conceptual understanding. In this way, students' questions have the potential in facilitating the design of teaching strategies according to students' needs and motivations (Chin 2007; Maskill and Pedrosa de Jesus 1997; Hadzigeorgiou 1999).

Students could also benefit from the questioning processes, triggering their thinking, scaffolded, and even extended, by a particular teacher question, or perhaps a question formulated by one of their peers (Chin et al. 2002; Pedrosa de Jesus and Moreira 2009). King (1994), for example, states that peer questioning promotes the social construction of learning since it might raise and solve socio-cognitive conflicts. During peer-interaction different perspectives emerge and are integrated. The formulation of one student question might lead to the formulation of a hypothesis, the establishment of predictions and reflections by other students.

Questions can lead to learning even when formulated by ourselves, since self-questioning might be helpful in organizing ideas and managing tasks. Indeed, from the students' perspective, the process of formulating a question requires attention on the contents and the selection of the main ideas, resulting in learning by itself (King 1994; Schmidt 1993).

The vast importance of questioning for the quality of teaching and learning processes seems to be rooted on its relation to the process of understanding itself:

> '*Questioning is one of the thinking processing skills which is structurally embedded in the thinking operations of critical thinking, creative thinking and problem solving. It consists of the smaller micro-thinking skills of recall, comprehension, application, analysis, synthesis and evaluation [...] Questions guide knowledge construction in the formation and changing of the cognitive networks or schemata.*' (Cuccio-Shirripa and Steiner 2000: 21).

Despite the benefits of asking (and answering) questions, our research work, developed in the context of higher education, has revealed that teacher-student interactions and even student-student interactions are not as frequent as would be desirable. In fact, many lectures tend to be monologues interrupted by some teacher questions, the main purpose of which tends to be the checking of knowledge acquisition. During the majority of university lectures, moments of 'true' dialogue, where the teacher and students try to share meaning, tend to be rare. Dialogues initiated by a student (question) seem to be even less frequent. Generally speaking, students rarely intervene by self-incentive and often only when solicited by a teacher question. Usually, undergraduates tend to adopt a passive posture, perceiving lectures as 'information reception moments' (Pedrosa de Jesus and Moreira 2009). According to Graesser and Person (1994), the low frequency (and low quality) of student questions can be connected to several reasons such as: difficulties in identifying relevant information, social barriers and lack of communicative skills.

Therefore, it urges rethinking classroom strategies in order to promote *effective* and *active* inquiry environments, as intended by the recent curricular reforms in Portuguese higher education courses as a consequence of the so-called "Bologna Process". More than the carrying out of 'new obligatory procedures' in a 'top-down' policy, it aims to promote the design and implementation of student-centred approaches, requiring an alignment between the teaching strategies, the learning outcomes and particularly the assessment methods (Biggs 2007; Veiga and Amaral 2009). It is well described in literature that assessment plays a central role in learning, strongly influencing student behaviour and the adopted study methods. Therefore, assessment should act as a tool for meaningful learning (Black and William 1998), and every learning experience should be assessed and properly adapted to each of the teaching-learning contexts (Hofstein et al. 2005).

"Questions in Biology" – Main Purposes and Initial Design

This chapter aims at describing the process of designing and implementing a student-centred strategy called "Questions in Biology", developed during two consecutive academic years (2007/2008 and 2008/2009) in the context of a curricular unit dealing with topics of evolution. The initial design intentions and its later adaptations due to field work experiences will be highlighted. The study aimed to support, motivate, inspire and 'contaminate' other Biology academics to 'try' out similar strategies, towards the promotion of the scholarly teaching (Trigwell, Martin, Benjamin and Prosser 2000).

The discipline, Themes and Laboratories in Biology (TLB), has been taught since 2006 by the same professor, during the second semester, to approximately 70 first-year undergraduates of Biology and Biology/Geology Education at the Aveiro University (Portugal). Students are divided into two classes, each one having a two-hour lecture per week.

The design and implementation of this strategy was planned collaboratively with the professor and two science education researchers, following a co-researcher methodology (Macaro and Mutton 2002). During the entire semester the two researchers attended and observed all lectures (non-participant observation) and had regular informal meetings with the professor to discuss future development of the teaching-learning processes of the discipline. Data was also gathered by interviewing (semi-structured interviews) the professor at the end of the semester in order to evaluate the implementation process of the strategy. Some students were also selected for individual and group interviews, based on their expressed interest in participating in the strategy. With the informants' consent, all interviews and observed lectures were audio-taped and verbatim transcribed in order to allow further content and discourse analysis, respectively (Tuckmann 1999).

The intention was to design a strategy that could result in the creation of a complementary environment for the lectures, providing students more opportunities to question, argument and discuss topics on evolution privileging a more structured way rather than the traditional 'one- way' classroom oral discourse. The central focus should be on students creating meaning, confirming and sharing understanding by means of developing an interactive community of learners addressed to high levels of thinking and meaningful learning, where questioning should play a central role. Since the discipline is only lectured once a week, in two-hour sessions, it was decided to take advantages of the potentialities of the new information communication technologies using the Online Discussion Forum (ODF) communication tool hosted at the e-learning platform of the university (fig. 1). ODFs have been used extensively and are one of the most commonly used tools, since asynchronous text-based discussions present several advantages for students: more opportunities to interact with each other, more time to reflect, to think and to search for further information before making a contribution to the discussion (De Wever, Schellens, Valcke and Van Keer 2006). There is, indeed, some empirical evidence about computer-mediated text-based communication facilitating higher-order thinking (Garrison et al. 2000; Garrison and Cleveland-Innes 2005). The design and implementation of an asynchronous online discussion plays a significant impact on the nature of interactions and influences the way students approach their learning. It is well known that the number of interactions does not reflect the quality of discourse content. Evidence of higher-order learning are usually related to the context of the learning environment and to an effective teaching presence that encourages participation and triggers discussion. The discussion should not be teacher-centred. Instead, interactions should be facilitated in a sustained manner by moderating and shaping the direction of the discourse. If deep approaches to learning are intended to be achieved, which is one of the main goals of "Questions in Biology", then there should be a specific design goal, providing clear participation requirements and content expectations. Also, assessment should be congruent with the intended goals (Garrison and Cleveland-Innes 2005).

Fig. 1: Main entrance to the ODF "Questions in Biology"

Considering the importance that assessment plays in learning, it was decided that all the students' contributions would be assessed and that each student should post at least two questions/comments in order to assure that their participation would be considered for summative assessment purposes, counting for a maximum of 10% of their final grades (2 points out of 20).

Before implementing the strategy, the main goals were discussed and clarified with the students, as well as the assessment criteria which have resulted in the following descriptive dimensions:

I. 'Quality criteria' (1,5 values out of 2), concerning two factors:

   I.1. Cognitive level of students' reasoning

   Questions can be classified according to the mental processes required to answer them. For example, the ASI categorization system of Pedrosa de Jesus, Teixeira-Dias, Almeida and Watts (2006) considers three levels of questions: A) Acquisition (low cognitive level), S) Specialization and I) Integration (high level questions) (see table 1 for a more detailed description).

   Since one of the aims was assessing the 'quality' of students' reasoning it was decided to code students' post using the ASI system. However we have introduced a slight adaptation: since the entire message is fundamental to contextualize and understand the thinking processes of each student, the assessment unit was the entire message and not only the questions *per se*. It was also decided to assess all the students' post whether the question was explicit or not, because the inquiry disposition of the student might be embedded in the message without explicitly formulating a question.

I.2. Scientific accuracy
Students' posts were also classified according to scientific accuracy considering three levels: B1) scientifically incorrect; B2) with some scientific incorrectness; B3) scientifically correct.

Finally, quantitative criteria were defined in order to better translate the 'quality' of student posts into a discrete value. The combination of the *cognitive level* with the *scientific accuracy* of each post resulted in four different 'levels of quality' of the post: 0, I, II and III. The definition of these ordinal categories was based on concrete examples of students' posts for all possible combinations (for example A2B1), and also took into account the defined learning outcomes of the curricular unit, (see step 1 of fig. 2). For evaluation purposes, different 'weights' for each level of 'post quality' were defined (see step 2 of fig. 2).

II. Frequency criterion (0.5 points out of 2):
Besides 'quality', the number of students' posts was also a criterion for assessment. To ensure that participation would be considered for final assessment, each student had to post at least two questions/comments during the semester. Higher frequencies were considered positively: more than 5 posts – 0.5; 2 or 3 posts – 0.3; 2 posts (the minimum required) – 0.

The final mark was achieved by adding the value of qualitative assessment indicators (maximum of 1.5 points) to the value concerning the frequency of posts (maximum of 0.5 points).

In the following sections, particular aspects of the implementation of the strategy during 2007/2008 and 2008/2009 and some student's outputs will be described and discussed. Transcripts from the interviews with the teacher and some students will also be presented to complement and sustain discussion.

Table 1: The ASI questioning categorization System (Pedrosa de Jesus, Teixeira-Dias, Almeida and Watts 2006) with examples of questions from the "Questions in Biology" forum

Acquisition (A) level (low cognitive level)	**Specialization (S) level**	**Integration (I) level (high cognitive level)**
• Deal with relatively simple ideas, objects, processes, or concepts, which do not require evaluation, judgement, or drawing conclusions. • Aim to confirm explanations or to clarify conceptual issues. • Tend to be 'stick-to-the-facts' questions, whose answers mainly rely upon memory processes.	• Go beyond mere search for information; • Aim to establish relations and understand/ interpret the surrounding world. This is to both expand and test constructed knowledge. • These questions transcend the specific or detailed level of comprehension, in order to generalize into meaningful patterns.	• Reorganize concepts into novel patterns. Hypothesize new or different applications of principles learned. • Aim to reconcile different understandings, resolve conflicts, test circumstances, force issues, and track in and around complex ideas and their consequences. • May also include characteristics from the specialization phase.
Example: *"What is after all the difference between Lamarckism and Darwinism? Lamarckism considers that the most used organs will develop, while Darwinism considers that the natural selection only allows the survival of the fittest?"*	Example: *"Taking into account the Neo-Lamarckism theory, and without contesting Weissman's experiences, is it possible that the environment influences the genes and gives origin to a heritable characteristic?"*	Example: *"The bee after stinging dies. This happens because the sting remains inside the victim and the bee's viscera are pulled out. How can we explain this by the mechanism of natural selection? Shouldn't the bees have developed an evolutionary mechanism that allows them to survive, since the bee uses the sting to defend herself (= survive)?"*

Fig. 2: Translation of the "quality" criteria of students' posts into a quantitative mark for summative assessment purpose

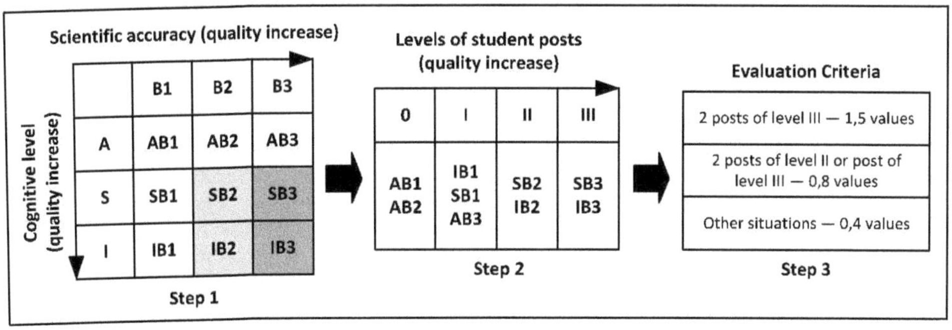

"Questions in Biology" – 1st Edition (2007/2008)

The purposes of the ODF as well as the assessment criteria were explained and clarified by the professor during the opening lecture of the semester. The "Questions in Biology" forum started with a message posted by the teacher (fig. 3), communicating explicitly a specific learning challenge, which represents a moment identified by Garrison and colleagues (2000) as a 'triggering event'. According to these authors, 'a critical role of the teacher is to initiate, shape and, in some cases, discard potentially distracting triggering events so that the focus remains on the attainment of intended educational outcomes' (Garrison et al. 2001: 10).

Fig. 3: First triggering event of "Questions in Biology" (2007/2008)

> Subject: (R)evolution in Biology
> Theodosius Dobzhansky (1900 - 1975, genetidst) published, in 1973, in the Journal *American Biology Teacher* (vol. 35, pp. 125 -129), a famous paper entitled "Nothing in Biology Makes Sense Except in Light of Evolution". What is the extend of this sentence? Does anyone want to comment it? (Article in attachment)

The ODF lasted nearly four months (the entire semester). Students questioned, discussed and argued about topics on evolution having posted a total of 222 posts, which represents a mean of 3.17 posts per student (fig. 4). A larger number was written near the end of the semester, namely during the first two weeks of June 2008. Despite the fact that the teacher enhanced in each lesson the importance of a continuous participation in the forum, the majority of students behaved strategically by posting the 2 required messages near the deadline in order to fulfil the assessment criteria. In spite of this, the global analysis of the quality of all posts during the semester revealed that 42% were at the highest level (Level III) and 13% were at Level II. Thirty per cent of the posts were assessed at Level I and 15% at Level 0. Since the majority of Level 0 and Level I posts were written during the final week of the ODF (76% and 65% respectively), it might be assumed that 'last-minute' posts tend to be of low quality, having the risk of jeopardizing deep learning. Some of the students interviewed confirmed this assumption:

> "When I realized that the deadline was so close I said to myself 'I need to do something!'... but I saw so many things written in the forum that I realized I wouldn't have time to read it all, so I have decided to choose 2 or 3 topics and answer to them... I wrote a brief message based on those topics (...) If this wouldn't have been included in the assessment, maybe I wouldn't have made an effort." (Filipe – fictional name)

> "Even though I was not able to make a comment with the objective of getting a higher grade, at least I have participated not to take the risk of losing everything, right? (...) this last comment was only to participate." (João – fictional name)

On the contrary, Level III posts achieved the highest frequency during the middle of the semester, namely during May, with 47 posts (51% of the total number of Level III posts). This evidence reinforces the fact that students need time to reflect, adapt and integrate knowledge, and that ODFs can promote high levels of thinking if proper conditions are created, such as giving students the opportunity to formulate several posts, and to improve the following ones by constructive teacher feedback, and also by reading posts from their peers.

Fig. 4: Frequency of students' post and level of quality during first edition (2007/2008)

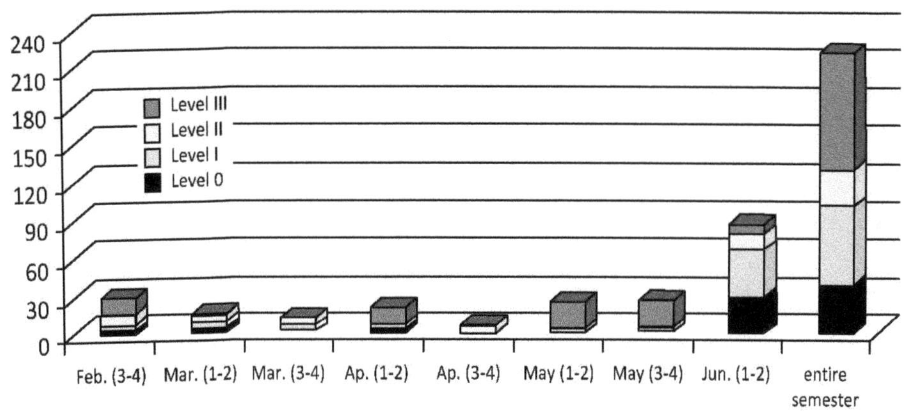

The fact of ODF being a written-based communication seems to also have a positive influence on the quality of students' intervention. Evidence from oral discourse analysis, based on transcripts from the audio taped lectures, showed that a total of 60 students' questions took place during the semester, corresponding to less than one question per student. A mean number that is, nevertheless, higher than in many other first year university disciplines. The large number of students' oral questions formulated during 'TLB' lectures was considered at a low cognitive level, corresponding to 53%, and only 13% were at the highest cognitive levels. These are not surprising results, since when it concerns oral discourse the number of students' questions tends to be too small and the cognitive level preferentially low (Maskill and Pedrosa de Jesus 1997). Indeed, according to Garrison and colleagues (2000), text-based communication provides time for reflection, for organizing and structuring ideas, for encouraging discipline and rigour in thinking and communication, being very closely connected to the development of critical thinking.

Also, comparing students' behaviour during lectures and the ODF, it was clear that many of the students that generally attended the lecture silently took the chance of presenting their opinion or doubts via the ODF. This is also an advantage of this alternative way to communicate, providing opportunities for students to interact with each other and the teacher, according to their preference of communication (written or orally):

> "At least while we are studying we have this facility, otherwise we would have to contact the teacher (…). Also I think it is useful for those persons who are shy and for those who haven't the capacity and time for formulating questions during classes." (Alexander – fictional name).

It is important to emphasize that the teacher, at any moment, didn't 'force' a specific topic. The most posted subjects were indeed raised by the students (not the teacher). For example: creationism/intelligent design; evolution theories vs religion; human evolution vs cultural evolution; evolution and homosexuality and specific examples of evolution – such as Platypus (*Ornihorunchus anatinus*). The intention was to allow students to discuss topics of their 'genuine' interest. However, whenever appropriate, the teacher enhanced particular important ideas raised by the students, challenging or giving feedback to them in order to stimulate participation and sometimes refocusing discussion (see fig. 5) concerning the contents and learning outcomes of the discipline.

Fig. 5: Example of a student-teacher interaction episode at the ODF

Maria: This discussion about human evolution is causing a lot of controversy not only in this forum but also in my own thinking. Sometimes, I just start agreeing with some of the opinions I read, but suddenly these ideas seem so controversial that I can't stop questioning myself. I have already referred to this, but I have to reinforce it again: is it plausible to talk about evolutionary mechanisms in humans?! Is it possible to apply to our own species this panoply of terms that humans themselves have idealized for describing other species? Don't you think that rationality could be a factor that makes us one great exception to evolution? Shouldn't we question ourselves about the evolution of human species when all terms created /used were idealized for other species?

Teacher: I'm writing this comment here because I consider that Maria has opened up a 'window' of discussion that nobody has taken to. To those that consider that comments are always around the same topic, we have here a good opportunity to 'deviate'. (Several students' post after)

Throughout the semester the teacher posted a total of 39 messages, which were deliberately reduced as we can see by the following statement from his interview:

> "I realized very soon that, after I have written a post on the forum, the discussion between the students tended to drop. They tended to post messages with the perspective of delivering a message to me. But that is not the point. It is THEIR discussion." (Teacher)

Indeed, the students initially clearly posted isolated messages addressed to the teacher, and the following posts tended to have no 'logic continuity' with the previously presented ideas. This means that the student can be actively engaged in the discussion, posting a lot of messages, but not interacting with his peers, he is instead constructing meaning individually. Students' engagement in monologues without connecting to other contributions is one of the challenges that the teacher faces while conducting discussion (Garrison and Cleveland-Innes 2005). In consequence, besides appealing several times for the importance of interaction between the students, the teacher started to structure his participation by integrating in one single post the ideas of several students, showing the 'expected' exercise of trying to connect others messages with own thoughts and not just 'dropping' an isolated idea.

The global evaluation of the "Questions in Biology" forum was very optimistic as it was considered successful in motivating students as well as supporting them in developing written questioning and argumentation. The teacher, at the end of the semester, also made a very positive balance focusing the potential of this kind of strategy in helping him to know the students better as well as their learning capabilities, which are normally only evidenced at the end of the semester during the typical written exam:

> "This kind of strategy is very time-consuming. But the final balance is very positive. It helps me to feel more confident about the students' final grades. I started to get to know the students better ... almost each student ..." (Teacher)

"Questions in Biology" – 2nd Edition (2008/2009)

After encouraging and stimulating results from the first edition, it was decided to continue using the strategy during the following academic year. Some improvements, based on acquired experience from the previous year, were introduced, namely:

i) Construction of a written document with important guidelines for students about the "Questions in Biology" forum:

   During the first year of implementation, students frequently had doubts, specifically about the assessment of their participation in the ODF. In order to better support students, and also to minimize the repetition of instructions, a document describing the main learning outcomes of the discipline and how they were related to the "Questions in Biology" goals, was written. Assessment criteria and specific guidelines on 'how to maximize' the quality of students' intervention using concrete examples of students previous posts were also integrated. This document was posted on the e-learning platform after being clarified and discussed with the students during the first lecture.

ii) Organization of the "Questions in Biology" forum into three smaller periods of time:

   As a result of the students' tendency to deviate from evolution as the main theme and to 'post' frequently isolated messages without any link to previously exposed arguments or repeating the same idea without adding new aspects of it, it was decided to try to 'orient' the discussion along specific themes by splitting the ODF into three thematic blocks (focusing on each discussion). Each thematic block was initiated by a teacher 'triggering event' related to the curricular contents being covered at that time during lectures (table 2). In order to ensure a continuous participation on the ODF, it was also decided that each student must post at least two comments/questions in each of the three blocks in order to ensure the inclusion of their participation in the final assessment (maximum of 2 points).

iii) Introduction of a formative assessment period:

Considering students' initial difficulties and doubts in understanding the purposes of the strategy, it was decided that the first block (three weeks) would be only considered for formative assessment purposes, giving them the time for adaptation to the new strategy. At the end of the first block, during a 'normal' lecture, the teacher, in collaboration with the researchers, commented on students' difficulties and gave some stimulating suggestions and constructive feedback, supported by exploring some students' selected posts.

iv) Extending the "Questions in Biology" strategy by implementing presential group discussions:

Data collected by interviewing some students at the end of the first year revealed that several colleagues, particularly those that tended to participate and discuss frequently during normal lectures, felt somehow 'harmed' by the fact that only written participation in the ODF was considered for assessment. Therefore it was decided to introduce two presential group discussions, aimed at giving students the opportunity to synthesize the main arguments/ideas that had been emerging during the lectures and the ODF discussion. Both were carried out during a 'normal lecture' lasting for two hours. The first discussion, with a formative purpose, was organized in order to match the end of the first thematic block. The second discussion moment was at the end of the semester after the third thematic block. It was then decided that only the second discussion moment would be considered for summative assessment purposes. Since it was the first time that this strategy was adopted, corresponding to an 'isolated' moment when compared with other tasks that students had to accomplish during the semester, the final discussion was proposed to account for 0.5 points out of 20. All students agreed. Therefore, the students' participation in the "Questions in Biology" could count for 2.5 points out of 20 on their final grade, instead of the 2 points like the previous edition. The assessment criteria for participation in the forum were maintained (1.5 points for quality and 0.5 points for frequency), since it was found useful by analyzing and assessing the discourse content, according to the learning goals.

Table 2: Initial triggering events of each thematic block (2008/2009)

Initial Triggering Event	Main topic of the Programme
Subject: Thematic block 1 The question is this – Is man an ape or an angel? My Lord, I am on the side of the angels. I repudiate with indignation and abhorrence these new fanged theories. Disraeli, Benjamin (1804-1881, British Statesman Prime Minister)	Block 1: Development of the scientific reasoning about evolution (February)
Subject: Is it possible to refute Evolution? Evolution, by natural selection, is a theory that explains the origin of adaptation, of complexity and diversity of all living beings. It is a theory based on group of explanations confirmed by observations and experiences that are already accepted as facts. The collection of evidences, that corroborate the formulated explanations, is abundant and diverse. Is it possible to observe evolution in nature? Is it possible to measure evolution in the lab? Who wants to follow these clues?	Block 2: Evidences of Evolution (March)
Subject: Evolution with(out) limits Ernest Mayer, in 1940, stated, during a debate, that a new species only will appear if a population, geographically isolated from his parental species, acquires, during the isolation period, characteristics that favour or maintain reproductive isolation, even when external barriers disappear. Is the geographic splitting of one species into two groups of individuals enough to observe the formation of a new species? What about the chromosomes? Where do they 'stand' in all of this? In other words, can we conclude that speciation is the same as Evolution?	Block 3: Speciation mechanisms (May)

As regards the online discussion forum, a total of 248 student posts were written throughout the semester: 43 posts during the initial formative period, 126 during the first summative discussion and 79 posts during the second summative discussion (fig. 6). Strategic students' posts, on the eve of the deadline, were again observed during the formative block. Therefore, the teacher used the formative period as an opportunity to give clear instruction to the students and to emphasize how their behaviour could have negative effects on the summative assessment of the posts: if this situation would repeat in the following blocks it would be very difficult to give feedback and to assess the posts at the appropriate time. The teacher's advice and recommendations had indeed some effect on the following blocks, but not as much as would have been desirable.

From the total number of students attending the curricular unit (N= 70), fourteen did not participate in the online forum. Thirty-nine participated during the entire summative period (both blocks) and 17 students only participated in the first summative challenge, having no participation at all during the second block. This corresponds to a mean of 2.25 messages per student during the first summative block and a mean of 2 messages during the second one.

As a consequence of the decrease in the number of students' participation, the number of students' posts also dropped to around 40%, dropping from 129 to 79 posts. Students' feedback from the interviews revealed that the nature of the theme was the main cause, since they recognized that they did not even know what to post on.

"For example, on the last theme, I was not able to comment until the day before because I thought it was too complex, and there was too many things to read (…) I didn't know what to write, it was a difficult theme to comment on." (Jorge – fictional name)

"On the last block there were less topics being addressed and none of them raised my interest … and to start a new topic, well, I didn't know where to start." (Manuel – fictional name)

On the other hand, the teacher also perceived that possibly some students did not understand that they should have contributed at least two posts in each block. This situation was discussed afterwards in the classroom, and it was decided to drop the criterion of 'obligation' to participate in each block. Therefore, for assessment purposes, the first and the second summative blocks were considered as just one. Frequencies of more than two posts per student were, like in the previous year, considered positive.

Concerning the quality of students' posts, the patterns used in the previous academic year were globally maintained (fig. 6): posts at Level III were the most frequent (51%, 63% and 31% for the total number of posts of each block respectively). Once more, Level III posts achieved its highest expression during the middle of the semester. On the contrary, lower level posts, namely Level 0 and Level I, were more frequent at the beginning and end of the semester (Level 0: 16% for block 1, 12% for block 2 and 20% for block 3; Level I: 16%for block 1; 13% for block 2 and 25% for block 3).

Fig. 6: Frequency of students' posts and level of quality during the second edition (2008/2009)

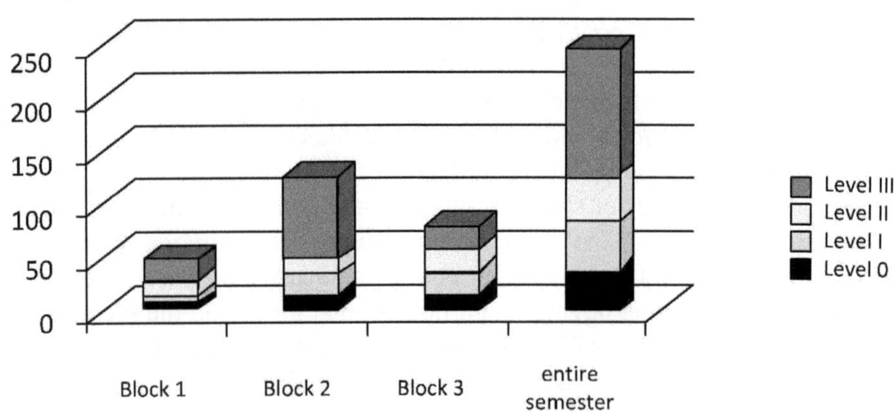

As mentioned previously, two presential group discussions were organized. During the summative group discussion, a mean of 21 students participated in each class. In 'normal' lectures, five to ten students tried to interact with the teacher, who constantly appealed and stimulated students' involvement in the classroom discussion. Also, student-student interactions, which were very rare during lectures, rose considerably during the group discussions addressed to the "Questions in Biology" forum. Other behaviour was also observed and considered very positive, proving students' engagement, such as: i) taking notes from what was being said, particularly statements from other peers, ii) using them to support and structure their talk, and iii) bringing into the discussion some references, such as books marked on specific pages. For example: "Climbing Mount Improbable" and "The Blind Watchmaker: Why the Evidence of Evolution Reveals a Universe without Design" by Richard Dawkins, or "The Red Queen" by Matt Ridley.

Since it was decided from the beginning to consider for assessment students' participation in the last summative discussion, a specific observation grid was designed in order to check for evidence of students' competences such as criticality and clarity of their interventions (fig. 7). During this summative discussion, the two science education researchers, acting as non-participant observers, filled in the observation grid for each student who participated in order to allow the teacher to fully concentrate and engage in the discussion. Afterwards, the researchers discussed and crossed their annotation, ranking the students on a scale from 0.1 points (level 1) to 0.5 points (level 3), according to the established criteria (fig. 7). This proposal was then discussed with the teacher.

As in the previous year, the final balance of the implementation of this strategy was considered to be very positive. Again it was possible to observe, during the semester, evidence of students' motivation and engagement in learning. The reflection on students' behaviour during the first year enabled the strategy to be 'fine-tuned' according to undergraduates' needs and characteristics. The teacher also considered "Questions in Biology" as better settled towards the desirable goals, and prized the entire *process* of designing and implementing the strategy, aiming to create a learning environment that provides stimulus and encouragement for students' development through small shifts in the adopted teaching practices. Indeed, like Teixeira-Dias et al. (2005) have stated:

> "(...) activity of 'fine' tuning is not so much a result of a once-for-all modification, but a relatively constant concern and evolution that requires almost permanent consideration of students' questions and feedback, and of student–teacher and student–student interactions." (Teixeira-Dias, Pedrosa de Jesus, Neri de Souza and Watts 2005: 1127).

Fig. 7: Levels of students' skills and competencies assessed during presential group discussion

Level	1	2	3
Relevance	Superficial opinion without any relevant arguments/ non-fitting arguments.	The students answers parcially to the topic in discussion.	The student express is a relevant opionion using valid arguments.
Criticality	The students doesn't show any criticality. He doesn't questions the intervention of his colleagues.	The students shows some criticality. Questions sometimes the intervention of colleagues on what concerns accuracy and relevance.	The student evidences clearly a spirit of criticality, questioning colleagues' intervention on what concerns accuracy and relevance.
Cognitive Level	The students participates in order to clarify, confirm ideas/information.	The students participates in order to unterstand, compare, relate the topics under discussion.	The student participate in order to explore/hypothetize arguments.
Clarity	Students' intervention is unclear. Ideas are mixed up. There is no evident logical sequence.	Ideas are presented in a clear, understandable mode.	Ideas are presented in a clear, understandable mode. Student's intervention clearly evidences an effort of organizing information.
Scientific accuracy	Scientifically incorrect student's intervention.	Students' Intervention with some scientific incorrectness.	Scientifically correct student's intervention.

Main Conclusions

The present context of Portuguese higher education asks, in the light of the so-called 'Bologna Process', for curricular and pedagogical reforms in order to promote the acquisition and development of competences such as critical questioning and argumentation. While curricular reforms ask for larger course tuning, pedagogical reforms might start in the classroom itself, with slight initial changes on teaching practices.

Within this chapter it was intended to describe the process of designing and implementing a strategy called "Questions in Biology", aimed at enhancing students' questioning about evolution topics, and, therefore, contributing towards active and in-depth learning of this theme. Although the strategy was implemented in the context of a discipline taught by one teacher, it was not a 'lonely' process. In fact, the design and also the implementation of the strategy resulted from collaborative work between the teacher and a group of researchers in science education. Following a co-researcher model (Macaro and Mutton 2002), which allows each participant in the process to benefit from the enterprise, the science education researchers had the opportunity to do research in natural teaching-learning contexts, and the teacher had the opportunity to reflect on data collected from his own lectures, obtaining also some support in the design and implementation of innovative strategies:

> "These kind of strategies are very important for the students, and exciting for the teacher. But, it is, honestly, impossible to do these things alone. There is no time and too much pressure." (Teacher)

The strategy "Questions in Biology" started to be an online discussion forum where the teacher initiated the discussion by communicating a specific learning challenge, and students interacted mainly with each other, discussing specific ideas related to evolution. During the second year, the strategy was implemented again, but with some adaptations, improvements and innovations according to the acquired experience and the students' outcomes from the first edition.

Focusing on the quality of students' learning, evidence for the positive impact of the strategy described were observed, such as the increase of teacher-student and student-student interactions, both orally (group discussion) and written (ODF). Particularly the high frequencies of Level III posts on the discussion forum provide evidence that students are able to think at higher cognitive levels when proper conditions are given. Through the ODF, students had the opportunity to share and construct collaborative knowledge, also training written argumentation. This allowed the teacher to get to know better the learning abilities of each student, even before the final exam. Finally, some evidence of promoting study methods was also observed. In order to participate in (oral and written) discussions with their *own* ideas and their *own* knowledge, several students started to seek information in books or other references besides the indicated course books.

It is believed that the success of the "Questions in Biology" is rooted on the conscious effort of aligning the adopted teaching methods with the real learning outcomes and particularly with the assessment modes, since evaluation influences student behaviour and adopted study methods:

> "I think assessment has influenced us … We had to pose at least two questions but I ended up making more … The first question I asked I felt obliged to do so … But, then the other questions came out spontaneously because we were interacting, and then it became natural." (Anna – fictional name)

Formative assessment moments, such as continuous teacher feedback in order to stimulate and sometimes focus students' reasoning, was also revealed as being very important for students' learning.

It was realized that the discipline, through the implementation of "Questions in Biology", stood out due to the creation of a 'complementary environment' that allowed students to 'reveal' other interests and abilities besides the 'memorization of information':

> "I liked very much this discipline, because here we were assessed by other skills, besides writing well at the exam ... I know that I am better talking. We discussed a lot in this discipline. In the other disciplines this didn't happen ... unfortunately" (Peter – fictional name)

Considering the above, it is our opinion that one possible way of successfully implementing innovative pedagogical approaches is to focus and reflect on its own dynamic characteristics. It is not positive, or realistic, to perceive 'pedagogical reforms' as once-for-all modifications that are perfect during their first trial. Collaborative work between university teachers and researchers in science education might be one step forward to enhance quality in higher education, promoting the scholarship of teaching and learning.

References

Biggs, J. (2007), Teaching for quality at University. 3rd ed., 360 p., Open University Press, Philadelphia, USA.

Black, P. and William, D. (1998), Assessment and Classroom Learning. Assessment in Education, 5 (1), pp. 7-73.

Chin, C. (2001), Learning in Science: what do students' question tell us about their thinking? Educational Journal, 29 (2), pp. 85-103.

– (2007), Teacher Questioning in Science Classrooms: approaches that stimulate productive thinking. Journal of Research in Science Education, 44 (6), pp. 815-843.

Chin, C., Brown, D.E., and Bruce, B.C. (2002), Students-generated questions: a meaningful aspect of learning in science. International Journal of Science Education, 24 (5), pp. 521-549.

Chin, C. and Osborne, J. (2008), Students questions: a potential resource for teaching and learning science. Studies in Science Education, 44 (1), pp. 1-39.

Cuccio-Shirripa, S. and Steiner, H. (2000), Enhancement and analysis of science questions level middle school students. Journal for Research in Scientific Teaching, 37 (2), pp. 210-224.

De Wever, B., Schellens, T., Valcke, M. and Van Keer, H. (2006), Content analysis schemes to analyze transcripts of online asynchronous discussion groups: a review. Computers and Education, 46, pp. 6-28.

Garrison, D.R., Anderson, T. and Archer, W. (2000), Critical Inquiry in a Text-based Environment: Computer Conferencing in Higher Education. The Internet and Higher Education, 2 (2-3), pp. 87-105.

– (2001), Critical thinking, cognitive presence and computer conferencing in distance education. American Journal of Distance Education, 15 (1), pp. 7-23.

Garrison, D.R. and Cleveland-Innes, M. (2005), Facilitating Cognitive Presence in Online Learning: Interaction Is Not Enough. American Journal of Distance Education, 19 (3), pp. 133-148.

Graesser, A.C. and Person, N.K. (1994), Question asking during tutoring. American Educational Research Journal, 31, pp. 107-137.

Hadzigeorgiou, Y. (1999), On problem situations and science learning. School Science Review, 81 (294), pp. 43-48.

Hofstein, A., Navon, O., Kipnis, M., and Mamlok-Naaman, R. (2005), Developing students' ability to ask more and better questions resulting from inquiry-type chemistry laboratories. Journal of Research in Science Teaching, 42 (7), pp. 791-806.

King, A. (1994), Guiding knowledge construction in the classroom. Effects of teaching children how to question and how to explain. American Educational Research Journal, 31 (2), pp. 338-368

Macaro, E. and Mutton, T. (2002), Developing language teachers through a co-researcher model. Language Learning Journal, 25 (1), pp. 27-39.

Maskill, R. and Pedrosa de Jesus, H. (1997), Asking Model Questions. Education in Chemistry, 34 (5), pp. 132-134.

Pedrosa de Jesus, M.H., Almeida, P., Teixeira-Dias, J.J. and Watts, M. (2006), Students' questions: building a bridge between Kolb's learning styles and approaches to learning. Education + Training, 48 (2-3), pp. 97-111.

Pedrosa de Jesus, M.H. and Moreira, A.C. (2009), The role of students' question in aligning teaching, learning and assessment: A case study from undergraduate sciences. Assessment and Evaluation in Higher Education, 34 (2), pp. 193-208.

Saadé, R.G. and Huang, Q. (2009), Meaningful Learning in Discussion Forums: Towards Discourse Analysis. Issues in Informing Science and Information Technology, 6, pp. 87-100.

Schmidt, H.G. (1993), Foundations of problem-based learning: some explanatory notes: Medical Education, 27 (5), pp. 422-432.

Teixeira-Dias, J.J.C., Pedrosa de Jesus, H., Neri de Souza, F. and Watts, M. (2005), Teaching for quality learning in chemistry. International Journal of Science Education, 27 (9), pp.1123-1137.

Trigwell, K., Martin, E., Bejamin, J. and Prosser, M. (2000), Scholarship of Teaching: A Model. Higher Education Research and Develoment, 19 (2), pp. 155-168.

Veiga, A. and Amaral, A. (2009), Survey on the implementation of the Bologna process in Portugal, Higher Education, 57, pp. 57-69.

Tuckman, B.W. (1999), Conducting Educational Research. 608 p. Wadsworth Publishing, New York, USA (ISBN-13: 978-0155054776).
Watts, M., Gould, G. and Alsop, S. (1997), Questions of Understanding: Categorizing Pupils' Questions in Science. School Science Review, 79, pp. 57-63.
White, R. and Gunstone, R. (1992), Probing Understanding. 196 p. Falmer Press, London, UK (ISBN-13: 978-0750700481).

Chapter 12

Field activities within an Environmental Sciences Program: a B-Learning Case Study

Carla Padrel Oliveira, Ana Paula Martinho, Sandra Caeiro, Filomena Amador, Fernando Caetano

Abstract

A reflection on the use of fieldwork in environmental science programmes is conducted in this paper. It is generally agreed that for fieldwork to be effective, it must serve a clear purpose in the curriculum. For a good curriculum design, a fieldwork course must complement, enhance or extend an existing part of the curriculum or fulfil some major objective of the curriculum as a whole. Thus a successful integration of fieldwork within an undergraduate degree programme is particularly important. For an environmental science course delivered through online methodology, it becomes evident that special attention should be devoted to the design and implementation of such a course. If we consider the more general definition where the field is seen as the location, outside the classroom setting, where learning takes place, then fieldwork is the set of activities that will facilitate students' learning. It is generally agreed that a field course is much more difficult to integrate than a set of activities closely related to a specific subject. However, in a distance-learning university it is not plausible that students perform this type of activity very often. Therefore, the field courses are designed as a stand-alone module. The current work analyzes the design of the course entitled Fieldwork II which is integrated in the undergraduate degree in Environmental Sciences at Universidade Aberta. Many issues have to be taken into account when designing such a course that encompasses legislative, pedagogical and logistical matters. An overview of the research work that has been developed on the pedagogical value of fieldwork is given. Taking into account the specific learning methodology adopted at Universidade Aberta, we also present a fieldwork model for environmental sciences in a context of a blended learning (b-learning) undergraduate programme. The organizational effectiveness of the model and students' participation was assessed in two scholar years of a fieldwork course at Universidade Aberta. Also, the project assignment synopses were assessed in the context of education for sustainability. Finally, it should be stressed that fieldwork should be regarded as a form of learning which exploits the unique characteristics of the field environment to improve the student learning experience.

Introduction

The crucial role of higher education in providing opportunities for social learning and, in particular, introducing the concepts of sustainable development has been reported by UNESCO at the Johannesburg World Summit for Sustainable Development. An improved environment and quality of life can only be achieved through active and knowledgeable citizens as well as informed decision-makers making the right choices within interrelated economic, social and environmental issues the world faces today (UNESCO 2002). The UNESCO report identified the need for an education that questions current models and called for "deeper, more ambitious ways of thinking about education, one that retains a commitment to critical analysis while fostering creativity and innovation". This interpretation of education has been promoted by environmental educators: Saul (2000) who requires culturally critical perspectives and Huckle (1996) who argues that only through asking socially critical questions can we progress towards a sustainable future.

Environmental studies appeared as an academic topic in the late 1960s when these challenges became a concern of society in general and coincided with scientific disciplines raising awareness of ecological imbalance. As a result, higher education institutions responded with two different approaches: an "environmentalization" of the disciplines and an "interdisciplinarization" of the environment (Bursztyn 2008). The first approach, which is an adaptation of the common disciplinary approach, introduces new subjects and new content in the existing courses. The second approach requires a new interdisciplinary programme between existing groups of disciplines.

Interdisciplinarity now plays a major role in the debate concerning the future of higher education and, consequently, the future of universities. To face these challenges, European countries devised a reform plan, known as Bologna Process, which aims to create a higher education area on a continental scale, while maintaining national and regional differences. A central issue for the Bologna Process was to simplify and also unify the higher education systems with three central pillars of mobility, employability and interdisciplinarity.

Environmental sciences are the study of the relationship between man and the natural world, and it provides an integrated, quantitative, and interdisciplinary approach to the study of environmental matters. This is more than a new science because it requires dealing with problems that cannot be effectively treated in isolation as it might have been with the mature disciplines often referred to as natural and social sciences. Indeed, environmental sciences must be a general approach, uniting concepts and skills.

Since 2007, the Universidade Aberta (UAb) has offered a 1st cycle degree in Environmental Sciences, according to Bologna Process principles. UAb is a distance-learning university where the pedagogical model is based on e-learning and relies on the use of online communication tools; the model promotes interaction between students and teachers and is deeply focused on students as individuals who actively build their own knowledge. The goal of our course is as follows: (1) provide students with the scientific principles, concepts, and methodologies required to understand the interrelationships of the natural world; (2) identify and analyze environmental problems both natural and human-made; (3) evaluate the relative risks associated with each problem; and (4) examine alternative solutions for resolving or preventing them.

The 1st cycle degree in environmental science (ES) includes a face-to-face field component, which requires the students to be present at a certain location for a specified time. This can be seen as a form of blended learning: e-learning complemented with instructor-led training and other live formats (Bersin 2004). This particular degree follows a blended learning methodology (b-learning) where the fieldwork permits use of the academic disciplines with real-world problems, encouraging the acquisition of generic and subject-specific skills, which will be valuable for future employment.

In this particular case, blended learning is driven by the following factors: (1) that field work is considered to be an essential component of environmental science; (2) proper training in environmental studies is driven by coupling "book-learning" to field work; and (3) that a "local perspective" is a useful teaching approach, enabling students to observe how concepts are relevant to their local environment (Bersin 2004).

Recently there has been a significant reduction in fieldwork conducted as part of a traditional science course. It seems that this pattern has several possible causes, including an over- prescriptive science curriculum, financial constraints, changing of professional values and bureaucratic complexities. However, for field sciences such as environmental, the laboratory is the "field"; anywhere else the experience becomes second-hand, out of context and relatively meaningless (Fisher 2001).

Fieldwork courses within an environmental science programme are important because they engage students and provide a first-hand connection with society and sustainability. This concept is supported by Besenyei et al. (2004) who states that there are a variety of matters that students can learn from their fieldwork experience. These can vary from practical aptitudes to the development of identification skills. Fieldwork provides a wider experience of the environment as a whole and the opportunity to relate theory with practice. According to Besenyei et al. (2004), it is not possible to replace fieldwork with any other learning methodology and it is impossible to gain a degree or qualification in environmental sciences without fieldwork. Fuller (2006) also supports that fieldwork helps to improve the understanding of a specific subject and consequently enhances learning skills. The work undertaken by Lima et al. (2010) also showed that it is possible to create field trips in an informal setting encouraging group discussion and team work that provides informed citizens.

In this paper, we reflect on the use of fieldwork in the UAb's environmental science programme, delivered through distance-learning methodology, which will be described. We also present our analysis of the results obtained from the Fieldwork II course for the scholar years of 2008/09 and 2009/10, as well as the registered interaction of the students in the appropriate forum. A final topic will be devoted to the discussion of the work developed towards the promotion of skills and competences for sustainable development within the course.

A Fieldwork Model for an Environmental Sciences Degree

Environmental sciences is a domain with poorly defined borders that combines many scientific disciplines. This in itself requires integration of teachers from both social and natural sciences in the same teaching project. Our experience has shown there are numerous obstacles to be overcome to achieve an interdisciplinary collaboration in a fieldwork course, and these are summarized in this section. We also present a reflective analysis, which will also be described, that has formed the basis for the development of an intervention model.

Field trips can provide teachers with different educational and epistemological functions such as the following: illustrative, inductive, motivating, training (skills training) and investigative (Compiani and Carneiro 1993). The illustrative function reinforces concepts already mentioned while the investigative function sequentially guides the process of observation and interpretation, thus students can solve a problem posed by a teacher. Motivating tours are intended to raise students' interests to specific questions that can be the subject of future study. Skills training is concerned with acquisition of practical skills, for example, operation of a compass. The investigative function promotes student autonomy, establishing a hypothesis, structure observations and interpretation, deciding on the most appropriate strategies to validate them and report their findings. These activities are complementary to the online component and this particular approach is more often utilized for an adult audience. In those cases there is a greater need to integrate different disciplines, and therefore the investigative component increases throughout the degree.

The integration of knowledge within environmental sciences is "generally situated in a dualistic intellectual paradigm in which 'nature' is perceived as being separate from human 'culture'" (Strang 2009). The term "environment" is associated with ecology rather than with socio-cultural and economical contexts.

This is of course the fundamental misconception regarding sustainable development. We need to reconceptualize human relationship with nature, accepting an influence in both directions and, at the same time, understanding that any distinction or classification are purely heuristic.

Before suggesting some guidelines for interdisciplinary collaboration between teachers it is necessary to identify the matters that hinder this work, and these are as follows:

- The term fieldwork assumes different meanings depending on the discipline. It is important to introduce an epistemological component to ensure the terms used have the same meaning for each teacher irrespective of the discipline;
- Development of the curriculum requires determining the subjects to be included, their association with the acquisition of skills, the appropriate sequence, required learning resources, and the assessment and evaluation methods to ensure the course objectives are achieved;
- Higher education occupies a central position in the way in which future generations learn to cope with the complexities of globalization. At this time universities need to teach the skills required to be successful in the labour market. Universities must also cultivate in students, faculty and staff a positive attitude towards cultural diversity regarding understanding of how people contribute to a better life in a safer world. It follows that significant and close collaboration is expected between academic staff. To do so requires conditions that promote and facilitate the internal communication so that at least a basic understanding of the principles and methods of 'Sustainability Science' are achieved. This position means all disciplinary identities are retained while the expertise of each is assimilated into one;
- Not less important are the issues related to the distinct "profile" of each of the teachers which requires coordination in assigning different tasks to different teachers.

So far we have identified matters that may deter interdisciplinary collaboration. However, the importance of interconnecting the goals, competences and skills developed by students in a fieldwork course must also be emphasized. The pedagogical value of fieldwork is recognized by several scientific disciplines where it is already part of core *curriculum* for it provides students with concrete phenomena and materials and positive cognitive, psychomotor, and interpersonal skills (Orion 1993). These matters are particularly relevant to the teaching of environmental sciences where a main goal is the development of holistic models; these include the interrelationships between environmental factors and processes, analyzing different case studies and developing a project based on problem-solving methods. Students are given autonomy to establish a hypothesis, structure the observations and interpretations, decide on the appropriate strategies to validate and discuss their findings.

On a theoretical level Giere (1988; 1992) proposes a model to explain the evolution of scientific knowledge that is of particular relevance to activities within fieldwork, and these are as follows: 1 – Identify the aspects that will be studied, 2 – Identify the model to represent the world; 3 – Identify the information obtained by observation, 4 – Identify the model predictions, and indicate what information can be obtained if the model explains reality; 5 – Determine if the observations are consistent with the predictions and 6 – Determine whether other models provide similar predictions. If we apply this approach in a pedagogical context, then we ensure students identify pertinent problems, examine assumptions, evaluate how to obtain data, verify the results and compare them with those determined from a model and apply the model to new problems.

We share Giere's perspective and extended it to include fieldwork approach advocated by Orion (1993). Orion (1993) highlights the following matters: i) the role played by fieldwork in the learning process; ii) the existence of different types of fieldwork and iii) the importance to an organization. This model follows a constructivist approach and conveys the need to establish a relationship between new information and the cognitive structure of students. The main determinants of meaningful learning are the students' predisposition to learn and the existence of material and strategies that establishes "bridges" within students' cognitive structures. Fieldwork then needs to combine concrete learning experiences with higher levels of the cognitive process. Orion (1993) states "the main role of the field trip in the learning process is the direct experience with concrete phenomena and materials and is the direct sensoriomotor experience that can facilitate the construction of abstract concepts, enhancing meaningful learning".

B-Learning Field Course Design

For the design of the fieldwork curricular units and activities, we have adopted the methodology proposed by Orion (2003). This approach includes a preparation unit that seeks to minimize the impact of adopting a different learning environment. In this case, students receive information in advance to permit integration of knowledge and understanding obtained from different cases. These materials can include maps, photographs and a booklet with the objectives, teaching method, identification of the different stops and its purpose; we will assume the student reads and absorbs the material provided in advance. These procedures reduce the element of surprise that may give rise to disruption of the student's observation and reasoning. The field trip is the central part of the course and is guided from modules in an illustrative manner. During the course, students are encouraged to discuss their observations with peers. According to Orion (ibid.), the final module of the course is the most complex for both students and teachers alike because they are required to integrate knowledge acquired in the field with hierarchical networks of information and to develop abstract concepts with a view to eventually providing a meaningful learning experience.

The model of fieldwork adopted within the ES degree is shown in Figure 1. Three phases were considered: previous to the opening of the course, a second stage corresponding to the teaching period by the teacher's team and the last phase focused on the evaluation processes.

Fig.1: Fieldwork course design model

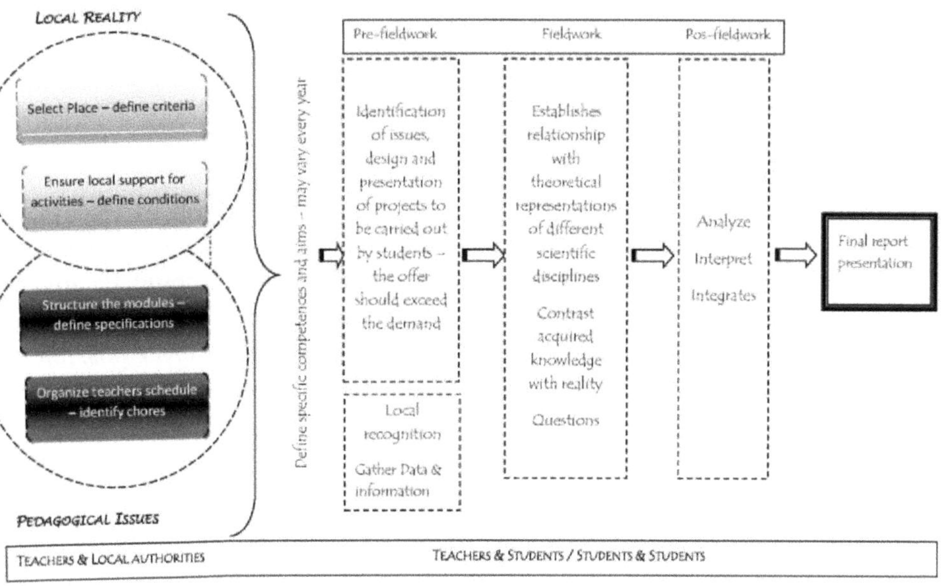

The starting point for the development of such curricular unit in each scholar year results from the junction between the local reality (socio-economic, cultural and environmental) of the selected place where the field trip will take place and the pedagogical issues that derive from the proposed model, as shown in Figure 1. One of the most critical aspects in this whole process is the selection of the region to conduct the face-to-face activities. From our past experience a set of criteria must be met and taken into account when choosing a region. These issues are as follows: i) ability to involve the local institutions and the local community; ii) good support concerning logistics, namely housing and transport; iii) existence of a diversity of attractive places with relevance to the context of an environmental sciences degree. It should be stressed that the involvement of the local community is a crucial point if we want to promote deep experiential learning as well as social learning. Furthermore, from the viewpoint of transport, it should be an accessible place and transportation should be provided during the activities. Accommodation, too, should be found in a way that the whole group can be housed in a single place, increasing the levels of interaction between students and between students and teachers. In addition, access to meeting spaces should be taken into account. Finally, costs also play an important role.

Simultaneously, it becomes necessary to organize the three modules that actually form the course. This requires the distribution of the different tasks among the teachers involved which entails a broader set of functions and competences such as: i) facilitators' skills in order to conduct online discussions and keep them within the defined themes and issues; ii) knowledge of the local reality and the important topics to be able to pose the appropriate questions that promote the involvement of students; iii) recognize the need for facilitating communication between the different players and identify the appropriate way to overcome those difficulties; iv) be aware of individual needs or behaviour to avoid creating any troubles within the group.

A further matter that has been privileged is the definition of the research themes for the final project. All participating teachers should be involved in the identification of research questions, and special attention should be given to those issues that are directly related to local needs or to the local community. Several of these examples will be given in this paper. Whenever possible, the projects should be developed in close liaison with the community who could also be involved in the final assessment.

Finally, it is also our purpose to contribute to the local development of those communities promoting the concept of sustainability in its different components, namely environmental and social.

Environmental Sciences Programme: the Pedagogical Model

The UAb has reorientated its educational and training activities from an "industrial" distance education model constructed around a learning package that favours the interaction between the student and the materials, to a model focused on the development of competences utilizing information technology. This transition was made possible by technological development and conforms to the European Higher Education Space, in which students are expected to take greater responsibility for their own learning and also requires teachers to emphasize the acquisition and development of core competences.

Interaction is seen as central to an educational experience, and the focus on interaction in online learning emerges from the potential of technology to support sustained educational communication. Internet technologies provide a communication through asynchronous interaction (Garrison and Anderson 2003). The purpose of an educational experience, whether it is online, face-to-face, or a mixture of both, is to structure the educational experience to achieve the desired learning outcome. In this context, interaction must be structured and systematic. Some have argued that it is valuable and even necessary in higher education to create a community of inquiry where interaction and reflection are sustained. Interaction in such an environment goes beyond social interaction and the simple exchange of information and must include various combinations of interaction among content, teachers, and students (Moore 1989; Garrison and Cleveland-Innes 2005).

At UAb, pedagogical strategies have been developed that engage students as active participants. The model includes principles permitting teaching and learning activities for each curricular unit to be performed at a distance, in a virtual learning environment, using an e-learning platform. An introductory module, held totally online, is intended to familiarize the students with the virtual environment and the e-learning tools as well as promoting the acquisition of online communication and online social skills.

The undergraduate courses follow a pedagogical model (Pereira et al. 2007) specifically designed for online learning at UAb. This learner-centered model, where students are responsible for knowledge building, is based on the flexibility of access without temporal or spatial constraints. It also relies on diversified student-teacher and student-student interaction as well as students and learning resources in a social context. The implementation of these principles requires three vital elements in the learning process:

i) the student attends a *virtual class* where the learning activities for each curricular unit take place. In this space, accessed both by students and teachers, the interaction occurs through two types of forums: those that are moderated by students and those moderated by teachers. The former are workspaces for the class in which students interact about the subjects they are studying while the latter are intended to permit students to ask questions and assist overcoming difficulties that have not otherwise been solved through the discussion between themselves;

ii) The *curricular unit plan* (PUC) is a document presented by the teacher at the beginning of the curricular unit intended to guide the students' learning process. The guide contains information on the objectives of the curricular unit, its subjects and contents to study, the competences to be developed, the organization of the learning process, the learning resources, what is expected of a student, what a student can expect from the teacher (or tutor), and the assessment criteria;

iii) The *learning card* is a personal device that allows students to visualize at any time the marks obtained in the continuous assessment activities (e-folios) during the semester. These e-folios are complemented by a p-folio which is performed in a face-to-face location. The final grade of the curricular unit will be the sum of the marks obtained in both the e-folios and in the p-folio.

The model previously described (fig. 1) has been used for the design of the current course and the three modules proposed are as follows: the first part is preparatory and an online asynchronous module where students are exposed to materials and information necessary for the field trip; the second module is the field trip, which provides the necessary stimuli for higher levels of cognitive learning; the third and final module is devoted to the elaboration of the project that will congregate the theoretical knowledge obtained from other curricular units.

Environmental Sciences (ES) Programme: an Interdisciplinary Approach

The UAb has developed an undergraduate *b*-learning programme in Environmental Sciences directed at an adult audience (over 21 years) who are typically working individuals seeking professional development. The purpose of the course is to promote and develop a set of professional skills and competences within environmental sciences. The first two years are composed of mandatory curricular units of Science and Environmental Technology (40 ECTS), Biological Sciences (22 ECTS), Earth Science (22 ECTS), Mathematics (12 ECTS), Chemistry (12 ECTS), Physics (6 ECTS) and Legal Sciences (6 ECTS), for a total of 20 compulsory curricular units. This structure provides the student with a broad based curricula. In the final year the student can then study one of three minors that are as follows: Natural Heritage, Environmental Health, and Environmental Management and Sustainability. Each of these courses comprises a 60 European Credit Transfer System (ECTS)[1].

---

1 More information about the study plan of the ES programme and its course description is available at *http://www.univ-ab.pt/guiainformativo/eng_planoestudos1.php?curso =10&ma =3.*

The whole programme has been designed and developed to promote integration of scientific disciplines with economics, earth sciences, biology and technology. This leads to the concept of sustainability science, which is an integrative science, a science that sets out to break down the barriers that divide the traditional sciences (Martens 2006). The fundamental principle of sustainable development is to manage both personal and social responsibility between generations and global issues (Juárez-Nájera et al. 2006). In the last 20 years it has become evident that science and technology education need to supply the knowledge and tools to permit transition to a sustainable environment. A pragmatic definition of 'Sustainability Science' was provided by Carter (2007) who stated that "sustainable science is a field defined by the problems it addresses rather than by the disciplines it employs".

Bursztyn (2008) has also considered the educational issues of sustainability and concluded universities should introduce environmental issues and relate these to the need for sustainable development within the framework of traditional courses. Bursztyn also advocated an interdisciplinary approach that is complementary rather than competing with discipline- orientated education. Academic communities and university departments are hierarchical discipline-orientated organizations. Juárez-Nájera et al. (2006) argue the education of sustainability requires a new academic and professional culture that revolves around the concept of sustainability rather than the particular knowledge obtained studying a subject. Solely on this basis can sustainability claim the status of a traditional science: hence the term sustainable science.

The Bologna Process has changed European universities. However, these changes have not addressed the educational requirements of society operating within the concept of sustainable education. In this regime, educators need to be conscious of the qualities required of future scientists devoted to sustainability: they have analytical insight, problem-solving competences, and the skills to present their findings both orally and in written form (Martens 2006). Today's students will be the citizens of tomorrow in all fields of life and work.

A preliminary study was conducted in the first year of the ES programme to examine the *curriculum* and evaluate how both teachers and students use the role of sustainability science (Amador et al. 2008). This study shows additional information on sustainability science and the education of sustainable development is required within the study plan of the programme. This recommendation resulted in continuous improvement to provide better sustainability education.

Fieldwork II

In Fieldwork II students are exposed to a variety of local economic activities related to agriculture, industry, production or tourism within a particular region of Portugal. This selection is based on the following requirements: the existence of activities, relationship with local and global environmental matters and the availability of local support for the organization. The final goal is the preparation of a report whose assessment will be based on written documentation produced as well as an oral presentation followed by a discussion with an examination panel. By the end of this curricular unit, students will be expected to be able to do the following:

1. Demonstrate a broad and interdisciplinary understanding of the problem posed;
2. Acquire data and organize information clearly and coherently, using specialized vocabulary as appropriate;
3. Write structured and reasoned arguments and develop teamwork skills;
4. Reflect on their own learning and elicit constructive feedback;
5. Develop oral and written communication skills using online and text-based resources;
6. Manage time effectively and prioritize learning tasks to achieve pre-determined outcomes;
7. Undertake independent study and work to deadlines.

The assessment rules and criteria, with which students are acquainted from the start of the course, are shown in Table 1.

Table 1: Assessment criteria of Fieldwork II course

	Activity	Criteria	Marks
Pre fieldwork	on-line asynchronous	Establish frequent interactions with teachers and other colleagues in the corresponding forum, raising relevant questions, helping to prepare face-to-face activities	2/20
Fieldwork	Residential week	• Follow the activities also showing interest, curiosity, placing relevant questions; • Comply with the pre-established timetable. • Interact with colleagues, teachers and other intervenient, revealing teamwork abilities.	4/20

	Activity	Criteria	Marks
Post fieldwork	Written project on-line asynchronous	• Demonstrates the relevance and interest of project. • Scientific rigor, appropriate vocabulary and adequate organization. • Creativity of the approaches taken	10/20
	Oral Presentation face to face	• Communication skills • Clear seminar, developing a coherent argument.	4/20

This course is organized in 3 distinct modules with specific aims and sequentially presented during the semester, as follows:

a) Pre-Fieldwork

An online asynchronous module that provides the necessary information, both from the logistics and scientific viewpoint, required for the activities intended for the residential week. The complete schedule for the residential period is given, and students are encouraged to provide additional information on the subjects as well as on the places to be visited. Following the work of Besenyei et al. (2004), it is also our understanding that the incorporation of an initial preparation period dedicated to the background information enforces what is going to be learned during the field trip.

As shown in Table 1, the interaction and relevant information provided to prepare the face-to-face activities is one of the assessment criteria. Table 2 shows the schedule for the residential week that took place in the last two academic years of 2009 and 2010 in two different regions. In the second year it was thought more efficient to have theme days with subsequent activities and visits.

Table 2: Schedule activities for the residential 5 days

	Day 1	Day 2	Day 3	Day 4	Day 5
2008/2009	*Seminar* Towards sustainability: environment, health and education	Organic farming; rural tourism	Integrated solid waste management systems	Water treatment plants; Electricity production	Food industry equipments; Waste water treatment plants
2009/2010	*Seminar* Sustainability: environment and health	*Cork* From the cork oak to cork	*Water* – Bottled water company; – Water treatment plant; – Waste management system	*Sustainable agriculture* Organic farming; Livestock and pasture	*Nutrition* Small food industry: mushrooms; Wine production

It should be noted that the first day of the residential week is devoted to a seminar with both academic presentations on the different subjects as well as reports of local entrepreneurs from different sectors. Since this is an open-day event an active participation of representatives from local industries as well as local intervenients is also expected.

Another very important purpose of this preparatory module is the choice of the themes proposed for the final project or report. The aim of the project is to promote and integrate knowledge acquired on the other courses to solve or to develop a practical approach to a specific problem. A list of different projects is published and it is expected that students interact with each other and with the teachers providing information and questioning the themes and the proposed projects.

Table 3 shows some of the projects taken in the past two years, grouped by main topic, which is within the 3 areas of the corresponding minor: Natural Heritage, Environmental Health and Environmental Management and Sustainability.

Although the projects are individual, some of the themes can be chosen by more than one student, provided they are adjusted to different realities. The final choice of the projects by the student takes place in the week before the residential period.

Table 3: List of Projects available to students in the last two scholar years

Environmental Health	**Natural Heritage**	**Environmental Management and Sustainability**
Environmental impact of meat consumption	Educating young people for the conservation of geodiversity	Waste management
Organic agriculture as a strategy for the control and reduction of health risks	Identification of threats to geodiversity	Turning waste into raw material
Traditional products with protected names: its influence in reducing the perception of food risk	Promotion of biodiversity	Waste water treatment
Food safety at home	Science, Society and Environment – formal education projects for young population	Biodiesel production
Challenges of nutritional information and health promotion	Municipal Markets: its importance in the revitalization of villages	Integration of local land-use management tool: potentials and conflicts
Household food waste and environmental impact	Video Production within Science, Society and Environment relationship	Environmental impact assessment in your area
		Carbon Economy – contribution of cork oak forests
		Alternative energy – solar energy
		Sustainability of an organization
		Life cycle analysis of cork products: potential for eco-label

b) Fieldwork

This module concerns the residential period which refers to the 5 consecutive days that students are together. It is expected that the students are actively involved in all the activities and discussions, bearing in mind that all the information was given and also time should have been spared in the preparation of the visits. A space for discussion is also provided in the virtual learning environment but not much asynchronous interaction is expected to happen there.

As shown in Table 2, the schedule is rather intense, especially if we consider that bus trips have to be taken quite often in order to get from one place to another. The whole period is organized to enable the following skills: observation, interpretation, investigation and critical analysis. Therefore informal face-to-face discussion spaces are also organized so that students can interact with the teachers as well as with their peers.

c) Post-Fieldwork

This final module occurs online with tutorial orientation focusing on the documentation of the final project. The interaction is mainly between the teacher and the student, developed in a more individual way, since it is mainly concerned with project documentation. The teacher will provide material for a specific project with the expectation that the student will perform a critical literature survey on the theme and organize the final written report. Within this block there is a general interaction *forum* where students can share resources. The main purpose of the individual space is twofold: provide the student with a *private tutorial* where he can exchange views on the project with the teacher and also to provide the teacher with an easy and simple way to follow the progress of the work obligatory for final assessment.

Assessment of Students' Participation and Learning Outcomes

Where appropriate, participation was divided into categories defined by types of discussion. For the scholar year of 2008/09 and 2009/10, the analyses were performed separately to permit a comparison. In the scholar year of 2009/10 only blocks 1 and 2 were analyzed owing to the absence of data; at the time this work was conducted the projects were incomplete and the interactions under development.

As previously mentioned, the ES programme started in 2007/08 and therefore the course Fieldwork II had its first edition in the scholar year of 2008/09. From a total of 111 students in 2007, 21 enrolled into the second year. According to the UAb regulations, only students that complete 60% of the first year curricular units can proceed to the second year of the undergraduate degree. From the total of students, who went through to the second year, 12 enrolled in this course. For the 2009/2010 academic year, the number of students who took the course doubled to 24 students from a total number of 72 students who registered for the second year. Fieldwork II had enrolment of about 30% of the total number of students. This low subscription rate can be ascribed to the requirement to spend 5 consecutive days at a location and the additional cost associated with accommodation and transport.

In the scholar year of 2008/09, in the preparation forum (that is the pre-fieldwork module), the student questions concerning the field trip were about 20% of the total forum participation. The other contributions were related to information about visiting places (25% of the total forum participation) and organizational questions concerning the fieldwork week which contributed to about 50% of the forum participation. In this forum the course coordinator published four posts that were related to the organization of the field work.

In the scholar year of 2009/10, in the same preparation forum, the number of interactions was higher. About 12% of total participants were concerned with locations and field trip information. The other contributions concerned information about the places to be visited (40% of total forum participation); only 3% of the questions posed were regarding fieldwork organization. The pre-fieldwork module contained a project choice step in the second year the course was run and, as expected, this generated a significant number of student-to-student interaction (76 entries representing 46% of total forum participation). This option was not available during the first year of the course. There was also a considerable increase in the posts published by teachers, and these were mainly related to the list of proposed themes and project selection (about 71% of total entries by teachers).

In scholar years 2008 and 2009 the registered contributions in the forum available within the fieldwork module were, as expected, very few and not worth considering. This is not surprising, bearing in mind that the students are all in a face-to-face environment with the teachers during this time, and little attention is therefore paid to the asynchronous communication environment.

Although definitive data is not yet available for the current scholar year of 2009/2010 similar results are expected to those obtained for 2008/09. So far, the forum with the most contributions is the project elaboration theme (post-fieldwork module), with a total of 300 participations. These *fora* were individually organized for each of the projects where only the student, the corresponding supervisor teacher(s) and the coordinator of the course had access. A pattern has been identified and is illustrated in Figure 2. In general, the forum discussions concern bibliographic support rather than work schedules and organization and content of the written report. Matters concerning the oral presentation and power point presentations have increased. Nevertheless, 39% of the participation in the forum concerned the scientific content of the written report and 19% addressed organization or the work.

Fig. 2: Discussion topics in the project elaboration forum in the scholar year of 2008/09

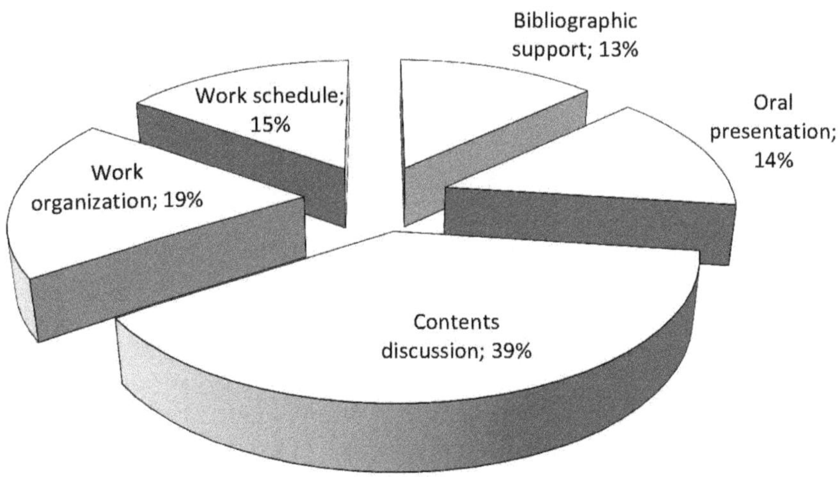

The contributions are, as shown in Figure 3, divided equally between teacher and students, showing that there is an effective interaction and supervision by the teachers.

Fig. 3: Students and teachers participations by topics discussed in the project elaboration forum in the scholar year of 2008/09

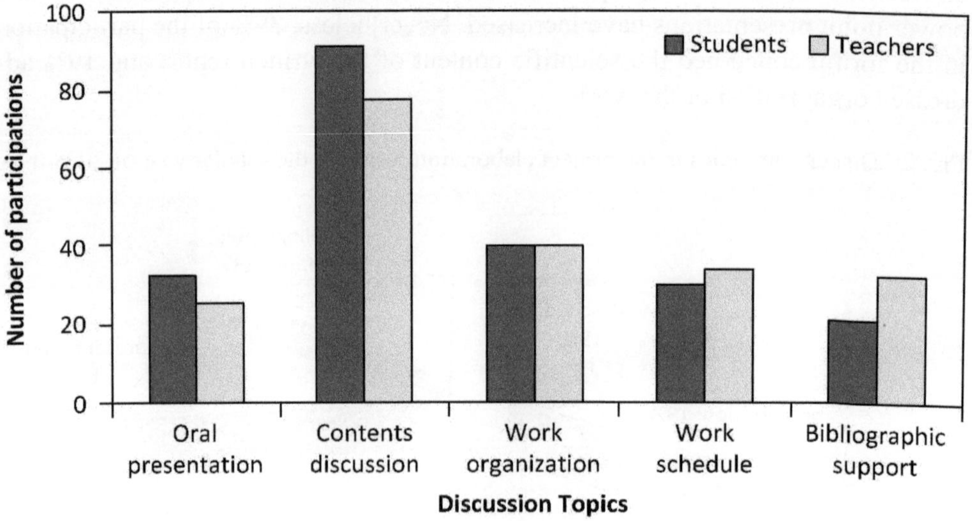

A further analysis of the different *fora* showed the following: the projects concerning specific and concrete situations were responsible for the highest participation. Figure 4 shows the different types of interactions that took place in each project. Project 7, devoted to the development of a waste water treatment system for a particular village, and 8, concerning a biodiesel production system, registered the highest number of participants mostly in relation to content. From the discussions in this forum and the posts published by students we conclude some students contacted the teacher through synchronous *Skype*™ meetings that were not considered in this analysis.

Fig. 4: Division by issues discussed in the forum of each one of the projects for the scholar year of 2008/2009

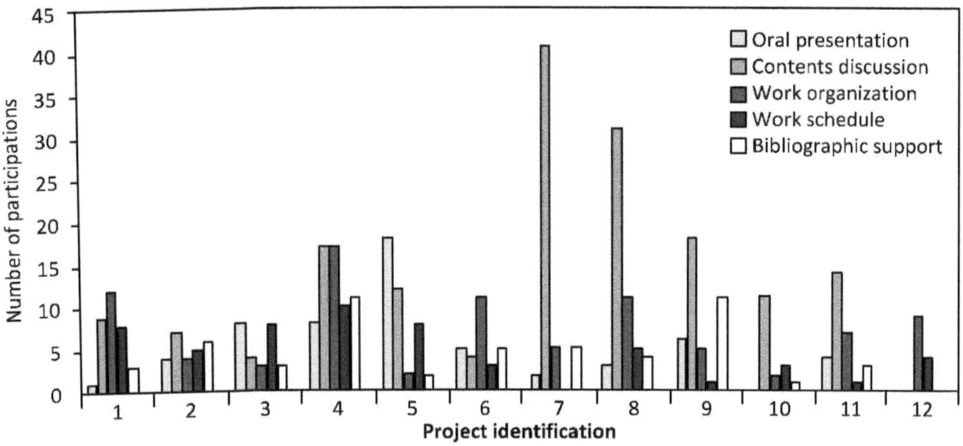

1. Environmental impact of meat consumption; 2. Organic agriculture: a strategy for control and reduction of health risks; 3. Integrated waste management; 4. Recycling: transforming waste into raw material; 5. Municipal markets: its significance in the revitalization of villages; 6. Traditional products with protected names: its influence in reducing the perception of food risk; 7. Waste water treatment in a production unit; 8. Biodiesel production; 9. Educating young people for the conservation of Geo-diversity; 10. Integration of local land-use management tool: potentials and conflicts; 11. Identification of threats to Geo-diversity; 12. Biodiversity promotion

In the post-fieldwork module a general *forum* can also be found where little participation related with organization of the projects supervising or assessment of the projects were also registered.

The final presentation of the project is done orally, in a synchronous face-to-face environment, which in some cases involves the use of video conference. The majority of the students showed contentment with the organization of the whole curricular unit as well as with the project supervision, through discussions conducted with peers in the informal virtual spaces (the *café* space). The majority of the testimonies reinforced the importance of contact with local communities with real problems where the application of the theoretical content learned previously is evident. According to Curtis and Mahon (2010), fieldwork activities and projects enhance student learning and provide a valuable real application of their classroom coursework. Students support this view because they found the project useful in building skills for future employment, which will also be valuable to their personal lives and growth as individuals. These results are in accordance with the work carried out by Besenyei et al. (2004) where face-to-face fieldwork effectiveness was tested within environmental science. Besenyei et al. (2004)

concluded that fieldwork provides a valuable learning experience. Accordingly, some of these experiences could be measured by learning outcomes but a large number are beyond measure and can be classified as life experiences. Students considered they had developed key learning skills through fieldwork and that their curiosity regarding environmental matters had increased significantly. It seems that the benefit is far more important than the cost of running these courses. Manzanal et al. (1999) conducted a study to evaluate the relationship between ecology fieldwork and students' attitudes towards environmental protection. Manzanal et al. (1999) concluded fieldwork clarifies concepts and intervenes directly in the development of more favourable attitudes toward the defense of the ecosystem. Both components are seen when judgements for the resolution of problems, which otherwise negatively affect the ecosystem, as well as actions and solutions are determined which should be adopted.

Undergraduates often find little connection between academic studies and their future career. Fieldwork courses and projects can be an important stage for students to get acquainted with the real world. This is particularly important in applied and interdisciplinary educational programmes such as environmental sciences. This is further emphasized when the course is taught for working students through a distance-learning process.

Evaluation is a key element of design for all teaching activities and is particularly important for fieldwork given the relatively high cost of provision. Therefore continuous work needs to be carried out in order to confirm the educational value of the fieldwork course and to evaluate to what extent this activity is achieving the desired learning outcomes. We will also continue to assess the effect that fieldwork is having on students' confidence and more simply whether they enjoyed it.

Assessment of the Project Assignment of the Curricular Unit Fieldwork II within Sustainability Education

To analyze the project assignments and their relevance within sustainability education, we adopted the methodology reported by Bardin, 1977: "A set of communication analysis techniques aiming at obtaining, by means of systematic and objective procedures for description of message content, indicators (quantitative or not) which will allow the inference of knowledge related to the conditions of production/reproduction of such messages." This analysis has three parts: description where the characteristics of the text are identified; interpretation that corresponds to the meaning granted to these characteristics; and inference which represents that explicit transfer from description to interpretation. These inferences seek to clarify the basis of the message or the consequences that the message might cause.

These principles were combined with the categories listed in Table 4, for the 30 project proposals listed in Table 3, to analyze Fieldwork II. Each proposal consists of a title, a synopsis as well as a list of the main objectives of the work to be developed, and this was presented at the commencement of the unit.

In the present study the categories are not exclusive in view of the fact that they can be concerned with different features of the concept of sustainability. Therefore the same text can be included in more than one category. In an attempt to diminish the inevitable subjectivity, the project classification was performed based on phases (registration units) of the proposals. Furthermore, the categorization was carried out by two of the authors and the average was also taken in order to avoid bias.

Table 4: Number of proposals by categories within education for sustainability

	Category/Indicators	**Nº of proposals (average)**
Competence objectives	Application of knowledge	8
	Acquisition of specific competences for sustainable development	13
Level of coverage	Respond to societal needs and demands at a regional and local level	13

Table 4 shows for each of the categories defined previously that it is possible to identify trends which, in some cases, might be related to the characteristics of the curricular unit. Since there are several projects related to local problems, the student report seeks to develop strategies for change to respond to societal needs.

Furthermore, it should be stressed that it is inevitable that the teachers involved in such a course not only have a very different background as well as a different understanding and interpretation of the sustainability concept: some with a weak concept of sustainability and others more linked with the values of nature conservation, closer to a concept of strong sustainability. However, Fieldwork II is the "time and the place" where students can apply their theoretical knowledge previously acquired as well as broaden horizons by being in direct contact with reality and real-life situations.

Conclusions

In this work we presented a fieldwork model to environmental sciences in a context of a blended-learning undergraduate programme. The organization effectiveness of the model and the students' participation was assessed in two scholar years of a fieldwork course at Universidade Aberta, Portugal. The project assignment synopses were also assessed in the context of education for sustainability.

According to the teacher's perception of student's participation during the whole programme, students were very satisfied with this course in terms of its organization, contents and assessment. Also, the course allowed them to get to know each other better, to know the teachers and to build a student's community and a learning community. The analysis of the student's participation in the asynchronous forums within the virtual learning space will be a valuable help to rethink the relevance of some of the forum and also to redefine their purpose even if the interaction students-teacher was always present and profitable.

Although the current work is developed within a formal learning environment, the contents of curricular unit Fieldwork II also include field trips which are intended for the students to gain a better sense of the real world environment and processes, as well as to promote the exploitation of the local features of the place where the outdoor activities happen. In this context there is also a non-formal side to the learning process which will contribute to the student's education for sustainability and integrated knowledge.

Fieldwork is frequently sidelined as a learning strategy by educational institutions because of factors such as time to cover comprehensive curricula, financial constraints, legal issues and commitment by teachers (Gerber and Kim Chuan 2000). In spite of the fact that in e-learning systems environmental virtual field laboratory being a growing reality (e.g. Barak and Nater 2002; Ramasundaram et al. 2005), it is our opinion that face-to-face fieldwork will always be required and will have a crucial role for the reasons explained throughout this work. The new virtual technologies are important but not sufficient, because they do not encourage the development of key learning skills, attitudes and values towards environmental conservation and sustainability development at the same level as face-to-face fieldwork.

Nevertheless, the discussion of the results needs a comprehensive analysis in order to be able to fully support our interpretations. A more detailed content analysis will be conducted into project assignments developed by the students and questionnaires and surveys to appraise the course aims, organization and assessment and evaluate students' perception about competent objectives and coverage level within sustainability concerns. With these new guidelines, we wish to promote the development of competences and skills – in both lecturers and students – that will lead to the implementation of problem-based learning methodology as well as proposals for final projects related to local issues and problems, presenting innovative solutions for different case studies in order to produce graduates with new competences for sustainability development practices.

Further work concerning the evaluation of field activities will be performed, namely in relation to the quality of the learning environment of outdoor science activities. The use of purpose-designed techniques such as that described by Orion et al. (1997) will be considered. In this case the method evaluates environmental interaction, integration, student cohesiveness, teacher supportiveness, and preparation and organization and material environment. Additional questionnaires, interviews and supplemental observation will be applied.

References

Amador, F., Caeiro, S., Oliveira, C., Caetano, P., Bacelar-Nicolau, P., Azeiteiro, U.M. and Martinho, A.P. (2008), Curricular development on higher education and sustainable science: A b-learning course in environmental science. In papers proceeding EMSU 2008 International Conference. Environmental Management for Sustainable Universities. A New Knowledge Culture. Barcelona. 15 to 17 October, pp. 1-12 (available at: www.emsu.org).

Barak, P. and Nater, E. (2002), The virtual museum of minerals and molecules (avalilable at: http://www.solis.wisc.edu/virtual_museum/index.html).

Bardin L. (1977), Análise de conteúdo. Edição 70, Lisbon.

Besenyei, L., Watkin, G. and Oliver, K. (2004), An evaluation of the educational effectiveness of fieldwork within environmental science awards at the University of Wolverhampton. University of Wolverhampton. Learning and Teaching Projects 2003/2004, pp. 63-70.

Bersin, J. (2004), The Blended Learning Book. Best Practices, Proven Methodologies and Lessons Learned, Pfeiffer.

Bursztyn, M. (2008), Sustainability Science and the University: Towards Interdisciplinarity, CID Graduate Student and Research Fellow Working Paper no. 24.

Cachapuz, A., Praia, J. and Jorge, M. (2001), Ciência, Educação em Ciência e Ensino das Ciências. Ministério da Educação, Lisbon.

Carter, L. (2007), Sociocultural Influences on Science Education: Innovation for Contemporary Times". Science Education, 92, pp. 165-181.

Compiani, M. and Carneiro, C. (1993), "Os papéis didácticos das excursões geológicas". In Enseñanza de las Ciencias de la Tierra, 1.2, pp. 90-97.

Curtis, K. and Mahon, J. (2010), Using Extension Fieldwork to Incorporate Experiential Learning into University Coursework. Journal of Extension 48 (2), pp. 1-8.

Fishera, J.A. (2001), The demise of fieldwork as an integral part of science education in United Kingdom schools: a victim of cultural change and political pressure? Pedagogy, Culture and Society, 9 (1), pp. 75-96.

Fuller, I.C. (2006) "What is the Value of Fieldwork? Answers from New Zealand Using Two Contrasting Undergraduate Physical Geography Field Trips". New Zealand Geographer 62, pp. 215-220.

Garrison, D.R. and Anderson, T. (2003), E-Learning in the 21st Century. A Framework for Research and Practice, RoutledgeFalmer.

Garrison, D.R. and Cleveland-Innes, M. (2005), Facilitating Cognitive Presence in Online Learning: Interaction Is Not Enough. The American Journal of Distance Education 19 (3), pp. 133-148.

Gerber, R. and Kim Chuan, G. (eds.) (2000), Fieldwork in Geography: Reflections, Perspectives and Actions. Kluwer Academic Publishers, Dordrecht.

Giere, R.N. (1988), Explaining Science. A Cognitive Approach. University of Chicago Press, Chicago and London.

– (1992), "Introduction: Cognitive Models of Science". In Giere, R., Cognitive Models of Science, vol. 15 "Minnesota Studies in the Philosophy of Science", xiii-xxv. University of Minnesota Press, Minneapolis.

Gold, J., Jenkins, A., Lee, R., Monk, J., Riley, J., Shepherd, I. and Unwin, D. (1991), Teaching Geography in Higher Education: A Manual of Good Practice. Blackwell, Oxford.

González, W.J. (eds.) (2002), Diversidad de la explicación científica. Barcelona Ariel.

Huckle, J. (1996), Teacher Education. In Huckle, J. and Sterling, S. (eds.), Education for Sustainability. Earthscan, London.

Juárez-Nájera, M., Dieleman, H. and Turpin-Marion, S. (2006), Sustainabilty in Mexican Higher Education: towards a new academic and professional culture. Journal of Cleaner Production 14, pp. 1028-1038.

Kempa, R.F. and Orion, N. (1996), Students' Perception of Co-operative Learning in Earth Science Fieldwork. Research in Science and Technological Education, 14 (1), pp. 33-41.

Kent, M., Gilbertsone, D. and Hunt, C. (1997), Fieldwork in geography teaching: a critical review of the literature and approaches. Journal of Geography in Higher Education, 21, pp. 313-332.

Kuhn, T. (1990), La estructura de las revoluciones cientificas. Fondo de Cultura Económica, México.

Lima, A., Vasconcelos, C., Félix, N., Barros, J. and Mendonça, A. (2010), Field trip activity in an ancient gold mine: scientific literacy in informal education. Public Understanding of Science, 19 (3), pp. 322-334.

Manzanal, R.M., Rodríguez, M.R.L. and Casal Jiménez, M.C. (1999), Relationship between Ecology Fieldwork and Student Attitudes toward Environmental Protection. Journal of Research Science Teaching, 36 (4), pp. 431-453.

Martens, P. (2006), Sustainability: science or fiction? Sustainability: Science, Practise, & Policy, 2 (1), pp. 36-41.

Mathews, M.R. (2003), Thomas Kuhn's Impact on Science Education: What Lessons Can Be Learned? Science Education, 88(1), pp. 90-118.

Moore, M.G. (1989), Three types of interaction. The American Journal of Distance Education 3 (2), pp. 1-6.

Orion, N. (1993), A Model for the Development and Implementation of Field Trips as an Integral Part at Science Curriculum. School Science and Mathematics 93(6), pp. 325-331.

Orion, N., Hofstein, A., Tamir, P. and Giddings, G.J. (1997), Development and validation of an instrument for assessing the learning environment of outdoor science activities. Science Education, 81, pp. 161-171.

Pereira, A., Mendes, A.Q., Morgado, L., Amante, L. and Bidarra, J. (2007), Modelo Pedagógico Virtual da Universidade Aberta para uma universidade de futuro. Universidade Aberta, Lisbon.

Popper, K. (1988), Conocimiento objectivo. Tecnos, Madrid.

– (1990), La logica de la investigación cientifica. Tecnos, Madrid.

Ramasundaram, V., Grunwald, S., Mangeot, A., Comerford, N.B. and Bliss, C.M. (2005), Development of an environmental virtual field laboratory. Computer and Education 45, pp. 21-24.

Saul, D. (2000), Expanding environmental education: thinking critically, thinking culturally. Journal of Environmental Education, 31(2), pp. 5-7.

Strang, V. (2009), Integrating the social and natural sciences in environmental research: a discussion paper. Environmental Development Sustainability, vol. 11, pp. 1-18.

UNESCO Report (2002), Education for All. Is the World on Track?

Chapter 13

Online Learning for Sustainability: The Student Perception in an Environmental Science Post-Graduation

Paula Bacelar-Nicolau, Ana Paula Martinho, Filomena Amador, Sandra Caeiro, Ulisses Miranda Azeiteiro

Abstract

The importance of a set of parameters on student satisfaction and on their perception of the learning effectiveness of an *MSc* online-learning programme in Environmental Citizenship and Participation at Universidade Aberta, Portugal, was assessed throughout its 4 editions. The majority of students showed a very high level of motivation and satisfaction with the course. Also, most students had the perception of having reached an affective learning outcome of values, attitudes and behaviour, which is a central element of education for sustainability. The main factors identified as contributors to student satisfaction and to their perception of the learning effectiveness were: (i) high quality and level of communication, both student-student and student-teacher, (ii) teachers' expertise and motivating skills, and (iii) pedagogical aspects, such as the depth of coverage of the subjects under study, receiving feedback in time, the didactic materials, the course structure, and the learning activities.

Introduction

One of the most important goals in achieving sustainable development is, along with social and individual understanding, developing environmental sensitivity. The last decades, especially the latter, have shown a change in the characteristic behaviour of advanced markets from markets which demand competitive prices to those which, in addition to price competitiveness, value other criteria such as quality, service and environmental awareness (Cloquell-Ballester et al. 2008). Educational programmes in environmental science are then receiving a growing interest by the working-student population seeking professional development in this field, but who have no available time or a university nearby to attend face-to-face courses.

Environmental science courses and curricula aim to train students in holistic thinking, which includes taking into account the interrelationships between environmental factors and processes (Ramasundaram et al. 2005). Laboratory exercises, field observations and field trips are then an important part of many environmental science courses. However, field observations and field trips can be constrained because of distance, time, expense, scale, safety, or complexity of real-world environments. Hence *online* education in the environmental sciences can be a complicated issue. Nevertheless, the use of up-to-date simulation models which allow the testing of multiple scenarios and their outcomes, videos, and other information and communication technologies, or intensive face-to-face activities given, in a short period of time, allow for the existence of these courses. Several authors have been working in developing new tools of e-learning in the environmental science field (e.g. Ramasundaram et al. 2005; Eneroth 2000).

E-learning has a role to play in this endeavour, as it allows global access to environmental education, independent of time and place, once the courses are available on the Internet. Additionally, *e*-learning has the potential of reducing the environmental impacts from face-to-face education. Potential resource savings can be found in the use of time, energy and waste (Eneroth 2000; Garrison 2000).

The increased use of internet-improved technologies, and web-based applications, as well as the increasing effectiveness of e-learning have become practical and theoretically important issues in both educational engineering and information system fields (Lee and Lee 2008). Research has strong evidence for the effectiveness of distance learning *per se* as a method of delivery (Moore and Thompson 1997). In distance-learning programmes, objectives and learning outcomes should be the driving forces (Tham and Werner 2005), and the effectiveness of measure becomes "how much and how well is this being absorbed?" (Kirkpatrick 1998). When designing and evaluating *e*-learning in higher education, essential aspects to take into account are the fact that distance learners are dynamic individuals that often change in response to both educational and life experiences (Gibson 1992), and that distance learners are a diverse group (Holmberg 1995; Altbach 2000) with different learning styles (Billings 1993).

Learning environmental satisfaction requires quality assessments of interaction, information, and system. Lee and Lee (2008) suggested the importance of qualitative assessment and interaction as well as system quality through e-learning. According to Cloquell-Ballester et al. (2008), it is important not to limit the evaluation of environmental education to a student satisfaction survey. It is important to develop specific studies about indicators for the real impact of environmental education on the socio-economic environment. The essence of quality education, in any form, is to ensure that learning objectives are achieved efficiently and effectively, without sacrificing the standards of the educator and his or her institution

(Tham and Werner 2005). Therefore, the evaluation process seeks to confirm whether the learning objectives and outcomes have been implemented and achieved effectively (Odini 2000). According to a study by Paecher et al. (2010), the main aspects which contribute to the students' learning achievements and course satisfaction are their achievement goals and the instructor's support and expertise. These are critical components and competences that educational institutions should provide students in order for them to be better able to face society (Gardiner 1994).

The aim of this paper is to assess the importance of a set of parameters on student satisfaction and on their perception of the learning effectiveness of an *MSc* online learning programme in Environmental Citizenship and Participation at UAb. A set of data was analyzed using a triangulation methodology to assess the importance of parameters such as student motivation, student-content interaction, student-student interaction, student-teacher interaction, type of learning activities, type/regime of evaluation, and student environmental citizenship attitudes and behaviour. Preliminary results of this research were published earlier (Bacelar-Nicolau et al. 2009).

The MSc Programme in Environmental Citizenship and Participation

In recent years the Universidade Aberta (UAb), Portugal, has implemented an online teaching and learning system based on a pedagogical model (Pereira et al. 2007) and using the learning management system (LMS) Moodle. This aims to provide students with equal access to learning resources and communication channels with their teachers, fellow students, and academic and administrative support services. The learning communities interact with secretarial/administrative staff, coordinating programme team, and teachers in separate, individual virtual spaces; a virtual *Café* space is also available for social interactions.

The *MSc* programme in Environmental Citizenship and Participation offered at the UAb had its first edition in 2006/2007, and has since witnessed four new course editions. It is a formal course, organized according to the European Credit Transfer and Accumulation System (ECTS). Curricular units are delivered *via* e-learning except for one curricular unit that includes an intensive one-day face-to-face lecture during the second semester. The *MSc* lasts for three semesters, the two initial semesters being dedicated to the curricular units (60 ECTS) (table 1) and the third semester being dedicated to planning, developing, writing and defending a dissertation or project (40 ECTS). The semester is defined as a period of 20 weeks, the 4 final weeks being committed to the final evaluation.

This programme is designed to prepare students for working in environmental policy-making as well as in improving their environmental citizenship, participation and planning abilities. The programme is designed to help governmental workers, public and private environmental advisors, members of environmental non-governmental organizations, teachers, researchers and individuals involved in environmental practices, policies, planning, teaching, participation and citizenship. Subjects, methodologies and case studies are drawn from areas of environmental sciences and social sciences in a sustainability perspective.

Table 1: The Curricular Plan of the *MSc* in Environmental Citizenship and Participation offered at the Universidade Aberta (UAb, Portugal)

1st Semester – Fundamentals (select for a sum of 30 ECTS)

Curricular units (optative)	**ECTS**
Land Use Management	6
Pollution and Resources	6
Waste Management	6
Instruments for Environmental Management	6
Biodiversity, Geodiversity and Conservation	6
Food Consumption and Environment	6
Environmental Impact Assessment	6
Promotion of Health and Environment	6
Sustainable Management of Marine Resources	6
Elements for the Analysis of the Social Conjuncture	6
European Virtual Seminar for Sustainable Development (*)	

2nd Semester – Conceptual, Practical Methodologies and Techniques (30 ECTS)

Curricular Units (obligatory)	**ECTS**
Ethics and Environmental Citizenship	6
Sustainable Policies	6
Participation and Interactive Methods in Environmental Decision-Making(**)	6
Methodologies of Social Intervention	6
Projects and Methodology in Environmental Citizenship	6

(*) This CU is coordinated by the Open University of the Netherlands and has students from different universities that also offer this course.
(**) curricular unit which includes a one-day intensive face-to-face lecture.

Methodology and Research Instruments

The study, herein presented, was conducted between June 2007 and January 2010, during the four editions of the *MSc* online learning programe in Environmental Citizenship and Participation, to a group of 78 students.

In methodological terms, two confidential questionnaire surveys, semi-structured personal interviews and observation grids for characterization of the curricular units were used to obtain a set of data that was analyzed using a triangulation strategy. Even though this was an exploratory study there was a concern to approach it both in a qualitative as well as a quantitative manner. This was undertaken in order to gain a deeper insight into the nature and the type of parameters involved in student satisfaction and in their perception of the learning effectiveness of the programme.

The first questionnaire survey contained 25 questions, designed according to appropriate questionnaire criteria of clarity and objectivity (Quivy and Campenhoudt 2006) to characterize the individuals (i) in demographic terms, (ii) from their evaluation of e-learning experiences, and (iii) from their evaluation of the course (structure, pedagogical resources, learning achievements, among others.). In section (ii) students evaluated statements related to the learning process on a 5-point scale ranging: "I agree completely", "I agree", "I do not agree nor disagree", "I disagree" to "I disagree completely". In section (iii) students assessed various aspects of the course on a 5-point scale: "insufficient", "sufficient", "good", "very good" and also "not applicable".

The second questionnaire survey was applied at the end of the second semester and was composed of 29 questions using the same criteria of clarity and objectivity. It aimed to evaluate (i) student's knowledge acquisition on key concepts of environmental sciences (e.g. water, waste, land use, air, natural conservation, soil and sustainable consumption), (ii) student's sensing of personal attitudinal changes in specific environmental areas, (iii) student's sensing of personal behavioural changes in specific environmental areas and (iv) student's course global achievements relating to their expectations (acquisition of professional competences, public participation, etc.). In all sections of this questionnaire students evaluated statements in a 5-point scale ranging, as above, from "I agree completely" to "I disagree completely".

The semi-structured interviews were carried out by a researcher on a convenience sample (4 students who completed the course) two weeks before the end of the course. Each interview lasted for ca. 1hr, and aimed to clarify trends and issues that were brought up by the questionnaires, as well as to characterize main weaknesses and strengths of the *MSc* programme. Informal observations of the master programme coordinators were also used with the same purpose as the semi-structured interviews.

Each virtual curricular unit was characterized, by the respective teacher, using a predefined observation grid (Bacelar-Nicolau et al. 2009), for content organization, pedagogical instruments, learning activities (e-activities), synchronous/asynchronous communication, support materials, and type of assessment.

Results and Discussion

Curricular Unit Characterization

The virtual space of each curricular unit, as characterized by the observation grid, had a number of essential structural and organizational elements: a learning contract, indicating the structural and organizational aspects of the CU (including a plan of learning activities, study materials, timetables, how/where/when students are expected to communicate with staff and other students in the context of learning activities, evaluation process), a notification forum for unit announcements, a discussion forum for student queries. Optional support elements *e.g.* electronic resources were available in most CU.

Units were similarly organized into topics, each topic being developed during one to three weeks, depending on the subject area. The LMS Moodle activities included lessons, mini-tests, glossaries, discussion *fora* and blogs. Each topic was generally associated with one learning activity. E-activities used were discussion *fora* (mostly involving group work, role-play, and case-study analysis), written assignments and blogs. Communication was mostly asynchronous. Support materials included e-books, textbooks, research papers, internet sites, study texts prepared by the teachers, power point presentations, and video clips.

Student Satisfaction Perception of the Learning Effectiveness

The questionnaire surveys had an overall response rate of 50% which is a good result for an online voluntary survey (Cook, Heath and Thompson 2000). Students enrolled in the course were 63% female and 37% male. Most students, 48%, were between 47 and 50 years old; 30% were 31 to 40 years old; 10% were 29 years old or younger, and 10% were 51 and older. Their main motivation for enrolling in the course was interest in the scientific area (35%) and flexibility of learning relating to time and place (31%), but some needed specific training in the area (13%) and some needed professional promotion (13%) or for other reason (1%). The majority of students, 70%, had not previously enrolled in e-learning courses; only 17% had attended one course in e-learning and 7% had attended two e-learning courses.

Even though most students had not previously taken e-learning courses, by the end of the 1st semester all students agreed that an e-learning course could be as good as a face-to-face course, and 79% agreed that the use of information and communication technologies (ICT) could efficiently replace face-to-face lectures. However, 25% of students still preferred enrolling in a face-to-face course, and 17% had no opinion on whether they preferred one or the other type of learning regime. The online regime was also considered by 74% of students to be able to enrich the learning methodology used by the teacher (10% had no opinion, 17% disagreed). Nevertheless, 17% of students thought that the e-learning regime weakened the importance of the teacher's role, although most (80%) disagreed with this statement, and some (3%) had no opinion.

During the course, 83% of students used the Internet daily, and 17% used the Internet 3 to 4 times per week. 40% of students spent more than 20 h/week connected to the Internet with activities relating to the course, 27% spent between 15 and 20 h/week, 27% between 8 and 14 h/week and only 7% spent 7h/week or less on the Internet.

Also, by the end of the 1st semester, 90% of students thought that overall competences gained with the course were good or very good (10% thought they were sufficient). The quality of the pedagogical materials was rated good to very good by 87% of the students (sufficient by 13%), and 80% considered that the proposed learning activities contributed well or very well to the learning objectives of the course (17% of students thought they were sufficient). Among the diverse pedagogical materials used in the course, students preferred, from a higher to lower degree of preference: Moodle's lessons and websites, texts prepared by teachers, power point presentations and scientific papers or textbooks. Concerning the learning activities, 82% of students felt that they were correctly assessed by the individual learning activities; 17% had not formed an opinion. Among the various learning activities used by teachers, students rated, from high to low, the written assignments (group and individual), followed by Moodle's lessons, discussion *fora* or Moodle's mini-tests, blogging activity, synchronous discussions and, finally, the use of glossaries. A teacher's involvement and teacher's competence were considered good and very good by 84% and 90% of the students respectively, and sufficient by 7%. Students rated the different types of assessment used in each curricular unit, from high to low, as follows: formative evaluation, group written assignments, online final exam, and finally, synchronous discussion of written assignments (e.g. via Skype).

All students agreed that the interaction between colleagues was an important factor in the e-learning process, and this was rated as good or very good by 90% of the *MSc* students, sufficient by 3%, and insufficient by 3%. Still, 37% of students thought that the *online* environment was emotionally poor and incapable of generating strong relationships among people, while 53% of the students thought that the *online* environment generated strong relationships among people; 7% had not formed an opinion. In agreement with this, to some extent, moments of synchronous communication in the e-learning course were dispensable for 23% of the students, but indispensable for most, 60%; 17% had not formed an opinion.

At the end of the curricular year most students (82 to 100%, depending on the subject) perceived they had acquired knowledge in the main areas of the *MSc* course e.g. sustainable use of water, sustainable use of solid residues, land management, air quality, management and soil conservation, nature conservation, sustainable consumption). As regards their attitudinal changes in those main areas, 82 to 100% of students perceived changes in their personal attitude, although only 76 to 89% had effectively changed their attitude in those environmental areas. On completion of the course, all students agreed that they intended to alter other people's attitude and behaviour in environmental subjects.

Also, on conclusion of the course, all students perceived that they had improved in their professional competences and the tools developed.

Most students, 64%, felt that the main obstacles in an e-learning course were ICT-related rather than pedagogically-related; 17% had not formed opinion; and 20% disagreed. This was particularly true for students experiencing e-learning for the first time, as has been reported by other authors (Tyler-Smith 2006). For these students, a greater effort in motivating and giving ICT support had to be given by the course coordinators, and also by teachers, particularly during the first three months of the course.

These results emphasize the growing acceptance of online distance learning, particularly in environmental education and in sustainability issues, as reported by other authors (Cloquell-Ballaster et al. 2008; Palmer and Holt 2010; Paechter et al. 2010), but the need to further improve through e.g. ICT methodologies, motivating skills, which are directed to different learning and intelligence types (Lee and Lee 2008; Ramasundaram et al. 2005).

Synchronous moments in the course were used as an attempt to bridge the distance felt by students. This was done in two distinct modes: (a) synchronous moments on the LMS, in the context of two curricular units, (b) face-to-face one-day sessions, between students and teachers, at 3 distinct periods: opening session of the *MSc* /1st semester; beginning of the 2nd semester; and one-day intensive session (half-way through the second semester), in the context of the CU *Projects and Methodology in Environmental Citizenship.* All of these sessions were transmitted, in real time, over the Internet for those students who could not be present. These synchronous moments were highly valued by the students. The UC characterization and the student interviews clarified that the synchronous moments on the LMS were orientated for the UC's final assessment, as oral presentations of individual or group assignments, and that these moments were cherished by the students. These results contrasted with those from the initial study, were synchronous moments were considered dispensable by the students (Bacelar-Nicolau et al. 2009); for the earlier edition of the *MSc*, synchronous moments were orientated towards the clarification of study matters, and this was valued less in the learning process. Moreover, the 3 days of face-to-face sessions were important for bridging the distance gap and for strengthening the e-learning community, a fact that was also evident from the increasing number of students who generally attended these sessions throughout the year. In fact, the b-learning regime, even though limited to a minimum of face-to-face sessions, has the potential of strengthening the virtual learning community, as has been pointed out by other authors (Tyler-Smith 2006; Moore and Gilmartin 2010).

The results of the personal interviews and the informal observations of the programme coordinators emphasized the importance of the "high communication level", both student-student and student-teacher, the "depth of coverage of the subjects under study" and the "effectiveness of the e-learning process" for the high motivation and *MSc* programme satisfaction felt by the students. Student-student interaction was considered to be "fundamental" because "it was a part of the asynchronous learning process" and also because "of the emotional support it gave when times were harder". The student-teacher interaction was also considered "fundamental", and in some cases the "fellowship bond, when distance between students and teachers had disappeared, contributed enormously to the learning progress". The teachers' motivating role was also pointed out by students to be a "fundamental pedagogical factor".

Feelings of declining motivation and dissatisfaction with the master programme were transmitted to the coordinators by students who, later, failed to complete the first year of the course. The main reasons for those feelings and for the subsequent student's dropping out were: (i) time conflicts between study, and work and/or family, (ii) ICT problems, (iii) "overload by too much to study, and multiple conversations to, and (iv) "solitude feeling that decreased motivation". Other reasons for dropping out, such as instructional design of the course/programme, or the feeling that students had learned what they needed or wanted (Tyler-Smith 2006) were not referred to during the 4 editions of the *MSc* programme. In fact, the drop-out percentage in the *MSc* programme declined from an initial 50% rate to 37% throughout its 4 editions (data not shown), values that are in the lower limit of the drop-out range referred to in Tyler-Smith (2006). The declining drop-out rate was attributed to (i) the implementation of a 2-week "introducing e-learning course", tutored by the programme coordinators, prior to the beginning of all *MSc* programmes, and (ii) to extra-support, both ICT-related and emotional, from programme coordinators and also from some unit teachers.

Conclusion

E-learning is becoming increasingly widespread in the context of higher education, responding to an increasing need to provide more flexible, time and/or place-independent study pathways in the face of growing trends towards part-time employment and student mobility, and also because e-learning potentially caters for student's different paces of learning and different learning styles. The use of e-learning in the education of various science courses, traditionally taught in face-to-face lectures, is also becoming more widely implemented and accepted (e.g. Mohanna 2007; Moore and Gilmartin 2010). Moreover, the new set of expectations of the younger "net generation" may cause in "those who have grown up with interactive technology" a discomfort with the information transmission approach of traditional lectures. The younger generation of "students expect a relevant and engaging learning approach" (Garrison and Vaughan 2008).

The large majority of students of the *MSc* programme in Environmental Citizenship and Participation at UAb showed a very high level of motivation and satisfaction with the course. Also, most students had the perception of having reached an affective learning outcome of values, attitudes and behaviour which was suggested by Shepard (2008) to be a central element of education for sustainability. The main factors identified as contributors to student satisfaction and to their perception of the learning effectiveness are similar to those reported by other authors (Palmer and Holt 2010; Paechter et al. 2010) and were the (i) "high quality and level of communication", both student-student and student-teacher, (ii) teachers' expertise and motivating skills, and (iii) pedagogical aspects, such as the depth of coverage of the subjects under study, receiving feedback in time, the didactic materials, the course structure, and the learning activities. The e-activities introduced later, e.g. role-play, problem-based learning, group analysis of real-case studies, discussion *fora* or blog activities, all of which stimulate reflection and affective responses (Shepard 2008). One aspect which was emphasized both in the personal interviews and in the informal observation by the programme coordinators is the central role that the programme coordinator has for the success of the virtual learning community. This aspect has also been pointed out by Moura et al. (2010).

The early attrition felt in e-learning courses can, and has been reduced in our MSc programme, by the implementation of a 2-week "introducing e-learning course" prior to the beginning of all programmes and by extra-support (technical and e-motivating) from both master programme coordinators and UC teachers, until the e-learning community is affectively and effectively bound. A recent administrative decision that allows for the possibility of enrolment as part-time students, which in the *MSc* corresponds to a minimum of one CU out of five per semester, appears to result in a decrease in the drop-out rate, even though it may delay the necessary initial cohesion of the e-learning community. We believe that the research methodology of this study should move towards a further deepening of reflection on teaching practice from the perspective of integration between research and action. In the present case, the results obtained already led to modifications in successive years of the *MSc* programme.

As regards students' attitudinal changes in the main areas of environmental citizenship and participation, and although most students perceived changes in their personal attitude and behaviour, a new survey will be conducted sometime after the end of the programme to confirm the results presented here.

The complexity of variables affecting e-learning need further and continuous research, from which will result, in time, e.g. improvements in ICT, more understanding of online pedagogy, more effective course design, more learner-centred design and support, or more adequate motivating and teaching skills. Some of the difficulties still felt in e-learning may be an evolutionary issue, and also in time, be resolved by the growth of a learner population for whom online learning is as normal as the classroom has been for former generations of learners for the last one hundred and fifty years.

References

Altbach, P.G. (2000), The crisis in multinational higher education. International Higher Education, 21, pp. 3-5 (https://htmldbprod.bc.edu/pls/htmldb/f?p=2290:4:710304345725727::NO:RP,4:P0_CONTENT_ID:99898).

Bacelar-Nicolau, P., Caeiro, S., Martinho, A.P. and Azeiteiro, U.M., Amador, F. (2009), E-learning for the Environment. The Universidade Aberta (Portuguese Open Distance University) experience in the Environmental Sciences post-graduate courses. International Journal of Sustainability in Higher Education, 10, pp. 354-367 (doi: 10.1108/146763709 10990701).

Billings, D. (1993), Learning style preferences and distance education: a review of literature and implications. In ACSDE Research monograph no. 8 (part 2), pp.1-11. The American Center for the Study of Distance Education, Pennsylvania State University University Park, PA., USA (ISBN: 1-877780-10-3).

Cloquell-Ballester, V., Monterde-Díaz, R., Cloquell-Ballester, V. and Torres-Sibille, A. (2008), Environmental education for small- and medium-sized enterprises: Methodology and e-learning experience in the Valencian regions. Journal of Environmental Management, 87, pp. 507–520 (doi:10.1016/j.jenv man.2007.01.041).

Cook, C., Heath, F. and Thompson, R.L. (2000), A meta-analysis of response rates in web- or internet-based surveys. Educational and Psychological Measurement, 60, pp. 821-836 (doi: 10.1177/00131640021970934).

Eneroth, C. (2000), E-Learning for Environment. Improving e-Learning as a Tool for Cleaner Production Education. 236 p. Licentiate Dissertation. Lund University, Sweden (http://www.iiiee.lu.se/Publication.nsf/$webAll/0404209 3A1557698C1256C1A003BF290).

Gardiner, L.F. (1994), Redesigning higher education: producing dramatic gains in student learning. 283 p., ASHE-ERIC Higher Education Report Series, Volume 23-7. George Washington University, Washington DC, USA (ISBN-10: 187838063X).

Garrison, D.R. and Vaughan, N.D. (2008), Blended Learning in Higher Education: Framework, Principles and Guidelines. 272 p. Jossey Bass Higher and Adult Education Series, John Wiley & Sons, San Francisco, CA, USA (ISBN-13: 9780787987701).

Garrison, R. (2000), Theoretical Challenges for Distance Education in the 21st Century: A Shift from Structural to Transactional Issues. International Review of Research in Open and Distance Learning, 1, pp. 1-17 (http://www.irrodl.org/index.php/irrodl/article/view/2).

Gibson, C.C. (1992), Changing perceptions of learners and learning at a distance. In ACSDE Research monograph no. 4 (part 1), pp. 34-42. The American Center for the Study of Distance Education, Pennsylvania State University, University Park, PA., USA (ISBN: 1877780065).

Holmberg, B. (1995), Theory and practice of distance education. Routledge, New York.

Kirkpatrick, D.L. (1998), Evaluating training programmes: the four levels. 289 p. Berrett-Koehler Publishers, Inc. San Francisco, CA (ISBN-10: 1576750426).

Lee, J. and Lee, W. (2008), The relationship of e-Learner's self-regulatory efficacy and perception of e-Learning environmental quality. Computers in Human Behaviour, 24, pp. 32–47(doi:10.1016/j.chb.2006.12.001).

Mohanna, K. (2007), The use of elearning in medical education, Postgraduate Medical Journal, 83, pp. 211 (doi: 10.1136/pgmj.2007.058610).

Moore, G.M. and Thompson, M. (1997), The effects of distance learning. 112 p. (ACSDE Research monograph no.15), The American Center for the Study of Distance Education, Pennsylvania State University, University Park, PA., USA (ISBN-13: 978-1877780189).

Moore, N. and Gilmartin, M. (2010), Teaching for Better Learning: A Blended Learning Pilot Project with First-Year Geography Undergraduates. Journal of Geography in Higher Education, 34, pp. 327-344 (doi: 10.1080/03098265.2010.501552).

Moura, A.P., Cunha, L.M., Azeiteiro, U.M., Aires, L.A., Graça, P. and de Almeida, M.D. (2010), Food consumer science post-graduate courses: comparison of face-to-face versus online delivery systems. British Food Journal, 112, pp. 544-556 (doi: 10.1108/00070701011043781).

Odini, C. (2000), Using evaluation in training and development. International Journal of Manpower, 21, pp. 511-515.

Palmer, S. and Holt, D. (2010), Students' perceptions of the value of the elements of an online learning environment: looking back in moving forward.

Interactive Learning Environments, 18, pp. 135-151 (doi: 10.1080/09539960 802364592).

Paechter, M., Maier, B., Daniel Macher, D. (2010), Students' expectations of and experiences in e-learning: Their relation to learning achievements and course satisfaction. Computers and Education, 54, pp. 222-229 (doi:10.1016/ j.compedu.2009.08.005).

Pereira, A., Mendes, A.Q., Morgado, L., Amante, L. and Bidarra, J. (2007), Virtual Pedagogical Model: Model of Universidade Aberta for a future university. Universidade Aberta, Portugal (http://hdl.handle.net/10400.2/1295).

Quivy, R. and Campenhoudt, L. (2006), Manuel de Recherche en Sciences Sociales. 256 p. Dunod, Paris, France (ISBN: 9782100500390).

Ramasundaram, V., Grunwald, S., Mangeot, A., Comerford, N. and Bliss, C. (2005), Development of an environmental virtual field laboratory. Computers and Education, 45, pp. 21-34 (doi: 10.1016/j.compedu.2004.03.002).

Shepard, K. (2008), Higher education for sustainability: seeking affective outcomes. Higher education for sustainability, 9, pp. 87-98 (doi: 10.1108/ 14676370810842201).

Tham, C.M. and Werner, J.M. (2005), Designing and evaluating e-learning in higher education: a review and recommendations. Journal of Leadership and Organizational Studies, 11, pp. 15-25 (doi: 10.1177/107179190501100203).

Tyler-Smith, K. (2006), Early Attrition among First Time e-Learners: A Review of Factors that Contribute to Drop-out, Withdrawal and Non-completion Rates of Adult Learners undertaking eLearning Programmes. Journal of Online Learning and Teaching, 2(2) (http://jolt.merlot.org/Vol2_No2_TylerSmith.htm).

Chapter 14

Agronomists, Civil Engineers and Architects: What Role in Earth Sustainability?

Ana Marta Gonçalves, Sónia Costa, Daniela de Figueiredo, Fernando Gonçalves, Miguel Ângelo Pardal, Ulisses Miranda Azeiteiro

Abstract

Sustainability is an actual and current concept discussed in political, economic, social, educational and environmental areas, involving ethical aspects of human behaviour. Man and the environment have been in equilibrium for a long time. However, during the last century, the development of science and new technologies, overuse of non-renewable energies and industrial globalization allowed the exponential growth of the human population which resulted in an increasing consumption of resources, environmental pollution and loss of biodiversity. Thus, in 1987, the Brundtland Commission launched the first definition of sustainability as: "*to suppress the needs of present generations without affecting the ability of future generations to suppress theirs*". A sustainable development may be followed by citizens in their everyday lives and by workers in their jobs in order to achieve human and environmental welfare without compromising future generations. Thus, training and awareness initiatives/activities may be developed and proposed to students and also to different sectors of economic activity contributing to the main decisions, like active employers and educators of future generations. Therefore, this chapter proposes a training session comprising different activities to agronomists, civil engineers, architects and also workers who are associated with both these sectors. The main aims of this work are: 1) to create awareness among professionals to environmental problems and 2) to propose the development of tools to be adopted in their daily life activities in order to contribute to a sustainable development.

Introduction

The usage of Earth's resources in an inadequate way induced an unsustainable condition. There is no part of the planet that is not suffering the influence of human activities. Climate change, imbalance of ecosystems, reduction of fertile soils, extinction of some species and depletion of natural resources are some of the profound impacts caused by mankind. These problems tend to intensify with the increase in the number of humans, mainly in cities, predicting to reach almost 5 billion out of a world total of 8.1 billion by 2030 (UN-HABITAT 2007). According to this scenario, the necessity of solving environmental and development problems is increasing, and the implementation of action plans integrating society, economy and environment is becoming crucial. The concept of sustainability was launched in 1987 in the report "Our Common Future" by the World Mission on Environment and Development (Klostermann and Cramer 2007). Sustainable development was defined by the Brundtland Commission as the "*development that meets the needs of the present, without compromising the ability of future generations to meet their own needs*" (WCED 1987). Still, in the context of the Earth's system, the most suitable and acceptable definition is referred to by Falkowski and Tchernov (2003) and Gallopin (2003), who consider sustainability as adaptive management with participatory decision-making and learning-by-doing being the mode of operation to strive for global stewardship. It is an intuitive and weighty concept with different meanings to different people, according to their advantage and purpose of study (Klostermann and Cramer 2007; Murtugudde 2009). This generated different areas in different domains of ecology, economy, society and education that defined and adopted principles of sustainability in their activities contributing to sustainable solutions in order to minimize environmental impacts (Alberti 1996; Xing et al. 2009). For instance, pressures from governments and non-governmental organizations in water issues concerning sustainable water management led to the creation of committees and forum commissions (the first World Water Forum was in Marrakech in 1997, and the World Commission on Water was initiated in 1998). One of the main goals of the Third World Water Forum in 2003 was to ensure the preservation of freshwater resources due to their importance to sustainable development, economic growth, social stability and poverty alleviation (*http://www.world.water-forum3.com*). Moreover, in some countries (such as UK, Canada and USA) in the building sector, there is an increasing requirement for companies to adopt principles of sustainability in their activities and policies to achieve sustainable solutions in urban development projects (Xing et al. 2009). New office buildings have an environmental assessment to evaluate impacts of construction and urban development to better support key decisions in different stages of the development projects. BREEAM, ENVEST, LEED, BEES and ATHENA are some institutes that develop methods/tools on environmental impacts assessment (Xing et al. 2009). Additionally, in the economic sector, a corporate sustainability index and sustainability indicators were established in industries and stock markets. These new metrics evaluate development sustainability based on the interrelationships between economic development, resource-energy consumption and environmental pollution (Xu et al. 2006). Industries are evaluated and rated yearly according to the relevant financial performance associated with social and environmental action plans of sustainability. One of the goals of an environmental market is to make available "ecological" products, meaning products that have been produced taking into account a reduction in the damage to the environment. Thus, the concept of "green consumer" came out (Ottman 1994). However, the development of all the economical activities referred to above have a major agent of transformation in common – technology. New technologies may be developed according to specific rules and respecting human freedom, without compromising the present and future generations. It is important to be an active citizen with a conscious behaviour, practising sustainable activities in daily work and everyday lives. The preservation of the environment should follow a lower anthropogenic perspective and a higher ecocentric view. In this new outlook, man is centered on the planet as his home ("oikos" = home in Greek), allowing the emergence of an ethic that is also studying the behaviour of man in relation to the global nature. Human beings come to better understand their role and responsibility to other living beings. In this environmental ethic, nature is seen as a place where man can take his resources without jeopardizing the preservation of life overall. In most cases, sustainable practices and activities are developed for students as they are seen as a future generation who has to acquire responsibility and become active citizens. Moreover, younger people are more receptive and conscientious to environmental problems and sustainable actions, making it easier to get feedback from them and achieve our goals. Nevertheless, it is also important not to forget the "older" citizens that are the key to most decisions made and are educators of the new generations. Sustainability is an ethical issue becoming necessary to educate the population by dissemination activities focusing on routine actions that must be put in practice everyday in order to preserve irreplaceable resources such as soils, energy resources, freshwater systems and biodiversity (Cabrita et al. 2003; Smith 2001). Furthermore, technical training is a component that must be applied in different sectors of the economy in order to achieve a sustainable development. Agronomists, civil engineers and architects were chosen to apply this activity to, as they are a huge group of professionals that can make a difference by acquiring sustainable attitudes in their work and everyday lives. Thus, they must be awarded for this theme in order to give them a range of choices to apply in their working routine. A variety and quantity of fertilizers and other chemicals are used in agriculture that could be a threat to the environment, having in some cases strong impacts on the services and goods provided by ecosystems (Gonçalves et al. 2008 In *Environmental Education and Sustainability*). A sustainable usage of natural resources by agronomists will avoid the erosion and overexploitation of soils, forest fires and water contamination that could induce the loss of biodiversity and the maintenance of populations. A sustainable agriculture[1] encompasses three domains – economic, social and environmental - the following being the main aims of this practice (Hoeflich et al. 1997; Granados e López 1996):

- to achieve and maintain fertile and productive soils with low costs (e.g. by recycling animals' waste);
- to fulfil the human needs;
- to promote an available economy;
- to give a better quality of life to society and farmers;
- to minimize the impacts in the environment, contributing to a better quality of the surrounding places;
- to promote as fast as possible profitable systems that can be used in the long term.

Merging traditional practices with new technologies allows the cropping and exploitation of different resources simultaneously in order to obtain products faster. Moreover, the use of computers and other electronic equipment to measure, for instance, the levels of nutrients in soils and cultures allows the usage of exact quantities of fertilizers and other chemicals with little or no damage to the environment. Furthermore, pest control could be performed using ecological forms (e.g. natural insecticides, traps) that could be easily prepared (see *http://attra.ncat.org/soils.html*). Sustainable agriculture is increasingly acceptable due to high productivity and quality of products. Furthermore, degrees and masters in sustainable agriculture or related areas are in the curricula of some international institutes (e.g. in Peru, Brazil, Portugal, USA, UK, Australia)[2]. Besides, there is a virtual network called RAS oriented to agronomists, farmers and other professionals that promotes the exchange of ideas and practices in order to promote this kind of agriculture (*http://www.rasnsw.com.au/default.htm*).

The construction industry represents one of the highest contributors to social and economic development, being one of the sectors with a profound impact on the environment. The application of the concept of sustainability to the housing sector requires a set of procedures that must be evaluated during construction work in order to link together environmental and architectural qualities. In several countries there are post-graduate courses addressed to civil engineers, architects, managers and promoters focused on techniques, principles and methodologies to

---

1 American Society of Agronomy in 1989 defined sustainable agriculture as that one promotes the quality of environment and resources; provides products to mankind needs; is economical practicable and promotes a better quality of life to farmers and society. In *http://www.soc.iastate.edu/sapp/soc235susag.pdf* and *http://sarasota.ifas.ufl.edu/AG/SusAgFAQ.pdf* (accessed on 31.1.2011).

2 *http://www.centerforsustainability.org/resources.php?root=79&category=82* (accessed on 31.1. 2011); *http://www.ehow.com/list_7282044_organic-agriculture-degrees.html* (accessed on 31.1.2011); *http://www.mastersportal.eu/students/browse/programme/20259/agroecology.html* (accessed on 31.1.2011).

evaluate sustainability of a building and thus to obtain the certification of environmental quality in their construction. Some tools have already been developed concerning this topic. For example, the international SBTOOL (Sustainable Building Tool) has the aim of recognizing, evaluating and certificating sustainable buildings. Thus, the construction of buildings may follow some requirements that can make the difference in sustainable development: 1. reduce energy consumption and water consumption; 2. change of conventional sources to renewable; 3. use of ecological materials; 4. waste management; 5. optimization of construction; 6. perform flexible spaces. All these parameters could be achieved using equipment available in the construction sector, as referred to in table 1.

Table 1: Requirements and building materials to a sustainable building construction

Requirements	**Building materials**
Reduction of energy consumption and change conventional sources to renewal	Devices for power generation
Optimization of construction	Prefabrication of buildings
Perform flexible spaces	Multifunctional spaces, with easy adaptation to the concept that a person wants to give to the space
Reduction of water consumption; waste management	Lodging of suitable equipment with the characteristics required
Usage of ecological material	Make construction of buildings with materials available on the market with low impact to the environment

Moreover, some policies have been applied in western countries in the recycling of construction waste instead of simply placing it in vacant lots to minimize environmental impacts (Linhares et al. 2007). Many countries have established criteria and procedures to manage residues, classifying them in several categories/classes (e.g. table 2). Furthermore, the application of the policy of the three Rs (Reduce, Recycle and Reuse) in residues management is essential due to the huge quantities produced everyday in urban centres.

Table 2: Classification of solid residues from the construction sector in Brazil

Class	**Description**	**Example**
A	Reusable or recyclable waste	Brick
B	Reusable or recyclable waste for other industries	Paper, plastic
C	Residues not recyclable	Plaster
D	Dangerous residues Contaminated residues	Paints, solvents From industries, clinical radiology

(adapted from Linhares et al. *2007)*

This chapter is organized as training sessions to workers related to agriculture (agronomists, farmers, foresters) and construction (civil engineers, architects, house-builders) sectors that have to develop activities of sustainability in their work. Thus, the main aims of this chapter are: 1. to create awareness among workers (e.g. civil engineers, architects, agronomists) to environmental problems; 2. to propose the development of tools that contribute to a sustainable world; 3. to adopt methodologies implementing sustainable activities (e.g. ways to spread these practices to workers); 4. to evaluate the impact of those methodologies/tools in a sustainable world.

Methodologies and Practical Activities

These activities are useful tools to be applied to students in order to develop and approach an issue about environmental education. Normally, these chapters have two components: theoretical and practical, giving the background and then applied in activities. Thus, students have the opportunity to evaluate their knowledge.

In future works it would be interesting to develop action plans/chapters applicable to other sectors and then apply it individually (to each sector without mixing workers from different sectors). In the next step, the same activity should be developed in groups of workers from different sectors. They have the opportunity to work in groups and exchange ideas. Finally, teachers should evaluate these two methodologies: 1. activities performed and developed by groups of workers from the same sector; 2. activities performed and developed by groups of workers from different sectors. The best methodology should be applied in future.

The topic of sustainability is introduced to students (e.g. agronomists, civil engineers and architects). Afterwards, some general ideas of the biggest challenges faced by developed human societies are highlighted. For example, increased poverty, social disintegration, ecological damage, failure of policies and the rise in population growth led to the over-exploitation of natural resources, causing an imbalance between economic and industrial growth and environmental preservation. Therefore, it becomes crucial to develop lifestyles, having in mind a common welfare, respecting natural resources and the dignity of mankind. The concept of sustainable development occurred firstly in an environmental perspective and later on, due to social changes, in social, economic and educational domains (Benneh et al. 1996; Giddens 2000; Meadows and Meadows 1993; Nattrass and Altomare 2002). The principles of each domain are referred to in tables 3 to 6 which can be projected by the teacher. However, students can be questioned about the principles as they are being referred to in order to know their opinion and how the concept of sustainability is structured in different areas of activity. Then, a questionnaire is given to each student to fill in, giving information to the teacher about "how sustainable" the farmer/builder is (Annex I).

Students are divided into groups of 4-5 elements to develop an activity (Annex II). It is proposed that each group refers to the principles/aims of a sustainable agriculture (to agronomists' groups) and a sustainable architecture (to civil engineers/architects' groups).

Table 3: Principles of a sustainable environment

• Restriction on the use of non-renewable energy
• Consumption of renewable energy may be done without overstepping capacity of regeneration
• Avoid discharges to the environment that are not degradable over time
• Avoid activities that can be a threat or dangerous to mankind and ecosystems

Table 4: Principles of a sustainable education

• Ensure basic conditions of welfare, justice, freedom and solidarity
• Transmit responsibility and social values
• Society may be attentive to changes and dynamism of systems in order to avoid risk situations
• Invest in a solid education system and provide professional goals and skills

Table 5: Principles of a sustainable economy

• Organize economic structures that are able to answer to human needs
• Capital preservation such as buildings, bridges, monuments
• Capital preservation by each generation which may be given to next generations
• Effective use of resources
• All economic services must be made using an open system
• Effective use of taxes paid by citizens and factories
• International businesses must perform without compromising future generations
• Costs and benefit services may be paid by those who enjoy it
• Stabilization of monetary value, preventing inflation

Table 6: Principles of a sustainable society

• Ensure the human rights of citizens
• Ensure security and justice by a judiciary system
• Citizens have the right to better conditions of life
• Promote the same rights and opportunities to citizens
• Ensure social protection according to citizens' needs
• Ensure conditions of social inclusion for people with lawsuits

Afterwards, students must refer to how each principle/aim is accomplished in a pragmatic way in their work. Finally, students must list some ways of encouraging their workers to follow a sustainable development. Each work may be shown to the class in order to debate their presentations. Agronomists and civil engineers/architects should refer to the main difficulties felt when following a sustainable development concept. In Annex III, the main difficulties of each sector are already referred to (agronomists and civil engineers/architects) in developing sustainable development. Finally, the teacher should propose to each group to make a poster with a general definition of sustainability and practices that should be adopted by a citizen in his/her everyday life to contribute to a sustainable world (Annex IV). Teacher and students may bring to an end the practice of sustainable activities in everyday life (in work and as citizens) in order to minimize environmental impacts and not endanger future generations by an overexploitation of resources.

As a general conclusion, students must be aware that some of the benefits of sustainability are: 1. fresh water available to the population; 2. low consumption of energy; 3. reduction of air pollution; 4. fertile soils; 5. avoidance of an overexploitation of resources; 6. preservation of biodiversity and ecosystems.

References

Alberti, M. (1996), Measuring urban sustainability. Environmental Impact Assessment Review 16, pp. 381-424.

Benneh, G., Morgan, W. and Uitto, J. (1996), Sustaining the Future-Economic, Social and Environmental Change in Sub-Saharan Africa. UNUP.

Cabrita, A.R., Branco, J.P. and Mourão, J. (2003), Qualidade, quantidade e sustentabilidade habitacional. LNEC.

Falkowski, P. and Tchernov, D. (2003), Human footprints in the ecological landscape. In Earth System Analysis for Sustainability. Edited by Schellnhuber, H.J., Crutzen, P.J., Clark, W.C., Claussen, M. and Held, H., Cambridge, M.A., pp. 211-226. The MIT Press.

Gallopin, G.C. (2003), What kind of system of science (and technology) is needed to support the quest for sustainable development? In Earth System Analysis for Sustainability. Edited by Schellnhuber, H.J., Crutzen, P.J., Clark, W.C., Claussen, M., Held, H. and Cambridge, M.A., pp. 367-386. The MIT Press.

Giddens, A. (2000), O Mundo na Era da Globalização. Editorial Presença, Lisbon.

Gonçalves, A.M.M., Antunes, S.C., Pereira, R. and Gonçalves, F. (2008): The input of nutrients in freshwater ecosystems – A simulation of the eutrophication process. In Azeiteiro, U.M., Gonçalves, F., Pereira, R., Pereira, M.J., Leal Filho, W. and Morgado, F. (eds.), Environmental Education, Communication and Sustainability, pp. 209-231. Peter Lang GmbH, Frankfurt am Main, Germany (ISBN 978-3-631-55406-7).

Granados, D. and López, G. (1996), Agroecología. Universidade Autônoma Chapingo, México.

Hoeflich, E.C.E., Cano Cano, G., Cuevas, R.A.G., Martínez, E.V. (1997), Ciencia Ambiental y Desarrollo Sostenible. International Thomson Editores, México.

Klostermann, J.E.M. and Cramer, J. (2007), Social construction of sustainability in water companies in the Dutch coastal zone. Journal of Cleaner Production 15, pp. 1573-1584.

Linhares, S.P., Ferreira, J.A. and Ritter, E. (2007), Avaliação da implantação da Resolução nº307/2002 do CONAMA sobre gerenciamento dos resíduos de construção civil. Estudos Tecnológicos em Engenharia 3(3), pp. 176-194.

Meadows, D., Meadows, D. and Randers, J. (1993), Beyond the Limits: Confronting Global Collapse, Envisioning a Sustainable Future. Green Publishing Company, Chelsea.

Murtugudde, R. (2009), Regional Earth System prediction: a decision-making tool for sustainability? Environmental Sustainability 1, pp. 37-45.

Nattrass, B. and Altomare, M. (2002), Dancing with the tiger: learning sustainability step by natural step. New Society Publishers, British Columbia.

Ottman, J.A. (1994), Marketing verde: desafios e oportunidades para a nova era do marketing. Makron Books, São Paulo.

Smith, P.F. (2001), Architecture in a climate of change. Architectural Press.

UN-HABITAT (2007), Urbanization: A turning point in history. In The Millennium Development Goals and Urban Sustainability. UN-HABITAT.

WCED (1987), Our common future, report of the World Commission on Environment and Development. Oxford University Press, Oxford.

Xing, Y., Horner, R.M.W., El-Haram, M.A. and Bebbington, J. (2009), A framework model for assessing sustainability impacts of urban development. Accounting Forum 33, pp. 209-224.

Xu, F.-L., Zhao, S.-S., Dawson, R.W., Hao, J.-Y., Zhang, Y. and Tao, S. (2006), A triangle model of evaluating the sustainability status and trends of economic development. Ecological Modelling 195, pp. 327-337.

http://www.world.water-forum3.com (accessed on 20th October 2010).

http://www.centerforsustainability.org/resources.php?root=79andcategory=82 (accessed on 31st January 2011).

http://www.ehow.com/list_7282044_organic-agriculture-degrees.html (accessed on 31st January 2011).

http://www.mastersportal.eu/students/browse/programme/20259/agroecology.html (accessed on 31st January 2011).

http://www.soc.iastate.edu/sapp/soc235susag.pdf (accessed on 31st January 2011).

http://sarasota.ifas.ufl.edu/AG/SusAgFAQ.pdf (accessed on 31st January 2011).

http://attra.ncat.org/soils.html (accessed on 31st January 2011).

http://www.rasnsw.com.au/default.htm (accessed on 1st February 2011).

Annex I

NAME ______________________________
DATE ______________ CAREER ______________

Below there are some questions about actions, opinions that may contribute to a sustainable development. Please answer to the questionnaire referring if you agree/do not agree and do/do not do.

In italic are the answers expected by a sustainable citizen/worker.

HOW SUSTAINABLE ARE YOU?

1. Domestic waste:

1.1 is separated to be recycled. *I agree / I do.*

1.2 go to a common garbage bin. *I do not agree / I do not do.*

2. I go to work:

2.1 by bike, public transports. *I agree / I do.*

2.2 by car, motorcycle. *I do not agree / I do not do.*

3. The usage of renewable and non renewable energies:

3.1 The usage of non renewable and renewable energies must be done according mankind needs, independently of their capacity of regeneration. *I do not agree / I do not do.*

3.2 The usage of non renewable energy must be restricted and the consumption of renewable energy may be done without overstep their capacity of regeneration. *I agree / I do.*

4. In agriculture:

4.1 may be used fertilizers and chemical compounds in plantations and fossils fuels to agricultural machinery. *I do not agree / I do not do.*

4.2 may be used animals' waste as fertilizers and avoid or not use fossils fuels to agricultural machinery. *I agree / I do.*

5. In pest control:

5.1. may be used organic products that are more expensive than chemical compounds, still are harmless to environment. *I agree / I do.*

5.2 may be used compounds that have a rapid action independently their toxicity and impact to the environment. *I do not agree / I do not do.*

6. In architecture buildings:

6.1 may be chosen materials less expensive, independently their capacity of retain heat or have high consumption of energy. *I do not agree / I do not do.*

6.2 may be chosen materials with low impacts to the environment, if they are more expensive. *I agree / I do.*

Thank you for your collaboration!

Annex II

NAME ____

DATE ____________________ CAREER ____________________

1. In groups of 4-5 elements please fill the table above. You and your colleagues may report the principles/aims of a sustainable agriculture or sustainable architecture (according the sector your group represents). Furthermore, you have to refer how to achieve each goal.

In italic are the answers expected.
Answers by agronomists' groups:

Principles / aims	**Achieve the principles aims**
Achieve and maintain fertile and productive soils with low costs	– *Usage of animals' waste and organic compounds;* – *Crop rotation of a variety of products;* – *Utilize technological equipment to make measurements (to avoid the huge amount of products in systems) and to contribute to a efficient and fast planting (crop requiring to agricultural machinery);* – *Reduce or not use fertilizers and other toxic compounds;* – *Recycle wastes.*
Fill human needs	
Promote an available economy	
Give a better quality of life to society and farmers	
Minimize the impacts in the environment, contributing to a better quality of the surrounding places;	
Promote in a short time profitable systems that can be used in a long term.	

Answers by civil engineers/architects' groups:

Principles / aims	**Achieve the principles/ aims**
Reduction of energy consumption and change conventional sources by renewal	*Devices for power generation*
Optimization of construction	*Prefabrication of buildings*
Perform flexible spaces	*Multifunctional spaces, with a easy adaption to the concept that a person wants to give to the space*
Reduction of water consumption; waste management	*Lodging of suitable equipment with the characteristics required*
Usage of ecological material	*Make construction of buildings with materials available in the market with low impact to the environment*

As agronomist or builder what are, in your opinion, the best strategies to disclose the information by workers/citizens to contribute to a sustainable development?

- *Internet (informative sites where workers may have the possibility to make questions and interact with other workers – like RAS);*
- *Journals;*
- *Television publicity;*
- *Lectures;*
- *Talk directly to workers in the fields/ industries;*
- *Creation of agricultural/building institutions;*
- *Regulation of actions and usage of agricultural products/ building materials.*

Good work!

Annex III

NAME ____
DATE ____________________ CAREER ______________

In groups of 4-5 elements report the main difficulties to follow a sustainable agriculture or sustainable architecture (according the sector your group represents).

In italic are the answers expected.

- *Lack of information about the issue of sustainability;*
- *Restrict access to products with low impact or harmless to the environment (or not available in the market);*
- *Few products available to workers (more investigation work must be done in the sector in order to combine the needs of the sector with the environmental preservation);*
- *Price of products with no subsidies from the government;*
- *Policies adopted by governments are not an incentive to follow a sustainable development by workers.*

Annex IV

NAME ____
DATE ________________ CAREER ____________________

In this activity you and your group may have to make a poster entitled "Sustainable Citizens". You should give a general definition of sustainability and report some actions that be performed by a citizen in order to contribute to a sustainable development.

In italic are the answers expected.

Definition of Sustainability:

1. *Sustainability is an adaptive management with participatory decision-making and learning-by-doing being the mode of operation to strive for global stewardship.*

2. *Sustainability is to suppress the needs of present generations without affecting the ability of future generations to suppress theirs.*

Citizens' actions in their quotidian in order to contribute to sustainability:

- *Go to work in public transports, by bike or on walk;*
- *Separate the garbage (e.g. plastics and tins; paper / cardboard; glass);*
- *Minimize energy and water consumption;*
- *Do not make discharges / waste to the environment;*
- *Give preference to ecological products;*
- *Do not make an overexploitation of resources (resources may be used according their capacity of regeneration);*
- *Use renewable energy instead of non-renewable.*

Chapter 15

Developing Activities on Waste Management – an Approach to Achieve a Cleaner Environment

Sónia Costa, Ana Marta Gonçalves, Fernando Gonçalves, Ulisses Miranda Azeiteiro, Miguel Ângelo Pardal

Abstract

Achieving the joint goals of sustainability requires a continuous process of environmental education and effective change of behaviour. It involves the further development of conscious decisions in order to minimize the impacts of our actions on the environment. In this context, environmental education assumes an important role in the development of attitudes for protecting nature, especially when applied throughout the educational processes. Consequently, this paper is a contribution to teachers' approach in formal environmental education concerning the topic of waste management. As practical activities are a useful tool in the promotion of scientific knowledge and the interest of the students concerning environmental issues, this work will be developed performing different and progressive activities individually or in groups. The activities comprise tasks to be performed at the basic education levels (ages 8-12), but could be adjusted for other ages, encouraging the relationship student-environment. Under this approach, young citizens will have the opportunity to develop their abilities and know-how. They will act as potential environmental researchers and, at the same time, are responsible citizens contributing towards effective environmental protection.

Introduction

Sustainability is today one of the most widely used words, either in a scientific context, particularly in environmental sciences, or in society. Sustainable development is the "development that meets the needs of the present without compromising the ability of future generations to meet their own needs" (World Commission on Environment and Development 1987). Environmental awareness is an essential tool for the change in society ethical values and behaviour. In order to achieve this, citizens must be active participants in decision-making progresses, particularly those concerning the environment.

In little more than three decades, sustainability in education has emerged and evolved from marginal beginnings. It currently claims the necessity and the possibility of fundamental change in our collective view of the goals and nature of education and learning – a change, if made effective, would be critical to the quality of life of future generations (Stirling 2004). In this way, educational institutions around the world have a unique responsibility towards that goal. Basic education provided by schools is essential in the development of future conscientious citizens having a strong influence on the direction of society. In this way, environmental education (EE) is assuming increasing responsibility of younger generations, becoming an essential goal to promote sustainable development (Jacobi 2003). EE must provide opportunities to acquire knowledge, values, attitudes, and skills needed to protect the environment (Tal et al. 2004). The aim of EE in schools is to help students to increase their appreciation of the natural environment and to gain an understanding and motivation to help change the world.

The teacher/educator has the role of mediator in the construction of environmental references and should use them as instruments in the development of a social practice centred on the concept of nature. The EE could be situated in a vast context, the education for citizenship (Jacobi 2003). All educational institutions have an important responsibility in the further implementation of rules and should aim to be linked to internationally or nationally recognized developmental strategies. For this, they will need to have staff fully trained and competent in education for sustainable development and to provide all students with relevant opportunities and methods for learning about sustainable development (Haga Declaration 2000). Children can be empowered early on to take part in sustainability in discovering small ways to take individual and group action to make the difference. They need simple constructive opportunities and explore outdoors in order to develop respect for nature. When outdoors activities are not possible, reading books, watching educational programmes and developing practical activities in groups (e.g. exhibition of works to school community) may help students in experiencing the diversity of life on our planet.

It is widely accepted that learning in EE and science depends on previous knowledge and perceptions of the learner. Learning mainly occurs in a social context, when learners interact and share ideas, thoughts and actions. All these are in accordance with the constructivist view of learning that focuses on the learners, and recognizes that they have to be active in the process of learning (Tal et al. 2004). According to Putnam and Borko (1998), teachers require changes in their knowledge, beliefs and practice in order to achieve goals such as helping students to construct understanding, develop expertise and encourage the use of higher order of cognitive skills. Each teacher must include an environmental vision into the subjects during the class. Including EE as a routine in school activities is nothing more than taking the main focus of any activity in the environmental issue that is placed within the context being developed.

In 2002, the United Nations General Assembly adopted the Resolution A/RES/57/254 creating a United Nations Decade of Education for Sustainable

Development (DESD), spanning from 2005 to 2014 (International Implementation Scheme 2005). The United Nations Educational, Scientific and Cultural Organization (UNESCO) was requested to lead the Decade and develop a draft of International Implementation Scheme for the decade. It summarizes the goals and objectives of the decade in order to integrate the principles, values, and practices of sustainable development into all aspects of education and learning. This includes facilitating networking, linkages, exchange and interaction among stakeholders. This educational effort will encourage changes in behaviour which will create a more sustainable future in terms of environmental integrity, economic viability, and a fairness society for present and future generations. The basic vision of the DESD is a world where everyone has the opportunity to benefit from education and learn the values and lifestyles required for a sustainable future and for a positive societal transformation. In this context, all resources that help teachers to improve the conscience of students about environmental sustainability are important and useful. Consequently, the aim of the present work is to propose some suggestions of activities to be developed with younger students on sustainability, namely about waste management, being one more tool to help teachers in their function of environmental educators.

Background Information

Nowadays, solid waste management is a problem of major relevance for all societies. Waste represents an enormous loss of resources, both in the form of materials and energy. Excessive quantities of waste result from inefficient production processes, low durability of goods and unsustainable consumption patterns. Indeed, quantities of waste can be seen as an indicator of the material efficiency of society (O'Rourke and Clarke 2001). In this circumstance, the Directive 2006/12/EC established the legislative framework for the handling of waste. To implement the precautionary and the preventive action principles recommended by the EU legislation, it is necessary to set general environmental objectives for the management of waste.

All of us produce large quantities of solid waste. It is claimed that 93% of production material is never used on the final product and 80% of products are discarded after a single use (Stirling 2004). Half the waste comes from the manufacturing industry, construction and demolition activities, while municipal waste, mining waste and waste from other sources contribute each to about one sixth of the total (EEA 1999). To manage the waste in the best way, the member states of the European Union (EU) must establish a framework to prevent, reduce and, as far as is possible, eliminate from the outset the sources of pollution by adopting measures where recognized risks are eliminated (Directive 2008/98/EC). The same Directive will help EU members to move closer to "recycling societies", seeking to avoid waste generation and to use waste as a resource. In this way, measures aiming to ensure waste separation, collection and recycling are need. Environmental education is essential in some steps of that strategy (essentially waste prevention and reduction). Based on this objective and as a means to facilitate or improve its recovery potential, waste should be collected separately before undergoing recovery operations that deliver the best overall environmental outcome. In general, the production and recycling imply complex problems which involve cultural, political, marketing and social organization factors (Moreno and Pedrosa 2008). There is no single correct method to achieve proper waste management. Nevertheless, there are common needs that must be addressed to manage it better. These needs will include: adequate knowledge of the types of waste to be disposed of; how much there is; where it arises; who produces it and what happens to it (Rushbrook and Finnecy 1988). EU member states maintain different approaches to the collection of household waste and waste of a similar nature and composition, but the aim is the same: to move towards a European recycling society with a high level of resource efficiency (Directive 2008/98/EC). EU members must develop waste prevention programmes focusing on the key environmental impacts and taking into account the whole lifecycle of products and materials (Directive 2008/98/EC of the European Parliament and of the Council on waste). The management of waste requires an active and adequate contribution from the citizens. It requires mainly their intervention in its correct separation and in the judicious use of the infrastructures aimed at its placing, gathering and leading to adequate firms (Moreno and Pedrosa 2008). Moreover, finding acceptable strategies to cope with the waste problem is becoming quite a hard task, owing to the increasing awareness of environmental issues by population and authorities (Fiorucci et al. 2002).

Today, several common methods are used to dispose of garbage. They range from burning (incineration) to burying it in sanitary landfills. But these destinations of garbage are used the least. In particular, European legislation recommends the development of local integrated management plans which give priority to prevention, waste reduction and recovery, and allow using landfill only for the disposal of materials that cannot be recovered (Fiorucci et al. 2002). The goal is to find alternative destinations for the waste collected, which will include recycling some materials and reusing others as well. Nevertheless, landfilling is still the most common treatment route for waste in many countries. For instance, in the United States 33.8% of waste is recovered and recycled or composted, 11.9% is burned at combustion facilities, and the remaining 54.3% is disposed of in landfills (United States Environmental Protection Agency).

Suggested Activities

The social constructivism emphasizes interaction among learners that enable asking questions, sharing ideas, debating, arguing, concluding and collaborating while creating products and artefacts (Tal et al. 2004). Based on these assumptions, some suggestions of activities to develop with students of upper primary/junior secondary (ages 8-12, adjustable to other learning levels) are proposed. The activities must be preceded by a first phase of diagnosis of students' knowledge about diverse aspects related to waste. This could be done in a questionnaire format (Annex 1) or through a dialogue. It is very important to identify their difficulties and previous knowledge. This diagnosis will permit the teacher to readjust the methodology to be developed with the students. The same questionnaire must be used at the end of the activities to evaluate the efficiency of them in the change of students' and family behaviour, since students could be seen as a good way to raise awareness among adults about these subjects. They will have a strong influence in the development of an ecological mind on their relatives.

The activities proposed will comprise several waste management perspectives. It will also focus on the awareness of environmental problems caused by waste pollution. It is essential for students to understand why effective waste management is so important, and what is the role of citizens for the proper functioning of the ecosystems. Educating children about the importance of recycling provides a path to a cleaner environment and a greener future.

Activity 1 – Discussing Pollution

Most of the time the contact with a real situation increases the interest of students for specific problems. In this way, the teacher must raise questions concerning the main local environmental problems related with pollution to identify the knowledge and concerns of students about it. This initial phase must be conducted as an interactive discussion where the teacher must contextualize the activities. Holding class discussion will allow pupils to express their opinions. Moreover, a series of mental definitions of key concepts/issues need to be addressed on the discussion (e. g. waste, pollution, sustainability). It is important to clarify the students' doubts about the theme.

Activity 2 – Talking about Waste

The class must be started with a lecture on a reflexive text about "waste" chosen by the teacher which could be a newspaper item, a report or a story. The lecture must be followed by an interpretation about it and debate. Topics like identification of the problem and possible solutions must be discussed. A map approach like the one present in the Annex 2 can be used to achieve more consistent knowledge.

Activity 3 – Analyzing Waste

For many of us, home recycling is a pretty easy process. Our city provides us with a recycling bin and we have diverse information where to put it in. However, it is necessary to teach children how to do it and why it is important. This task is a way of confronting students with some important facts about it. To perform this activity, the teacher must exhibit solid garbage in front of students like plastic bottles, glass bottles and plastic cups (students may take to school solid garbage for this activity). Afterwards, students must choose one waste element. The next step is to divide the students in groups in accordance with the waste component. Each group will answer the following questions about their waste: decomposition time; impact caused by the production of the package; analysis of the label; the slogan and advertizing appeal and possibilities of reutilization of waste (Annex 3). The information needed by the students should be searched using several information resources.

At the end, each group will present their analysis and conclusions to the entire class, summarizing the information on the board.

Activity 4 – Reduce Waste to Protect the Environment

This activity will be useful for students to assume behaviours that will contribute to reduce the production of waste, discussing what each one can do to make a contribution.

One of the aims of waste management is to extract the maximum practical benefits from products and to generate the minimum amount of waste. Source reduction methods involve changes in manufacturing technology, raw material inputs, and product formulation. The term "pollution prevention" may refer to source reduction. Source reduction is typically measured by efficiencies and cutbacks in waste. Each of us has the duty to contribute towards reducing and reusing the materials. Waste, and how we choose to handle it, affects our world's environment. It is important to emphasize that all of us need a healthy environment. The waste we create has to be carefully controlled to be sure that it does not harm our environment and our health. Considering that waste management includes reducing, reusing and recycling, the present activity will start with a brainstorming on the board about the behaviour that each of us could make to reduce waste. In the context of the school, some behaviours highlighted by the students will be probably: rationalize paper use; make use of the paper as best as possible by writing on both sides; use recycled paper and use washable table napkins instead of paper napkins. In general, we must reuse all materials in their original form instead of throwing them away, or pass those materials on to others who could use them too. For instance, the scholar books could pass from one brother to another or to a neighbour, a cousin, etc.

Each sentence written on the board must be followed by a guided discussion about it. Subsequently, the teacher must manage the discussion to include some behaviours that students, and their parents, could adopt at home. Below are some examples.

- First and foremost, buy and use less. To buy only what we need and use all of what we buy.
- Start making wise package selections. We can reduce waste by selecting products that are not wasteful in their packaging. The package surrounding the product only needs to be designed to protect the product from damage, present the product information and keep it in good sanitary condition. The packaging is more often fun, elaborated, and consequently more expensive than necessary. It has marketing purposes to attract the consumer's attention. It usually adds little or no value to the product and could be considerably harmful to our environment by creating more waste or waste disposal difficulties. So, whenever we have choice, buying recyclable packaging is an excellent way to reduce packaging waste in our environment.
- Avoid disposable containers whenever possible. We can buy juice or water in large recyclable bottles or cans and then divide it up into smaller portions in reusable containers or travel mugs as we need it at home or to take with us. If students, for example, want to take juice or water when they go for a walk or go to school just take it along in their reusable sports bottle.
- Instead of buying bottled water, citizens could take city water that is usually just as healthy and cheaper.
- Buy concentrates rather than diluted products. The result is less waste for disposal when it is empty.

Activity 5 – How Can we Help the Recycling?

Ask students about how they can define recycling. Recycling occurs when we save and take reusable materials to places where they can be remade into either the same product or new products, rather than to just throw them in the bin. Making new items from recycled ones also takes less energy and other resources than making products from brand new materials. This reduces the need for "conventional" waste disposal, and consequently, reduces the energy usage, the air, water, and soil pollution.

After this introduction, groups of students must search on the Internet for information to find examples of products produced from others already used. They would be amazed what can be done with a recycled product. For instance, a recycled soda bottle can be made into T-shirts, combs or other plastic goods that can be used for many years. Paper products, too, can take on different forms. An old phone directory or colouring book might become one of the student's school books.

The next step focuses on identifying appropriate local sites to put waste for facilitating the recycling. The teacher could show several garbage items and images of recycling containers and will ask students to distribute each item by the appropriate container. The images must include not only glass, paper and metal garbage, but lamps, batteries and home appliances too. This activity is an opportunity to rectify incorrect students' knowledge about proper garbage separation. The goal is to help make recycling so easy and automatic that it blends into the flow of everyday life.

Activity 6 – Tell a Story Through Ecological Art

There are a number of activities for pupils that will educate them about recycling and its importance to the environment. Involving children in recycling schemes and other activities to improve the environment can be a lot of fun. In order to stimulate the students' creativity, ask them to tell some character names from stories and in small groups they have to elaborate a story about waste using some of the story elements referred. Each group, using available waste (magazines, newspapers, packaging boxes, cardboard boxes, cans and other materials), will construct puppets to represent each story character. Afterwards, they will make a theatre with the puppets to represent the story about waste.

Activity 7 – To Turn our Biodegradable Waste into Organic Fertilizer (set up a Compost Bin)

Compost is simply decomposed organic material. The organic material can be plant material or animal matter. While composting may seem mysterious or complicated, it is really a very simple and natural process that continuously occurs in nature, often without any assistance from mankind.[1] The students could set up a compost bin in the school to use the organic waste produced by them or use the remainder of plants from the school garden. The compost could be used to fertilize the garden or indoor plants. It is necessary to follow some steps to

1 *http://www.thegardenofoz.org/composting101.asp.*

efficiently produce an organic fertilizer, although it is an easy way of reusing organic waste.

First of all, it is necessary make an introductory explanation about the definition and importance of using compost. Advantages of composting include: contribution to proper waste disposal, conversion of organic waste into rich fertilizer, reduced dependence on inorganic fertilizers, improvement of the structure of making it easier for soil to hold and use the right amount of moisture and air and increases its ability to retain nutrients.[1] Composting is also based on renewable and locally available raw materials. Instead, synthetic fertilizers have disadvantages. They are derived from deteriorating petroleum resources and consume excessive energy during production. Excess chemicals also leach out of the soil, pollute waters and drive away or destroy helpful organisms (Marcelo 2007).

The first step to make a compost is to build a compost bin. Some outdoor space will be needed dedicated to the process. It is necessary to choose a site preferably near a source of water and in an open area, and it is important to ensure that the bin has enough openings to allow air to penetrate in so that the bacteria and fungi that do the composting can get enough oxygen. Compost works best with a mixture of easily available coarse and fine materials such as scrap lumber, chicken wire, bricks or concrete blocks.

The composting must be done by laying down a bottom layer of twigs, cornstalks, wood chips or other coarse material. It must be followed with a layer of high-nitrogen material such as manure or grass clippings. Then, it is necessary to add a layer of garden soil, more rough materials and more clippings, and so on. Sprinkle the materials with water while building up the pile is important. This sequence must be repeated. The pile must be turned every 3 or 4 days to move the fully composted materials out of the hot centre of the pile and replace it with partially composted material from the sides. The time required to develop finished compost also depends on the size of the pile and time of year or air temperature. It is important to control the composting to avoid unsafe use. Generally, the more compost is turned the faster it decomposes (Marcelo 2007). In Annex 4, a protocol shows how to make a composter that could be an example for an orientation.

This activity could be a good way of getting students to influence their parents on constructing a home composter to use organic residuals produced every day in their homes.

Activity 8 – Visit a Recycling Centre and a Sanitary Landfill

Students need constructive opportunities for active exploration outdoors to develop respect for the environment and to be near the reality. In this way, two study visits are proposed: to a recycling centre and to a landfill. These visits are important for students to apprehend the global idea of the whole cycle of waste and the different forms of recovery already addressed in previous activities. During both visits students must collect data through pictures, observation and notes. They will compile the collected data, pictures and knowledge into a short report or in the format of a poster that must be presented at school, for instance, in an exhibition on the theme of pollution.

– *Recycling centre*
  Visiting a recycling centre could be a good way for students to come into contact directly with the processes through which the garbage passes before being recycled. The recycling centres usually provide support for local authorities by means of selective collection and non-reusable packaging waste sorting programmes. It also guarantees the taking back, recovery and recycling of sorted waste under its contracts with manufacturers of packaging and packaging waste (paper and cardboard, glass, plastic, wood, steel, and aluminium). Additionally, recycling centres manage and deal with the final disposal of non-reusable packaging placed on the market after consumption, and guarantee to distributors that their non-reusable packaging is covered by an integrated system for the management of packaging waste.

– *Landfill*
  Historically, landfills have been the most common method of organized waste disposal and remain so in many places around the world. Students must visit a landfill to understand *in situ* how it works. They will be able to describe in a simple way its mechanisms. Discuss the possible hazards that might result from an improperly filled landfill. Emphasize right and wrong ways to construct a landfill.

Final Remarks

To achieve a cleaner and healthier environment all social groups need to collaborate in environmental campaigns or in daily routines. The students are a basilar group that must be encouraged to do it. The proposed activities will develop the students' knowledge, allowing them to deal with the problems of waste and the necessity of preserving our planet. Generally, it is difficult to promote the motivation of children towards issues that typically cause aversion, like waste. However, this theme worries society since it has direct implications on human health, on sustainable development and environmental protection, and it must be taken into consideration. EE in schools, along with media, will certainly ensure children are aware of the serious situation of natural resources. Moreover, students will feel that their behaviour and actions will be decisive for the future of our world.

References

Directive 2006/12/CE of the European Parliament and of the Council on waste of 5. April 2006 (available at http://eur-lex.europa.eu/LexUriServ/LexUriServ.do?uri=OJ:L:2006:114:0009:0009:EN:PDF).

Directive 2008/98/EC of the European Parliament and of the Council of 19 November 2008 on waste (available at http://eur-lex.europa.eu/LexUriServ/LexUriServ.do?uri=OJ:L:2008:312:0003:0030:en:PDF).

EEA – European Environment Agency (1999), Waste generation and management, Environment in the European Union at the turn of the century, pp. 203-236 (available at http://www.eea.europa.eu/publications/92-9157-202-0/ 3.7.pdf).

Fiorucci, P., Minciardi, R., Robba, M. and Sacile, R. (2003), Solid waste management in urban areas: Development and application of a decision support system. Resources, Conservation and Recycling, 37, pp. 301-328.

Haga Declaration (2000), 1st CBSS Ministers of Education meeting. Stockholm.

International Implementation Scheme (2005), United Nations Decade of Education for Sustainable Development (2005-2014). UNESCO Education Sector (available at http://unesdoc.unesco.org/images/0014/001486/148654e.pdf).

Jacobi, P. (2003), Educação ambiental, cidadania e sustentabilidade. Cadernos de pesquisa, 118, pp. 189-205 (available at http://www.scielo.br/pdf/cp/n118/16834.pdf).

Marcelo, R. (2007), Smart guide to household composting. Tao Shelter (available at http://www.tao-pilipinas.org/files/taoshelter/issue3/composting.pdf).

Moreno, J.S.M. and Pedrosa, A. (2003), Ecologic sustainability and individual and collective everyday practices – Views of the EQOFAR. In Azeiteiro, U.M., Gonçalves, F., Pereira, R., Leal Filho, W. and Morgado, F. (eds.), Science and Environmental Education, vol 27, pp. 333-348, Peter Lang, Frankfurt.

Nassour, A.C. (2003), A degradação do lixo urbano. Revista Eletrônica de Ciências, 18. São Carlos, Brazil (http://www.cdcc.usp.br/ciencia/artigos/art_18/lixourbano.html).

O'Rourke, J. and Clark, L. (2001), Social barriers to the recovery and reuse of recyclable construction waste material, In Dhir, K.R., Limbachiya, M.C. and Dyer, T.D. (eds.), Recycling and reuse of glass cullet, pp. 147-156. Thomas Telford Publishing, London.

Putnam, R.T. and Borko, H. (1997), Teacher learning: Implications of new views of cognition. In Biddle, B.J., Good, T.L. and Goodson, I.F. (eds.), International handbook of teachers and teaching, pp. 1223-1296. Kluwer, Dordrecht, The Netherlands.

Resolution A/RES/57/254. United Nations Decade of Education for Sustainable Development, UN documents gathering a body of global agreements (available at http://www.un-documents.net/a57r254.htm).

Rushbrook, P.E. and Finnecy, E.E. (1988), Planning for future waste management operations in developing countries. Waste Management and Research 6, pp. 1-21.

Sterling, S. (2004), An analysis of the development of sustainability education internationally: evolution, interpretation and transformative potential. In Blewitt, J. and Cullingford, C. (eds.), The Sustainability Curriculum: The Challenge for Higher Education, pp. 43-62. Earthscan, London.

Tal, R.T. (2004), Using a field trip to a wetland as a guide for conceptual understanding in environmental education – a case study of a pre-service teacher's research. Chemistry education: research and practice 2, pp. 127-142 (available at http://www.uoi.gr/cerp/2004_May/pdf/05Tal.pdf).

United States Environmental Protection Agency – Wastes – Non-Hazardous Waste – Municipal Solid Waste, USA (http://www.epa.gov/osw/nonhaz/ municipal/ – accessed in January 2010).

World Commission on Environment and Development (Brundtland Commission) (1987), Our Common Future, p. 43. Oxford University Press, Oxford (available at http://www.un-documents.net/wced-ocf.htm).

Annex 1

Eco-Questionnaire

Name____ Date__________________

Please answer this questionnaire in the most sincere and complete way you can.

1. Do you know what type of solid residuals must be separated?
   ___Yes ___No

2. Do you separate newspapers, plastic bottles and tins from other rubbish at home?
   ___Always ___Sometimes ___Never

3. Do you separate newspapers, plastic bottles and tins from other rubbish at school?
   ___Always ___Sometimes ___Never

4. Do you try to reduce the quantity of solid you produce?
   ___Yes ___No

5. Do you know all recycling bin colors?
   ___Yes ___No

6. In the area where you live are there waste recycling banks?
   ___Yes ___No

7. What king of garbage do you and your family produce in larger quantities?
   ___Glass ___Paper ___Metal ___Textiles ___Batteries ___Others

8. Do you try to reduce the quantity of solid waste you produce?
   ___Yes ___No

9. Are you and your family careful as to by products that can be reused?
   ___Yes ___No

10. Do you know where the appropriate local to deposit the old home appliances is?
    ___Yes ___No

11. Do you know the R's politic meaning?
    ___Yes ___No

12. Do you use recycled paper?
    ___Yes ___No

13. Do you know what a sanitary landfill is?
    ___Yes ___No
14. Do you know what a recycling center is?
    ___Yes ___No

15. Do you know what composting is?
    ___Yes ___No

Annex 2

Concept map about garbage

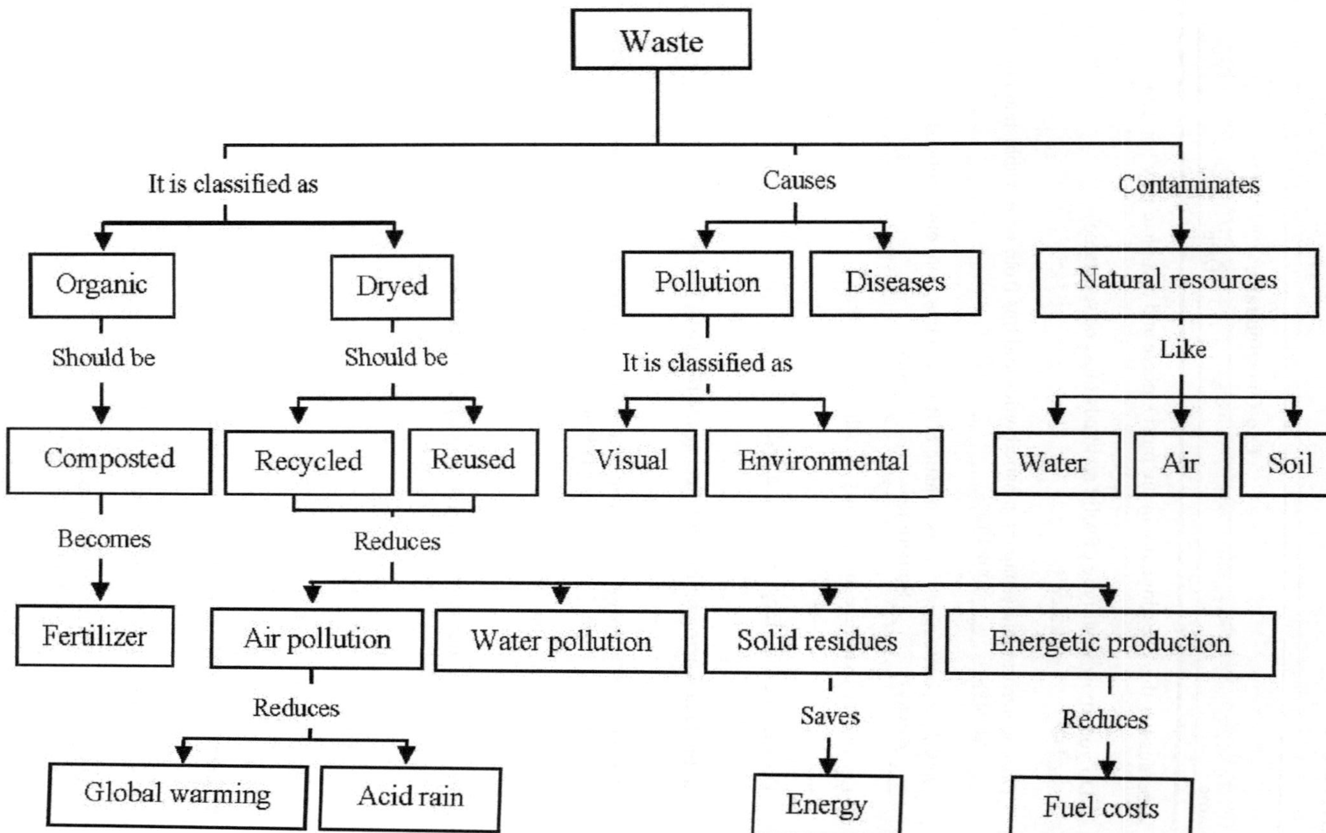

Annex 3

Analysing waste

Group name ______________________ Students name__________________________________

Garbage ____ Date_______________________

1. Search for information about the waste that you have chosen and fill the table.

Parameters of the waste analysis	**Information**
Decomposition time	
Impact caused by the package production	
Label information	
Slogan and advertising appeal	
Possibilities of reutilization	

Annex 4

How to make compost?

Material necessary to make compost

- Compost bin
- Fork aeration
- Water
- Thermometer

What residues can you place in the compost bin?

In the following table are indicated some residues that you produce in the school and in your house and which of them can you place in the composter.

Table I: Residues that can be placed in the compost bin.

Residues	**Place**	**Not place**	**Put in small quantities**
Vegetable scraps	X		
Fruit peelings	X		
Eggshells	X (crushed)		
Coffee grounds	X		
Leftover bread			X
Potato peels	X		
Cooked food scraps			X (cover with soil)
Leftover meat and fish		X	
Bones and spines		X	
Herbivorous excrements	X		
Grass clippings and leaves	X (cut in small pieces)		
Wood and ash			X
Ashes and cigarette butts		X	
Branches and shrubs	X (cut in small pieces)		
Straw and hay	X (cut in small pieces)		
Pine needles			X
Paper and paperboard			X (cut and wet)
Sawdust			X
Cork		X	

Steps to make a composter

1. Install the compost bin.
2. Prepare materials to put in the compost bin.
3. Control the following factors during the composting process.
   - temperature
   - humidity
   - aeration
4. The compost is ready after 4-12 months when presents the following appearance:
   - looks homogenous
   - texture similar to ground
   - brown colour

Chapter 16

Sustainability of Fisheries

Graham Pierce, Cristina Pita, Begoña Santos, Sónia Seixas

Abstract

This chapter reviews the concept of sustainability in fisheries, focussing on fisheries in Europe and paying particular attention to the human dimensions of fisheries. The particular problems presented by fisheries (related to the "Tragedy of the Commons") are introduced, followed by brief accounts of the importance of fisheries worldwide and of their history in Europe. We attempt to summarize the concepts embodied in fisheries management and governance and review the different dimensions (pillars) of sustainability in the context of fisheries: environmental, economic, social and institutional. We describe some current developments in management and governance of European fisheries, including the introduction of property rights, the role of ecological labelling and the concept of demand-led management, participation and co-management, marine protected areas and Integrated Marine Management. We advocate a system of governance under which more emphasis is placed on achieving the possible than in quantifying the unachievable, a system which delivers successful implementation of sustainability objectives based on holistic (and multidisciplinary) assessments of environmental, economic and social-cultural consequences of proposed actions and which is based on the full and active participation of all relevant stakeholders.

Introduction

In contrast to the terrestrial environment, provision of food from the sea remains largely based on what is effectively a hunter-gatherer system, albeit an increasing technologically sophisticated one. Unregulated marine fisheries provide a paradigmatic example of the "Tragedy of the Commons" (Hardin 1968), whereby individual fishers do not benefit in the long-term from refraining from catching more fish in the short-term, so long as other fishers choose not to show similar restraint.

It is generally accepted that heavy and poorly regulated exploitation has led to the depletion of a large proportion of the world's fish stocks and has caused some scientists to forecast the imminent end of marine fisheries as we currently know them (e.g. Worm et al. 2006). It is of course human nature to foresee disaster:

previous impending disasters, like the feared consequences of human overpopulation (a major focus of popular science in the 1960s), have not always materialized. In addition, at least in the northeast Atlantic, concern about the state of fish stocks is nothing new (see Sims and Southward 2006, who recall the debate occurring in the UK towards the end of the 19th century), and every generation judges the current perceived status of fish stocks against the shifting baseline of its own experience (see Pauly 1995). Nevertheless, and despite the cautious optimism expressed by some authors (e.g. Hilborn 2010; see below), there are profoundly worrying global trends. Zeller et al. (2010) show that a range of indices indicate global scale ecosystem damage due to fisheries, for example a shift from "unstressed" marine communities, dominated by species with large individual biomass but low abundance, towards "stressed" communities dominated by highly abundant species of small individual body size.

The International Council for the Exploration of the Sea (ICES), which is the main organization at European level producing advice on the state of the stocks exploited by the EU fleets, considers that many fished stocks in EU waters are currently overexploited (ICES 2010). In addition, wasteful practices such as discarding persist, to the frustration of both fishers and fishery scientists, among others. There may thus be better ways to use limited resources to achieve sustainable fishing. Indeed the present emphasis in Europe on developing an Ecosystem Approach to Fisheries may be counterproductive in the sense that it requires dedication of more resources to obtain the required biological and ecological information, potentially making it even harder to successfully balance competing objectives and ensure effective implementation. These are, of course, also not new ideas. For example, the EU project INCOFISH ("Integrating Multiple Demands on Coastal Zones with Emphasis on Aquatic Ecosystems and Fisheries") was based on the ideas that traditional assessment could be largely replaced by using simple indicators of stock status that could be readily understood by scientists, stakeholders and consumers, and that consumer power could be harnessed to promote sustainable harvesting.

In general, when speaking of fisheries, unless otherwise stated, we refer to capture rather than culture, to commercial fisheries rather than recreational fisheries, and to harvesting of fish, crustaceans, molluscs and some other marine invertebrates (including jellyfish, sea anemones, sea urchins and sea cucumbers). The term can also legitimately be extended to encompass the hunting of marine mammals (e.g. seals, whales) for human consumption although marine mammals are now protected in many countries.

There are many textbooks about fisheries (e.g. Cushing 1975; Hilborn and Walters 1992; Jennings et al. 2001; King 2007, to list just a few), and a huge literature on related topics in scientific journals. Here we will not attempt to cover all the biological and technical aspects of fish stock assessment and fishery management. Our aim is, rather, to provide a short overview of current issues, with an emphasis on the human dimensions of fisheries[1], and to highlight possible solutions to historically persistent problems. It should be self-evident that successful fishery management, the usual broad goal of which is "sustainable" fishing (i.e. meeting the needs of the present without compromising the needs in the future), depends on managing human behaviour. In this sense, while traditional fisheries science encompasses consideration of the economic benefits of fishing (e.g. based on the yield-per-recruit concept, Beverton and Holt 1957), in general the tendency to focus on the (biological) status of fished stocks has led to less consideration of balancing ecological, economic and social objectives, or the implementation of policies, both of which also require a strong scientific basis. Slightly more radically, there may be a case for developing demand-based fishery management rather than focusing only on regulating the supply of fish.

The Importance of Fisheries

The Food and Agriculture Organization (FAO) of the United Nations estimates that capture fisheries and aquaculture supplied the world with about 110 million (metric) tons of food fish in 2006 of which capture fishery production was about 92 million tonnes. Since the mid-1980s, landings of captured fish have been stable or decreasing, as many stocks were overexploited. In the current context of overall resource scarcity and decreasing catches, aquaculture production assumes a key importance. Aquaculture is a rapidly growing industry, which has experienced an enormous expansion in recent decades, and continues to be the fastest growing animal food-producing sector. It presently contributes about 47% of the marine food supply (FAO 2009), although this figure is misleading in that capture fishery landings include a significant proportion which is not used directly for human consumption, one important use of which is the production of feed for use in aquaculture and to feed other animals. It should also be noted that capture and culture fisheries are not completely separate entities. Apart from the use of wild caught fish in feed, the practice of "ranching" high value wild-caught species such as tuna and octopus (e.g. Volpe 2005; Chapela et al. 2006) is increasingly blurring the line.

Fish and fish products contribute to food security in many regions of the world, in some cases representing staple foods, or at least the main source of animal protein, and elsewhere representing a valuable supplement for diversified and nutritious diets. Overall, fish provided more than 2.9 billion people with at

1 Some basic concepts in fishery management are defined and briefly described in Boxes 1 and 2.

least 15% of their average per capita animal protein intake (FAO 2009). In addition, capture fisheries and aquaculture play, directly and indirectly, an essential role in the livelihoods of millions of people around the world. In 2006, an estimated 43.5 million people were directly engaged (part-time or full-time) in fish production, either by fishing or working in aquaculture. Furthermore, for every person directly employed in fisheries and aquaculture it is estimated that there could be four more employed in related activities (such as fish processing, marketing and services industries) (FAO 2009). In addition to their contribution to worldwide national economic activities and employment, fish and fishery products are important in generating foreign exchange: they are highly traded, with more than 37% (live weight equivalent) of total production entering international trade, as various food and feed products (FAO 2009).

The EU is the fourth largest seafood producer world-wide, contributing approximately 4.6% of global capture fisheries and aquaculture production (see table 1 for macroeconomic indicators). In 2006 European capture fisheries and aquaculture supplied the market with 6.4 million tons. Domestic demand for fish products within the EU is high, and per capita consumption of seafood is increasing. In addition, the fish catching, culture and processing sectors provide over 141 000 jobs in the EU (European Commission 2010).

Table 1: Macroeconomic indicators for fisheries in the world and the EU-27

	World	**EU-27**
Total fisheries (= catches + aquaculture)	143.6 million tonnes	6.44 million tons
Capture fisheries (weight)	92.0 million tonnes	5.14 million tons
Capture fisheries (value)	91.2 billion US$	8.2 billion EUR
Aquaculture (weight)	51.7 million tonnes	1.31 million tons
Aquaculture (value)	78.8 billion US$	3.2 billion EUR
Imports (value)	85.9 billion US$[1]	16.5 billion EUR[2]
Exports (value)	85.9 billion US$[1]	2.9 billion EUR[2]
Human consumption	110.4 million tonnes	11 million tons
Per capita annual consumption	16.7 kg	22.3 kg
Total people employed	43.5 million	141 thousand
Vessels (number)	2.1 million[3]	89 thousand
Vessels (gross tonnage)	–	1.93 million tons
Vessels (engine power)	–	7.06 million kW

Source: EUROSTAT (2009), European Commission (2010), FAO (2009). World data refer to 2006, EU-27 data refer to the 27 EU Member States, in 2007.

Notes: [1] Total value traded worldwide; [2] External trade of fisheries and aquaculture products (trade between the EU and third countries); [3] FAO obtained data on national fishing fleets from 97 countries (slightly fewer than half of those catching fish). Value represents number of vessels powered by engines.

Aquaculture is increasing in importance within the EU, now accounting for almost 20% of fish production. However, total EU seafood production has been decreasing for the past 20 years (European Commission 2010). For some time now, EU fisheries and aquaculture have not been able to meet domestic demand, resulting in the EU being the world's biggest importer of fishery products, at a time when many exploited fish stocks in European waters are at historically low levels. The environmental and socioeconomic sustainability of fisheries and aquaculture within the EU is thus a major current concern.

History of Fisheries Management in Europe

Concerns about the depletion of fishery resources are not a recent phenomenon: 120 years ago, the rise in fish trade and reports about the scarcity of fish from all around the United Kingdom's coasts strengthened widespread concern among fishing communities and scientists prior to the International Fisheries Exhibition in London in 1883 (Sims and Southward 2006). Indeed, fisheries have rarely been "sustainable". Rather, fishing has induced serial depletions, masked by improved technology, geographic expansion and exploitation of previously spurned species lower in the food web (Pauly et al. 2002). As previously mentioned, inevitably our perception of the status of fisheries is informed by our own experiences, i.e. the shifting baseline concept (Pauly 1995). In some areas traditional fisheries continue much as the previous generations might remember them; in others, the fishing has all but disappeared (compare pictures from different parts of Lisbon in the 1940s and in 2010 (fig. 1, 2).

Fig. 1: Postcard of a fishing village (Cascais, Portugal), (a) around 70 years ago. Private collection. (b) the same area in 2010

(a)

(b)

Fig. 2: Postcard of the fishing activity in Cais da Ribeira (Lisbon, Portugal), (a) around 70-80 years ago. Private collection. (b) the same area in 2010

(a)

(b)

In some parts of Europe, fisheries management has a long history. For example, in the Lofoten islands (northern Norway), laws have existed for more than 200 years to control fishing activity. However, fisheries management as we currently understand it is a relatively new concept (see boxes 1 and 2 on the evolution of concepts in fisheries management). As recently as the 1960s, marine fisheries in Europe were largely unregulated. The first common measures in the European fishing sector can be traced back to the early 1970s, when the then member states of the EU reached agreement on access rights for their fishing vessels and the establishment of a common market for fisheries products. The late 1970s and early 1980s saw the development of a common policy on fisheries and the assumption of the centralized authority, by the EU, for the regulation of commercial fisheries. This culminated in the establishment of the Common Fisheries Policy (CFP) in 1983.

Two decades later, in 2002, the CFP underwent a radical reform and several new concepts were introduced and/or implemented in an attempt to tackle the serious consequences of overfishing (see table 2 for a chronology). This reform saw the beginning of a more long-term approach to fisheries management, with the implementation of multi-annual recovery plans, and the introduction of restrictions on fishing effort and of a precautionary approach to management. It saw the end of aid for the modernization of the fleet and the establishment of measures to support the decommissioning of fishing vessels. It also facilitated the implementation of measures to allow for an increase in stakeholder participation in the decision-making process (with the subsequent creation of Regional Advisory Councils) and aimed to progressively implement an "ecosystem-based approach to fisheries management" (EAFM). Further reform is envisaged for 2012. The Green Paper, setting the way for the forthcoming reform of the CFP, indicates that the new policy will focus on addressing the deep-rooted problem of fleet overcapacity, improving the system of governance by giving more power to the fishing industry, developing a culture of compliance, adopting long-term management plans, adopting a regionalized management regime, and integrating fisheries management in the broader maritime policy context.

Table 2: Development of the Common Fisheries Policy (CFP)

Year	Description	Regulation
1976-1982	Set the rules for the CFP, with emphasis on conservation of fish stocks, allocation of fishing rights, regulation of fishing fleets	–
1983	Establishment of the CFP (main areas of activity: Conservation of stocks, vessels and installations, market controls, and external agreements with other nations).	Regulation 170/83
1992	First review of the CFP, resulting in minor amendments (measures to improve compliance, more restrictive regulations and improve monitoring)	Regulation 3760/1992
2002	Reform resulting in a new framework regulation (aimed at ensuring the sustainable development of fishing activities, and improve the basis of the decision-making process)	Regulation 2371/2002
2012	Forthcoming reform of the CFP	–

In order to ensure sustainable fisheries, the CFP introduced various "conservation measures", including rules for total allowable catches, limitation of fishing effort, and technical measures (related to gear and vessel characteristics). However, as in fisheries throughout the world, few of the measures introduced worked exactly as planned. In practice, the CFP has had very mixed success. Some fish stocks, especially those that have been subject to long-term management plans, have remained in a healthy state. For example, among stocks exploited by Scottish fleets, North Sea haddock, North Sea herring, Rockall haddock, and mackerel are all currently doing well. On the other hand, many stocks have been seriously overexploited, including North Sea cod (although biomass is increasing from a historic low point in 2006) and herring (which has gradually recovered since the mid-1970s) (ICES 2010). However, serious concerns remain about fleet overcapacity, exploitation of vulnerable species (e.g. deep-sea fish), wasteful practices such as discarding, environmental degradation (e.g. damage to the sea bed caused by trawling) and effects on non-target species including by-catch mortality of sharks and marine mammals. In short, there is a widely held (if not wholly correct) view that European fisheries are in crisis.

Box 1: The evolving concepts of fishery management in Europe:
A. Fisheries as biological systems

Here we present a very brief (and simplistic) overview of concepts in fishery management. Some of the recent terminology has its origins in the USA but the processes described are increasingly relevant to European fisheries, even if (paradoxically) we have yet to fully engage with the human dimensions of fisheries (see Fig. 3).

In its simplest current form, **fishery management** consists of **stock assessment** (i.e. evaluating the status of a fished population or stock[2], and setting **harvest rules** (e.g. catch quotas, fishing effort limits, specified gears, closed areas) designed to ensure a sustainable harvest. For the system to have any chance of working there must also be **monitoring** of compliance and **enforcement** of regulations, which is where management becomes subsumed within the concept of **governance**. Since this falls outside the remit of traditional fisheries science, we leave further discussion until Box 2.

Maximum Sustainable Yield (MSY) can be achieved by harvesting the stock so that abundance is maintained high enough to ensure good **recruitment** (i.e. a sufficient number of young fish are produced that can enter the fishery) but not so high that density-dependent regulation starts to restrict population growth. Needless to say, while mathematically easily defined, keeping a fishery at MSY is very hard to achieve in practice, since (a) the precise values will keep changing due to environmental variation, through its effects on environmental carrying capacity and, probably more importantly, on recruitment (i.e. the stock's ability to replace itself) and (b) once a fishery has grown sufficiently to achieve MSY it is very hard to reduce effort again when MSY is exceeded. Reflecting these difficulties, the current focus in EU fisheries is on achieving F_{MSY}, i.e. the fishing mortality required to achieve MSY.

Not all mature fish are equal. Taking into account the effect of body size and condition (and environmental factors) on the number and quality of eggs produced could lead to a better understanding of recruitment and a better estimate of the "optimum" stock size. Large, old females (**megaspawners**) may contribute a disproportionately large number of offspring and it may be useful to take special measures to protect them.

In many modern fisheries, stock abundance is so far below MSY that it becomes important to define "limit" or "precautionary" **reference points** for fished populations. If abundance falls below defined reference points, some kind of remedial management action should be triggered (e.g. fishery closure).

In practice, we never have perfect knowledge of fish abundance, we cannot perfectly control fish catching (if indeed we can control it at all), and the intrinsic variability of natural systems (**stochasticity**) makes us less certain (e.g. the correlation between the number of breeding fish and the number of young fish reared, the **stock-recruitment relationship**, may be very weak). According to the **Precautionary Approach** to management, catch limits should be set low enough to make stock failure (due to any of the above reasons) unlikely.

Many fisheries catch several different species, while the same species may be caught by more than one fishing fleet. In **multispecies** fisheries, we need to consider that optimizing catches of one species could result in too many of another species being caught. Managers must decide on the objective (is it to maximize total yield, to preserve the most vulnerable stock or to identify an acceptable trade-off?) and set harvest rules accordingly. It may be possible to define a "**multi-species MSY**". In **multi-fleet** fisheries, more complicated harvest rules may be needed to ensure that each fleet takes an equitable share of catches.

Fishing can have profound negative impacts on the marine ecosystem. Firstly and most obviously, overfishing of target species can reduce their abundance and disrupt trophic webs. Fishers also catch

2 Although we use "population" and "stock" interchangeably, the first strictly refers to a genetically distinct, reproductively isolated (and usually geographically discrete) grouping of conspecific animals, while the latter is a management unit, often the same as a population but sometimes (e.g. for political reasons) representing only a part of a population or even a mixture of several populations.

species which were not being targeted (**by-catch**), some of which may be vulnerable and/or have a high public profile (**charismatic megafauna**, e.g. dolphins, turtles). Fishers may throw overboard (**discard**) those species and sizes of fish that cannot legally be landed or which are of low value. Trawling can damage the seabed and benthic communities. The **Ecosystem Approach to Fishery Management (EAFM)** envisages use of **Ecosystem Health Indicators** to carry out an "ecosystem assessment" (including the usual stock assessment), possibly leading to the definition of an "**Ecosystem MSY**" for the target fish species (i.e. the MSY is modified to take into account effects on the ecosystem), presumably leading to modified harvest rules.

Even more than multi-species fisheries, EAFM presents a major challenge to fishery scientists and managers by increasing the amount of information needed, to permit a full assessment and define adequate harvest rules, by orders of magnitude. This has resulted in two very different (although not mutually exclusive) trends in methodology: (1) development of increasingly complex mathematical models, sometimes by combining different kinds of models (e.g. traditional fish population dynamics models can be combined with models of ecosystem trophic structure such as ECOPATH and linked at one end to physical oceanographic models and at the other end to bio-socio-economic models (creating "end2end" models), and (2) complex models are replaced by simple indicators which "capture" the pertinent features of the system. It should be also noted that increased model complexity comes at a price, not just in terms of the data requirements to estimate multiple parameters, but because the uncertainty in each one makes it increasingly unlikely that the final estimate will be either accurate or precise.

The foregoing account very much follows a traditional fishery biologist's worldview, focusing on the biological and ecological aspects of sustainability. **Harvest rules** came within the ambit of the fishery scientists only in that they helped to define the currency in which fish stock status is expressed (e.g. tons of fish which may be safely removed). In practice, fishery scientists are increasingly required to provide **dynamic stock assessments** (Hilborn and Walters 1992), in which the future trajectories of fished populations are predicted under different possible harvest rules. Thus, recommendations about harvest rules now fall within the competence of fishery scientists – even if the options to be evaluated are still specified by the client, normally the management/regulatory authority. More recently the fishing industry has also been involved in setting, for example, limits on Total Allowable Catches (TACs), signifying the increasing importance attached to socioeconomic objectives (e.g. market stabilization). Of course, some would argue that the industry has always been involved in setting quotas (and at levels which make sense only in terms of short-term profit), through successful lobbying. As Froese (2011) points out, an industry which in purely economic terms is less important than the sowing machine industry has always punched above its weight.

The increasing amount of information that must be processed and the inadvisability of conducting field experiments on real fisheries have led to the use of "**Management Strategy Evaluation**" (MSE), by which computer simulations (**scenario models**) are used to explore the interactions between the fish-fishery system and management (assessment + harvest rules), allowing much faster and less risky evaluation of options before they are implemented. This approach was developed by the International Whaling Commission for its "Revised Management Procedure" (see Cook 1995). A key component is understanding fisher behaviour in response to different stock status and harvest rule scenarios (Branch et al. 2006). This concept can be expanded to include consideration of effects of fishing (and other human impacts) on the ecosystem, leading to "**Integrated Ecosystem Assessment**" (IEA).

Under **adaptive management**, a feedback loop is introduced such that harvest rules are regularly modified in the light of new evidence about the state of the stock following implementation of existing rules.

Even under this more sophisticated worldview, economic, social and political considerations are simply things that happen to advice after it has left the scientists, and often involve the advice being modified or ignored. This is of course an oversimplification but the **International Council for the Exploration of the Sea** (ICES), the body responsible for assessing many of Europe's marine fish stocks, followed this kind of model for around 100 years (although it is currently undergoing extensive reform).

Box 2: The evolving concepts of fishery management in Europe: B. The human dimensions of fisheries

To state the obvious, fishery management is about managing people. Fish cannot be managed. The concept of the **three (or four) dimensions or pillars of sustainability** emphasizes that fishery management is not (or should not be) designed only to achieve biological or ecological sustainability: this represent just one of the pillars. Two of the other three, to be given equal weight, are social and economic sustainability. **Economic sustainability** captures the value of fishing as a source of income to the fishers, support industries (e.g. processors) and ultimately the contribution to national Gross Domestic Product (GDP). It also considers fishing costs (e.g. the capital invested in a boat, and the running costs – fuel, salaries, etc). As noted by Hilborn (2010), economic objectives are often fairly compatible with biological objectives – although there are circumstances under which the "rational" economic decision is to fish a stock to commercial extinction (e.g. the case of whaling, where it may make more sense in economic terms to fish a stock to extinction and live off the interest on earnings, since whale populations grow at too slow a pace to justify waiting for recovery). **Social sustainability** concerns societal benefits of fisheries, such as the provision of jobs, and a major protein source, and maintenance of local communities. As Hilborn notes in the article cited above, social objectives can be difficult to reconcile with economic or ecological objectives.

Some sources (e.g. O'Connor 2006; Valentin and Spangenberg 2000) define a fourth pillar, as **institutional** (or **cultural) sustainability**, which may be thought of as the sustainability of the human system in which the fishery takes place. In particular it is useful to consider this as including governance, i.e. the **implementation** of fishery management, including the monitoring of compliance and the means by which regulations are enforced. In this way it can be seen that the governance system is not a fixed constraint, it is something that can itself be optimized. Indeed, arguably this is the single most important pillar of sustainability: successful implementation of a sub-optimal environmental/biological objective is likely to be less environmentally damaging than an unenforceable but otherwise optimal solution.

Fishers are more likely to "buy-in" to management decisions in which they have participated, and more likely to be hostile to a **top-down** system in which they have no say. The move toward **fisher participation** and **co-management** is one of the key issues in the current reform of the CFP. Various local and regional bodies (e.g. **Regional Advisory Committees**) have given roles to fishers and other stakeholder groups (e.g. local communities, NGOs) in the decision-making process. In addition to allowing the consideration of non-biological objectives to become a legitimate part of governance, such arrangements can facilitate the inclusion of **fisher knowledge** into stock assessments.

In practice, current fishery governance structures tend to be more or less hierarchical, for example involving (moving from national or European to local or regional scales) (a) provision of advice, (b) strategic management decisions and (c) tactical management decisions, with different combinations of bodies involved at each level. There is of course a risk that such a hierarchy can perpetuate ingrained beliefs that environmental concerns should take precedence over social objectives or vice versa at particular scales.

The increased number of bodies participating in decision-making (each with their own sectoral objectives) and the need to explicitly incorporate objectives related to all four pillars of sustainability, including the need to optimize the management and regulatory (governance) system, has led to development (at least in the USA) of the "**Systems Approach**" to fisheries management (SA, or possibly SAFM), a generalization of the IEA/MSE approach. The key to reconciling different kinds of objectives is to have a common currency, normally the value in $ US. Clearly, issues remain about how social and ecosystem benefits should be valued in monetary terms. Under the SA, the cost of administering the management and regulatory system (governance costs) is included as a component of the model (something that, if done in Europe, might make the present governance system look impossibly cumbersome and costly in proportion to the size of the sector it governs), and it is possible to incorporate "hidden costs" (e.g., the risk of undesirable outcomes).

Results-based management does not involve specific harvest rules but rather passes the burden of proof to the fishers, i.e. requiring them to demonstrate that they are meeting specified targets for stock or environmental status (e.g. reducing by-catch of threatened species). The approach aims to provide incentives for fishers to devise mechanisms to achieve these goals, for example rewarding success by permitting increased fishing effort. This would normally be combined with co-management, such that all proposals are discussed and agreed on by fishers, scientists and regulators, and the success of the measures is regularly monitored. Current examples tend to involve local or regional scale fisheries.

Fishing is just one sector of human activity in the ocean. Others include aquaculture, oil and renewable energy extraction, marine transportation, tourism and marine conservation. Management and governance for sustainability thus becomes a multi-dimensional problem, enshrined in concepts such as **Integrated Coastal Zone Management** and **Marine Spatial Planning**.

Virtually every variant of management we have discussed above is **supply-side management**. However, **demand-side management** also offers options. **Eco-labelling** or **ecological certification** provides a mechanism to use consumer demand to promote sustainable fishing practices. The EU project INCOFISH proposed the slogan "Don't eat babies", and the "fish-ruler" tool, both to promote the use of simple harvest rules (and simple indicators of fish stock health) and to harness public opinion as a driver for sustainable fishing, by stigmatizing the landing of under-sized fish.

Sustainable Fisheries

What is Sustainable Fishing and why is it so Hard to Achieve?

In 1988 the FAO Council adopted its definition of sustainable development[3]. This was defined as

> "the management and conservation of the natural resource base, and the orientation of technological and institutional change in such a manner as to ensure the attainment and continued satisfaction of human needs for present and future generations. Such development conserves land, water, plant genetic resources, is environmentally non-degrading, technologically appropriate, economically viable and socially acceptable".

Following this, in 1995, FAO adopted a Code of Conduct for Responsible Fisheries, which offers an integrated comprehensive framework for the sustainable development of fisheries and aquaculture. More recently, it was agreed during the 2005 World Summit that sustainable development requires the reconciliation of environmental, social and economic demands, which became known as the "three pillars" of sustainability. Numerous policies and initiatives implemented in the EU over the last decade have been giving a new direction to fishery management. Concerns about sustainability, including environmental sustainability, are now at the heart of fishery management initiatives. Historical fishery management regimes are being gradually replaced by new regimes covering the eco-

---

3 Very similar concepts had been formulated previously, for example in the World Conservation Strategy (1980).

system as a whole. There have already been shifts in the nature of fishery management in Europe and further significant changes will follow in the near future (see also section 3 above). Table 3 shows some of the major drivers of current European policy in marine conservation which directly impact on fisheries management. Almost without exception, their aim is to increase sustainability.

Table 3: Some major developments in conservation and fisheries management measures over the last decade which impact on the fishing industry within the European Union (adapted from Pita 2010)

Year	Measure	Scope
2000	EU strategy for sustainable development	EU
2001	Biodiversity action plan for fisheries	EU
2002	Reform of the Common Fisheries Policy (CFP)	EU
2002	Johannesburg World Summit on Sustainable Development (WSSD)	International
2002	EU recommendation for Integrated Coastal Zone Management (ICZM)	EU
2003	OSPAR and HELCOM agreement	International
2004	Convention Biological Diversity (CBD)	International
2004	Establishment of the Regional Advisory Councils (RACs)	EU
2004	Introduction of requirement to monitor and mitigate cetacean by-catches in certain fisheries (although only applying to larger vessels)	EU
2004	Cod recovery plan adopted covering four stocks: (1) North Sea, Skagerrak and Eastern Channel, (2) Kattegat, (3) west of Scotland, and (4) Irish Sea. Effort limits impact other demersal stocks.	EU
2007	Proposal for an Integrated Maritime Policy (IMP)	EU
2008	Move from fisheries policy to maritime policy (EU Directorate General for Fisheries to DG MARE)	EU
2008	Marine Spatial Planning (MSP)	EU
2008	Marine Strategy Framework Directive (MSFD)	EU
2009	Green paper on the 2012 reform of the CFP	EU

However, commercial fishing is arguably particularly difficult to manage sustainably for several reasons. As noted above, marine fisheries are among the very few large-scale economic activities that could be described as "hunter-gatherer" systems. Fishing attracts active, individualistic persons who do not readily accept regulation of their activities. The sea is huge (covering 8 times the area of the world's land masses) and largely inaccessible and hostile to air-breathing animals. Much of the area of the world's oceans falls outside national boundaries and, even where national boundaries apply, in coastal waters, most fish species are highly mobile and are not confined by such boundaries, and ownership of these resources is (at least in practical if not legal terms) ill-defined. It is notable that in the EU, where most formerly national waters are now shared between many countries, fishery management has been less successful in achieving sustainability than in countries like Norway and Iceland that retain exclusive rights to their fisheries as far as the 200 NM limit.

As with exploitation of the great whales, when investment is high and there is little prospect of maintaining a viable fleet if it is inactive, it may make economic sense to keep fishing until stocks are commercially extinct and then live off the interest from the income thus generated (although in times of economic crisis, the low interest rates available may, paradoxically, favour conservation of exploited stocks). Several great whale species were brought close to actual biological extinction, an extreme outcome which is unlikely in most fish species.

History suggests that fisheries are often unsustainable in practice, with many well-known examples of failed fisheries (e.g. North Sea herring, Peruvian anchoveta, and Northwest Atlantic cod), in which the phases of fishery development (from pre-development to collapse, see Csirke and Sharp 1984) were apparently inevitably played out. Evidence of imminent stock collapse rarely resulted in appropriate remedial action: Pauly (2009a) refers to the "toxic triangle" of underreporting catches, ignoring scientific advice and blaming the environment. Yet, while some authors (e.g. Worm et al. 2006) see this as a harbinger of doom, others, notably Hilborn (2007, 2010) are more optimistic, pointing out that, at least in the first world, overfished stocks are usually able to recover given appropriate management (e.g. North Sea and other herring stocks).

In Europe, great advances have been made in fish stock assessment, taking into account the multi-species and multi-fleet nature of many fisheries, as well as impacts on non-target species and the ecosystem as a whole (see box 1 and fig. 3a, b) and with a longer-term perspective. This has been matched by increasing recognition of environmental issues by policy-makers. Harvest rules have moved from simple Total Allowable Catches (TACs) or quotas to a combination of quotas, effort and gear restrictions, and area closures. However, scientific evaluation of the economic and social dimensions of fisheries has lagged behind and, in any case, implementation has consistently failed: agreed harvest rules may contradict scientific advice (e.g. quotas may be set higher than recommended), may be inherently flawed (e.g. rules about discarding), and may be ignored by fishers and/or not enforced. In principle the latter issue may be addressed by incorporating fisher behaviour into management models (see, e.g. Wilen et al. 2002), i.e. fishers are viewed as part of the system to be managed, and/or by involving fishers as partners in the management/governance process.

European fisheries face several major challenges: reducing fishing mortality to sustainable levels, adjusting fleet size to match the resulting opportunities, eliminating discards and by-catches, and achieving a sustainable system of governance. Once fishers (and other stakeholders) are fully engaged in the process, and it is evident that social and economic objectives are being given full consideration, the prospects of fisher "buy-in" to the regulatory system should be much improved (see box 2, fig. 3).

Fig. 3: Different concepts of fishery management: a) a simple system from a fishery biologist's viewpoint; b) the ecosystem approach to fisheries management, incorporating multi-species and multi-fleet assessments; c) co-management and integrated marine management d) the three (or four) pillars of sustainability, with integrated ecosystem assessment and/or a systems approach.

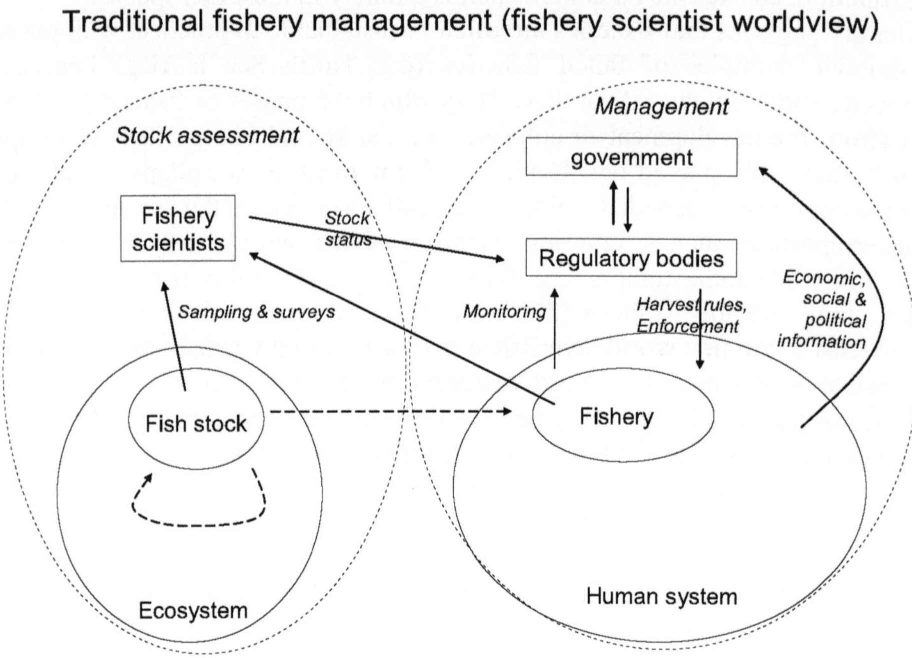

(a)

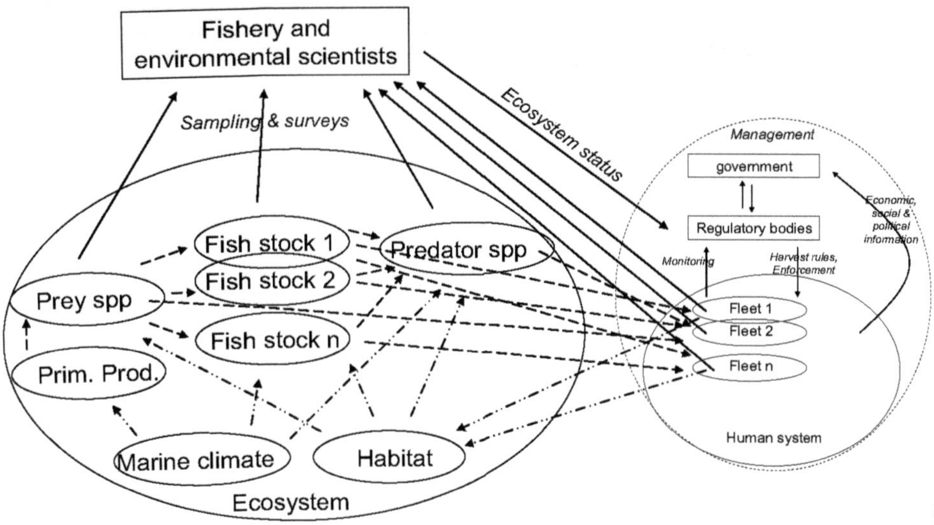
Fishery and environmental scientists
Sampling & surveys
Ecosystem status
Management
government
Regulatory bodies
Economic, social & political information
Monitoring
Harvest rules, Enforcement
Fish stock 1
Fish stock 2
Fish stock n
Predator spp
Prey spp
Prim. Prod.
Marine climate
Habitat
Ecosystem
Fleet 1
Fleet 2
Fleet n
Human system

(b)

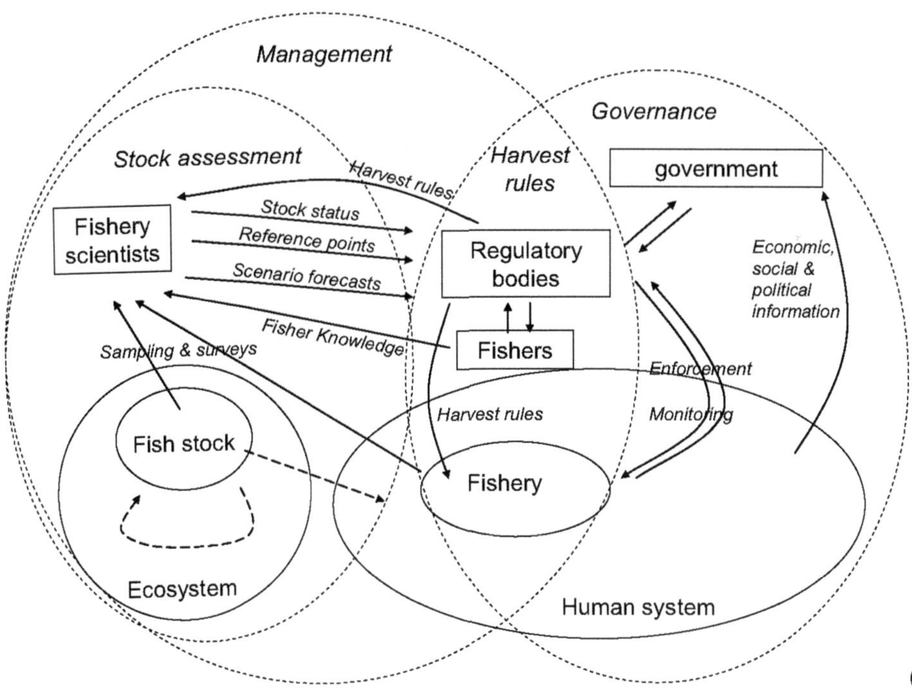
Management
Governance
Stock assessment
Harvest rules
Harvest rules
government
Fishery scientists
Stock status
Reference points
Scenario forecasts
Regulatory bodies
Fisher Knowledge
Fishers
Economic, social & political information
Sampling & surveys
Enforcement
Harvest rules
Monitoring
Fish stock
Fishery
Ecosystem
Human system

(c)

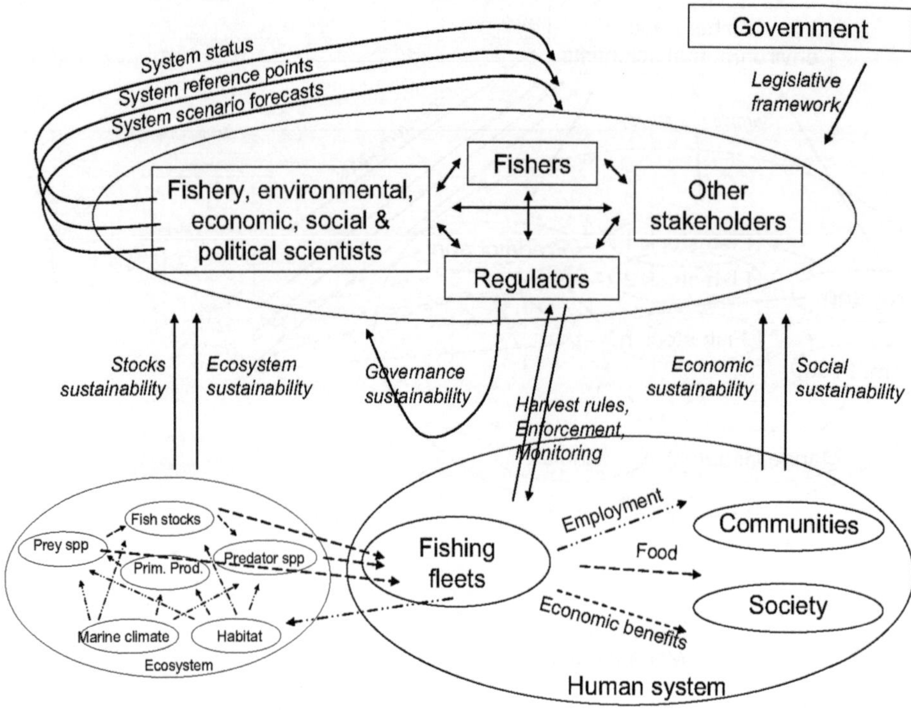

(d)

Environmental Sustainability of Fisheries

Here we briefly review (a) aspects of fishing activity that may have adverse environmental impacts and (b) other anthropogenic factors that may negatively affect fisheries.

Reflecting past policy to invest in fisheries, and "technological creep", European fisheries suffer from *overcapacity*, i.e. the fleet has the capacity to catch more fish each year than can be sustainably removed. It is estimated that over 80% of assessed fish stocks in EU waters are overfished, compared with approximately 25% globally (Pew Environment Group 2009). It is common for fishers to be able to catch their annual quota allocation of fish within a few months (although, to put this in perspective, there are fisheries elsewhere in the world that remain open for only a few hours a year). It is politically difficult to reduce fishing capacity, although some steps have been taken in this direction, and care is needed to ensure that vessels taken out of a fishery (and the associated fishing effort or catch quota) do not subsequently re-enter fishing elsewhere. Another option (albeit increasingly difficult, and clearly not without ecological risks) is to begin exploiting previously unexploited species. One such trend was the move into fishing deep-water fish species, a development doomed to become a very short-term solution by the low reproductive capacity of most such species and the lack of regulation while these fisheries were expanding, as was seen in Scotland (Royal Society of Edinburgh 2004).

Many types of fishing gear are unselective in the sense that they catch species and size-classes that are not being directly targeted (e.g. because they are not commercially valuable or the quotas have already been filled). Such catches are referred to as *by-catch*. By-catches of *charismatic megafauna* such as dolphins, turtles and sharks attract most public attention, while by-catches of commercially important species for which the fleet has no quota (or none remaining) may adversely impact on the state of the fished stocks. Under the current EU system, by-catches of species for which no quota remains must be discarded, i.e. thrown overboard. *Discarding* is almost universally perceived as wasteful (in Norwegian fisheries it is illegal), and enforced discarding of potentially valuable catch undermines fishers' respect for the regulatory system. However, fishers also discard by-catches of species with no commercial value and may discard less valuable species and size-classes if they anticipate that more valuable ones can be caught subsequently (so-called *high-grading*). Based on 1980s data, Alverson et al. (1994) estimated that between 17.9 and 39.5 million tons (average 27.0 million tons) of fish were discarded globally, each year, in commercial fisheries, almost 1/3 of which arise from shrimp trawl fisheries. A more recent estimate based on 1990s data (Kelleher 2005) indicates a much lower, but nevertheless substantial, annual total of 7 million tonnes. Aside from the obvious waste (most discarded fish are dead), such practices distort marine trophic links, increasing the numbers of scavenging species. For example seabird populations have benefitted from feeding on discards (Furness et al. 1988), as have benthic scavengers such as crabs (Catchpole et al. 2005, 2006).

Lost or abandoned fishing gear, including gill nets, trammel nets, longlines and cage traps, can continue to catch fish and other marine organisms, as well as representing a hazard for boat traffic. Recent studies in the northeast Atlantic verify that *ghost fishing* by abandoned gillnets is a significant cause of mortality which must be accounted for in fishery management (Large et al. 2009). Dumping of damaged or obsolete fishing gear could be reduced by educating fishers and establishing collection points in ports for unwanted gear. Matsuoka et al. (2005) proposed that such measures should be included in coastal zone development and management policy and that all fishing gears should carry labels to identify the owner.

Mobile fishing gears such as bottom trawls and dredges can cause significant *habitat damage*, something of which fishers have been aware since at least the 14th century (Jones 1992). Bottom trawling impacts on both the physical structure of the seabed substrate and the composition of benthic communities, with consequent effects on the epifauna – including demersal fish – which form part of the benthic food chain. In the English Channel, analysis of trends over 100 years indicated a loss of larger species from the benthic community (Capasso et al. 2010), and there is general concern about loss of biodiversity (e.g. Norse and Waitling 1998). Such damage may be of particular concern in vulnerable marine ecosystems, e.g. coral reefs (including cold-water corals). Bottom trawl nets used to catch groundfish and shrimps can drag along the seafloor, ripping up or scooping out inert material and animals. Dredges, used for capturing bivalves, not only capture everything that is bigger than the mesh of the net but also remove the upper layer of substrate, killing the animals living there and degrading the habitat. These impacts may be reduced by protecting a large proportion of the habitat of concern although, ultimately, the only way to completely avoid such damage is to use other kinds of fishing gear.

Fishing is obviously not the only human activity that can negatively affect marine ecosystems, and fisheries, in turn, may be negatively impacted by a range of other anthropogenic factors. It is estimated that some 10 billion tonnes of *ballast water* (used by ships to ensure their stability in the water) are transferred globally each year, and that 7,000 species are carried around in ballast water every day. Ballast water is responsible for dissemination of microorganisms and different life stages (e.g. eggs and larvae; cysts, spores or resting stages, adults) of plankton, plants and other animals. These newcomers have the potential to disrupt marine communities (Raaymakers 2002) and hence the fisheries that depend on them.

Pollution is another major issue in the world's oceans; indeed, consultation of Web of Science (in 2010) revealed over 41,000 published articles on marine pollution. The world's human population is largely concentrated in coastal areas and the sea has become a repository for sewage, agriculture run-off, industrial effluents, oil spills, garbage, etc. A wide variety of pollutants (e.g. heavy metals, pesticides, oil derivates and chemical fertilizers) are then incorporated into the marine food web with potentially harmful results. Some (e.g. persistent organic pollutants and heavy metals) can bioaccumulate and biomagnify through the food chain, ultimately reaching concentrations harmful to marine animals, especially in upper trophic levels, and posing a significant risk to humans who consume marine fishery products.

Heavy metals include mercury, cadmium and lead, which have no known biological function and can be harmful to marine animals at any concentration. One of the best known cases of the links between marine pollution and fisheries occurred in the Japanese town of Minimata in the early 1950s, where many of the human population (as well as cats and fish-eating birds) suffered severe neurological disorders due to eating fish and shellfish that had been contaminated by mercury in factory effluents (Kurland et al. 1960). High levels of mercury are routinely recorded in large predatory fish such as swordfish (e.g. Glover et al. 2010). Concentrations of cadmium in mussels from the Baltic and North Seas are above limits set by OSPAR (Rüdel et al. 2010) and lead is recorded in various marine animals, e.g. octopus (Seixas et al. 2005).

The *persistent organic pollutants (POPs)* include several categories of compounds of mainly anthropogenic origin, notably polychlorinated biphenyls (PCBs), pesticides (e.g. dichloro-diphenylethanes (DDT)) and, of more recent concern, the brominated flame retardants (e.g. hexabromocyclododecane (HBCD) and polybrominated diphenylethers (PBDEs)). In marine mammals, high PCB concentrations – which arise from the fish and cephalopods in their diet – have been linked with immune suppression and reproductive failure (e.g. in seals, Reijnders 1986). The production of PCBs has been banned for some decades, but high concentrations persist in many marine animals. The greatest risk to humans though, arises from eating marine mammals. The traditional human diet in Greenland, which includes seal muscle, liver, kidney and blubber, as well as whale blubber, results in high exposure to both POPs and heavy metals (Johansen et al. 2004). High PCB levels are also found in some fish, for example bluefin tuna (Ueno et al. 2002). HBCD and several of the most abundant PBDEs are biomagnified with increasing trophic position in the marine food web (Sørmo et al. 2009).

Anti-fouling paints, such as tributyl-tin (TBT), widely used to avoid problems of fouling in ships, deserve special attention due to their toxicity to aquatic life (with severe effects on the reproductive capacity of some aquatic organisms). TBT compounds have been included on the Red List of dangerous substances and their control is a priority action for the protection of the marine environment under the OSPAR and Helsinki conventions.

Chemical fertilizers can cause eutrophication, creating anoxic conditions and resulting in mass die-offs of fish. One example is the dead zone in Gulf of Mexico, provoked by chemical fertilizers from the Mississippi river (Turner et al. 2008).

Marine *oil spillages* (caused accidently from ships, oil platforms or pipelines, or intentionally from illegal oil releases from platforms or flushing of ships' tanks) can cause major environmental disasters, strongly impacting the fishing industry. Poly-aromatic hydrocarbons (PAHs) from oil result in contamination of fish and shellfish, potentially making them unsafe for human consumption and resulting in a distinctive taste (taint) (e.g. Whittle et al. 1997). In recent years, several major oil spills have led to temporary closures of fisheries, e.g. in Shetland (UK), Galicia (NW Spain) and, most recently, the Gulf of Mexico. In Galicia, losses in the fishery sector over the 12 months following the "Prestige" oil spill (2002) were estimated at around 76 million euros (Suris-Regueiro et al. 2007).

Social, Economic and Institutional Sustainability of Fisheries

Fisheries have linked social and ecological dynamics. Besides the important ecological and environmental aspects of fisheries, the activity has also *human dimensions* (i.e. the social, economic, cultural and institutional components) which need to be taken into consideration to achieve sustainable and effective fishery management.

Arguably, the more important failings of traditional fishery management in Europe, and elsewhere, relate directly to the lack of integration of the human dimensions into the fishery management process (Hilborn et al. 1995). In summarizing papers presented at the fourth World Fisheries Congress, Chuenpagdee and Bundy (2006) commented that although the human dimensions of fisheries were recognized, few studies involved stakeholders beyond their roles as objects to study. All fishery management measures have social and economic consequences. In addition, the performance of any fishery management measure (i.e. its successful implementation), depends (at least partly) on fishers' behaviour. For all these reasons, the failure to take the human dimensions of fisheries into consideration may undermine the success of any new fishery management measure. A governance system which does not achieve fisher buy-in is, in short, not institutionally sustainable.

A major contributor to overfishing (and consequent failure to achieve environmental sustainability) in EU waters, the *overcapacity* of the EU fleet, also has major socio-economic implications. Aside from the initial economic cost of subsidizing capacity building, reducing capacity creates the need (or at least the moral imperative) to provide financial assistance to prevent unemployment, both in the catching sector and those sectors dependent on it (e.g. traders, processors), and to mitigate consequent damage to coastal communities or loss of investment. As argued by Clark et al. (2005), the provision of subsidies is possibly the major contributor to unsustainable fisheries worldwide.

Hilborn (2010) observes that social sustainability is often incompatible with environmental and economic sustainability. The logic of market forces may lead to the increasing concentration of ownership of fishing rights to involve a smaller and smaller number of companies operating fewer and larger fishing vessels. This is clearly incompatible with retaining thriving coastal fishing communities, and some governments (e.g. Australia) have elected to accept the social costs of maximizing economic sustainability. Environmental sustainability increases as competition between fishers is reduced (and the tragedy of the commons no longer applies) while institutional sustainability improves as the system becomes less difficult and less costly to manage. One of the main contributors to lack of *institutional sustainability* (i.e. sustainable governance) has been lack of trust between fishers, scientists and managers/regulators, impeding the flow of reliable information and thus (through a positive feedback loop) contributing to inaccurate stock assessments, sub-optimal harvest rules and lack of buy-in by fishers, which, in turn, leads to further distrust. Over the last decade new institutional arrangements within EU fisheries have been aiming at improving the system of governance by increasing the fishing industry's participation in the fishery management decision-making process and, as such, aim to facilitate common understanding, establish trust between fishers and management bodies and researchers, increase fishers' accountability and responsibility, and contribute to more effective enforcement of rules and regulations by increasing the likelihood of compliance (Pita 2010).

There are inevitable differences in perception between fishers and fishery scientists. Again as noted by Hilborn (2010), fishers have more direct experience of real fish abundance than fishery scientists. However, fisher experience is limited to the areas where they fish. Since, broadly speaking, fishers aim to minimize costs and maximize landing value in order to maximize their profit, they are naturally drawn to easily accessible areas with high fish densities. Their observations in these areas of high density can result in an over-optimistic view of overall fish abundance.

There is always *uncertainty* associated with information on the status and trends in fished populations and on amounts of fish caught and landed (or discarded). This arises for numerous reasons, from the patchiness of fish distributions, and the relatively low sampling effort that can be applied by fishery-independent surveys of fish abundance or on-board monitoring, to the inaccurate reporting of their activities by fishers (which is generally more likely to involve under-reporting of catches than over-reporting – unless the aim is to build up a track record of fishing on a particular species). In turn, this creates a risk that even an apparently successfully implemented management action may have unexpected consequences. The environmental risks (principally the collapse of fished stocks) can be reduced by application of a *precautionary approach*, setting catch limits lower than might otherwise be optimal, especially if a stock is known to already be seriously depleted. However, uncertainty is one of the hardest concepts to communicate, since it means different things to scientists (for whom it is a statement of fact) and non-scientists (to whom it often signifies a reluctance to state the facts).

When talking about the economic and sociocultural sustainability of fisheries it is especially important to pay attention to *small-scale fisheries*. Although there is no universally agreed definition of this type of fishery, the European Fisheries Fund regulation defines small-scale coastal fishing as "fishing carried out by vessel less than 12 m in length and not using towed gear"). Such boats constitute the majority of the EU commercial fishing fleet (77% by number), even if their total catching power is relatively low. As is the case in many other regions of the world, these fisheries are important in terms of providing employment and income (especially in rural communities with limited employment opportunities), as well as maintaining traditional practices (Pita et al. 2010a).

To date, the CFP does not make a distinction between small-scale and large-scale fisheries in its system of governance. However, this could change since the Green Paper on the forthcoming reform of the CFP considers the possibility of adopting a differentiated fishing regime to protect small-scale coastal fleets. This would result in considerable changes in the way small-scale fisheries are managed, and several issues have already been identified as challenges associated with this specific treatment of small-scale fisheries, such as issues of preferential access, specific control measures and application of conservation measures.

Several problems surround management of small-scale fisheries within the EU; first, there is the problem of defining what constitutes small-scale fisheries (different member states have different definitions) and small-scale fisheries within the EU (as in other locations) are traditionally data-poor. Methodologies developed for large-scale finfish fisheries may be inappropriate for artisanal fisheries[4], which often exploit shellfish species with very different population dynamics and distribution patterns to finfish. For instance, Freire and Garcia-Allut (2000) argue that the overexploitation of shellfish in Galicia has resulted from a mismatch between management (i.e. assessment and harvest rules) and the biological and socioeconomic nature of these fisheries.

Social research has a key role to play in revealing how fishers may respond to specific management measures (e.g. the creation of protected areas) or different approaches to improve the system of fisheries governance (e.g. the implementation of co-management approaches). One of the most simple and obvious findings of such research is that fishers are more positive towards management measures about which they have been consulted (as opposed to the measure being handed down unannounced by government). By understanding how fishers are likely to respond, it should be possible to devise more efficient harvest rules and monitoring/regulatory systems and, as such, reduce uncertainty about the outcome of management measures. For example, under the Conservation Credits scheme, Scottish fishers are asked to notify the authorities of areas with high concentrations of cod, which (subject to confirmation) are then temporarily

4 However, methodologies developed for large-scale fisheries, for example use of log-books carried by fishing boats, have been successfully applied in assessment of recreational fisheries (ICES, 2009a)

closed to fishing. In return, the fishers are awarded additional fishing days (Anon. 2010).

Social science research in fisheries is still in its infancy when compared to the biological sciences. However, everything points to social sciences playing a more important role in fisheries management in the near future (see Symes and Hoefnagel 2010). Fishery advisory bodies, such as ICES, are starting to integrate economic and social considerations into their traditional scientific advice. ICES, in its revised Science Plan for 2009-2013, considered the understanding of the interactions of human activities with ecosystems as one of its high priority research topics (ICES 2009b). Recent developments in marine policy make it crucial to include social, economic, cultural, political and institutional information into fisheries management. These types of information have the potential to contribute to improve the realism of scenarios, better understand and/or predict the behavioural response of fishers, improve the buy-in of stakeholders, and develop suitable decision processes.

Recent Developments in Fisheries Management (the Way to Sustainable Fisheries)

Several measures have been implemented in recent years with the aim of achieving long-term sustainable management of fisheries resources; we give a brief overview of some of these measures, while also ending on a cautionary note.

Property Rights

As noted above, the absence of property rights in the sea has been one of the major stumbling blocks to delivering sustainable fishing. Competition between fishers, in the absence of property rights, creates perverse economic incentives and encourages expansion of fishing power, leading to excess fishing capacity (Clark 2006). Thus, creating a property rights system has been a major challenge for policy makers (Wilen 2000). Provision of property rights attempts to eliminate the common property problem by establishing private property rights over the fish stocks (Arnason 2002). Property rights in fisheries offer several advantages, such as security or quality of title, exclusivity, permanence (durability) and transferability, although there are both technical and social limitations (Arnason 2000).

Several types of property rights regimes have been employed in fisheries, including Territorial User Rights Fisheries (TURFs), individual catch quotas and community fishing rights (Arnason 2002). TURFs involve allocation of units of space to a state, private firms, cooperatives, fishers, etc. This approach has been successfully applied to fishing on benthic resources in Chile (see Gelcich et al. 2008). Use of individual catch quotas attempts to solve the common property problem not by conferring ownership of the fish stocks themselves but by allocating individual rights to harvest these stocks. Transferable and divisible catch quotas are usually referred to as *Individual Transferable Quotas* or ITQs (Arnason 2002).

Setting catch quotas does not itself define property rights but, for instance, ITQs potentially do so, at least when coupled with an appropriate level of license fee, and adequate monitoring and enforcement (Clark 2006). Several major fishing nations employ ITQs as their primary fisheries management system (e.g. Australia, Denmark, Canada and New Zealand), and many others employ ITQs in some of their fisheries (Arnason 2002). A global survey found that fisheries managed using ITQs, or a close variant, were half as likely to collapse when compared to other management systems (Costello et al. 2008).

Rights-based management (RBM) is discussed on the Green Paper on the forthcoming 2012 reform of the CFP (see COM (2009) 163 final). The Commission argues that transferable rights (individual or collective) to fishing may be a more efficient and less expensive way to reduce overcapacity and allocate more responsibility to the fishing industry. ITQs may offer the most flexibility within a private ownership system and may reduce the "race to fish". However rights-based fisheries management systems, such as ITQs, are not free from controversy, since this system may encourage discarding and high-grading, promote quota leasing, and facilitate privatization and the creation of monopolies (Burke and Brander 2000; COM (2007) 73; Ecotrust 2009). This process has clear social costs, since it will ultimately lead to the loss of vulnerable coastal communities where fishing is the primary source of income, as in the northeast of Scotland.

Eco-Labelling and Ecological Certification

The history of fisheries suggests that supply-led management of fishing has so far failed to deliver long-term sustainability and that management based on consumer demand could represent a viable alternative (Starkey et al. 2008). Some significant measures promoting sustainable seafood products at consumer level are already in place. These function primarily by raising public awareness of the issue and ensuring that it remains it on the agenda of governments and regional fisheries management organizations. A recent development in fisheries is the increase in the use of "eco-labelling". Eco-labelling is a market-based approach, which involves independent third-party certification, to enhance and reward responsible fishing practices (Gardiner and Kuperan-Viswanathan 2004; Roheim and Sutinen 2006). The growing importance of eco-labelling and organic certification in seafood marketing is a reflection of the increasing interest by retail chains in the topic, no doubt reflecting what appears to be a growing consumer, and media, interest in the issue.

Eco-labelling seeks to inform consumers about product attributes other than price, in particular environmental considerations, but potentially also ethical, social and health issues, e.g. "Dolphin safe" labelling with the aim of reducing the number of dolphin deaths in tuna fisheries in the Pacific. Surveys to date indicate consumer willingness to pay a price premium for labelled fish products (see Johnston et al. 2001; Wessells et al. 2001). For producers, the price premium achieved from eco-endorsed products may serve to recompense them for the extra effort required to generate products to the standards demanded by eco-labelling and organic product schemes. In addition, the seafood industry fears that without certification of sustainability it will be unable to sell its products to the world's major retailers (Roheim and Sutinen 2006). This presents a very real market incentive for sustainable fishing.

The Marine Stewardship Council (MSC) is currently the only eco-labelling organization which provides a world-wide eco-label for wild capture fisheries. The MSC eco-label was created in 1997 by Unilever and World Wildlife Fund (WWF) and the MSC has operated as an independent entity since 1999. Today, over 150 fisheries around the world are certified or undergoing assessment. The MSC eco-label is carried by more than 2,500 seafood products in 52 countries and the market for MSC certified sustainable seafood is estimated to be worth over $1.5 billion (MSC 2009). However MSC certification is not without its critics. The system came under fire in 2004 during certification of the Alaska pollock fishery, with two independent studies criticizing the MSC for lax application of standards.

There are various other eco-labelling schemes such as Friend of the Sea, Naturland and Marine Aquarium Council (Sainsbury 2010). Friend of the Sea provides an ecolabel called the Friend of the Sea Mark and applies to fisheries and aquaculture. Naturland was founded in 1982 for agriculture products, and in 2006 extended certification to marine capture fisheries with the ecolabel Naturland Wildfish. The Marine Aquarium Council certifies wild capture and subsequent treatment of fish for the ornamental aquarium trade.

Co-Management Arrangements

Intrinsically linked to the goal of achieving long-term sustainable management of fisheries resources is the need to ensure greater responsibility and accountability by all individuals involved in the harvesting, processing and marketing of fish. As such, measures have been developed for some years now to increase stakeholders' participation in the fisheries management decision-making process in Europe. The European Commission's (EC) first step for involving resource users in decision-making was the introduction of the Advisory Committee on Fisheries (ACF) in the early 1970s. This measure was reformed in 2000, to include new sectors and interest groups, and renamed the Advisory Committee on Fisheries and Aquaculture (ACFA). More recent measures have included the implementation of a network of Regional Advisory Councils (RACs). The RACs resulted from the 2002 reform of the CFP, meeting the aim of improving the system of governance by increasing the involvement of stakeholders, particularly fishers, in the decision-making process. The RACs have been in operation since 2004 and a total of seven exist within EU waters. Six of these cover specific areas (North Sea, Baltic Sea, Mediterranean Sea, North-western waters, South-western waters and distant water fisheries) while the seventh concerns pelagic stocks. The RACs' main task is to provide the EC with advice on fisheries management in their specific regions/stocks (Pita et al. 2010b). New developments in stakeholder involvement in management are expected with the forthcoming 2012 reform of the CFP. The EC Green Paper setting the rules for the reform identified the need to develop regional co-management arrangements. This will result in a new set of co-management arrangements which will change the face of EU's fisheries management (Pita 2010).

Co-management measures aim at the sharing of authority and responsibility between government and local communities in the management of natural resources. Co-management depends on the willingness and ability of fishers to participate. Since 2000, under the Scottish Shellfish Regulating Order scheme, local stakeholders in some regions of Scotland have been granted the responsibility to regulate and manage sustainable shellfish fisheries. Noble (2003) noted that, while this scheme was successfully applied in Shetland, in another region (Orkney Islands) it was unsuccessful due to opposition by the local fishers. Frangoudes et al. (2008) argue that government investment in training and improving organization of fishing activities, and the empowerment of women, have been key factors in achieving sustainable co-governance of shellfish gathering by Galician women.

Marine Protected Areas

As described above, past fishery management regimes in Europe are gradually being replaced by new management regimes covering the ecosystem as a whole, reflecting a global trend. In this context, Marine Protected Areas (MPAs) have been widely used world-wide as a tool for fishery management, to help to address overexploitation of some fish stocks. MPAs are used, for example, to create area-based restrictions on particular types of gear and/or fishing methods, thus allocating zones for different gear types. Alternatively, closed areas may be established with set time limits, as currently implemented in Scotland to aid the recovery of North Sea cod. These closures are named "Real time closures" and were introduced in 2007 as a voluntary system with the co-operation of the Scottish fishing industry (Anon. 2010; Pita 2010). The primary goals of MPAs are to protect critical habitat and biodiversity, to sustain or enhance fisheries by preventing spawning stock collapse, and to provide recruitment to fished areas. Possible advantages include protection of large mature females (megaspawners) which have disproportionately high fecundity compared to small fish. The increased egg output, and potentially increased recruitment, can benefit the fished part of the stock in adjacent areas. In addition, as protected stocks build up, reserves are predicted to supply local fisheries through density-dependent spillover (movement out of MPAs) of juveniles and adult fish into adjacent fishing grounds (e.g. Amargós et al. 2010).

In addition to MPAs created for fishery management purposes, international obligations and EU legislation have resulted in MPAs also being widely implemented for conservation purposes. This is mostly the result of commitments stemming from the World Summit on Sustainable Development (WSSD) and the United Nations Convention on Biological Diversity (CBD), for the establishment of a global network of MPAs by 2012. At the EU level, various agreements (e.g. OSPAR and HELCOM) reinforce the commitments for the establishment of a network of MPAs within European waters by 2010. This is further reinforced by environmental legislation, such as the Marine Strategy Framework Directive (MSFD), and the Habitats (92/43/EEC) and Birds (79/409/EEC) Directives (Pita 2010). Of special importance is the fact that, under the MSFD, a programme of measures must be implemented to maintain or achieve "Good Environmental Status" in the marine environment by 2020. The Directive specifically requires the implementation of network of MPAs. As such, MPAs will continue to be implemented within EU waters as part of the long-term conservation and fishery management plans.

Empirical evidence shows that no-take MPAs (i.e. areas fully closed to fisheries) can significantly increase species richness, biomass and density of fish although there is considerable uncertainty about the effect of no-take MPAs in temperate waters and several authors have questioned the value of MPAs as a means to address broad conservation and sustainable fisheries objectives (see Halpern and Warner 2002; Halpern 2003; Kaiser 2005; Jones 2006, 2007; Stewart et al. 2009). While there is general consensus amongst the scientific community on the use of no-take MPAs for restoring overexploited ecosystems, it is also suggested that the improvement of conventional fisheries management approaches can be more effective in promoting sustainable fish stock yields (see Hilborn et al. 2004,

2006; Kaiser 2004, 2005; Pauly et al. 2002; Roberts et al. 2005). There is the risk that the implementation of a no-take MPA may merely result in the displacement of fishing effort to other locations (and, worse, to new areas which were not intensely fished before). To avoid this, several authors argue for a combined use of both no-take MPAs and conventional fisheries management approaches (see Greenstreet et al. 2009; Hilborn et al. 2004, 2006; Kaiser 2004, 2005; Roberts et al. 2005; Pita 2010).

The issue of displacement of fishing effort due to the creation of no-take MPAs is yet to be resolved. A recent study by Greenstreet et al. (2009), in the North Sea, concluded that MPAs on their own are unlikely to achieve significant regional-scale ecosystem benefits, because local gains are largely negated by fishing effort displacement into the remainder of the North Sea. To add to the problem of displacement of effort, the whole question of how the socio-economic impact of MPAs will be measured, mitigated and dealt with is a seriously underdeveloped part of the research into MPA implementation in Europe.

High Seas Marine Protected Areas (HSMPAs) represent a potentially important tool for protecting biodiversity of the oceans. The first six HSMPAs were created in 2010, in the North Atlantic, covering a total area of 285 000 km^2, enclosing a series of seamounts and sections of the Mid-Atlantic Ridge that host a range of vulnerable deep-sea habitats and species. This was the first high seas network beyond the jurisdiction of national governments.

Integrated Management of the Marine Environment

The numerous policies and initiatives implemented in the EU over the last decade, such as the Integrated Maritime Policy (IMP), ICZM and MSFD, have been giving a new direction to the management of marine space, under which fishing is simply one of many human uses of the seas. Nowadays, the focus is on the integrated management of human activities in the marine environment, which is a reflection of the pressing need for environmental protection and the increasing competition for sea space. Such integrated approaches imply adoption of a common currency to facilitate comparison of the values of biological, economic and social outcomes of management (Pita 2010). They also imply some commonality of value systems and world views. This could be difficult to achieve in practice. For instance, Mee et al. (2008) pointed to the difficulty of defining "Good Environmental Status" due to different concepts of "good" in different European countries.

The IMP sets the broader vision for the integrated management of all maritime activities in the EU according to the economic, social, and environmental dimensions of sustainable development. Historically, coastal management and fisheries management evolved separately, addressing different needs. The result of this was that the fishing industry, until recently, had enjoyed a surprising level of "favoured status" in relation to its use of marine space (Symes 2007). However, the IMP will imply a less influential fishing industry, a new set of decision-making processes and procedures and the development of new management instruments. For instance, the IMP has made ecosystem-based fisheries management obligatory in the forthcoming 2012 reform of the CFP. The IMP has severe consequences for fisheries management and is possibly the most significant change to the institutional framework for fisheries governance since the implementation of the CFP in the 1970s (Symes 2007).

A Note of Caution!

Fisheries are a global activity and achieving sustainability in Europe may reduce sustainability elsewhere, not least because it will almost certainly require a substantial decrease in fishing activity in European waters. Europe's fishing fleets have long plied the world's seas and the lack of enforceable regulations in international waters presents a potentially tempting economic incentive to exploit such waters unsustainably, a phenomenon evident in the plight of bluefin tuna. Pauly (2009a) argues that the EU has taken advantage of weak governments in Northwest Africa to reach agreements that allow unsustainable fishing in their waters. It has also been argued that one reason for the rise of marine piracy in Somalia has been the decimation of regional fisheries, caused at least in part by foreign vessels exploiting the lack of fishery protection during the civil war (Dagne 2009).

Supermarket chains, currently the champion of eco-labelled fish products have, in fact, been largely responsible for driving consumer demand in the opposite direction over many years by selling cheap, poor quality fish, often imported frozen, and driving traditional fish shops out of business.

Against a background of rising global market demand for seafood, the future of world fisheries will be affected by fluctuations in the abundance of fish and other living marine resources, along with factors such as technical developments, scientific discoveries and economic progress. However, a critical issue will be how capture fishing and aquaculture activities are managed. Clearly, trying to predict future world fisheries (say) three decades from now is subject to many uncertainties (Bruinsma 2003). One of the main difficulties encountered in the prediction of the future of fisheries is that it depends on future political, social and economic conditions, prediction of which represents an even higher order challenge (Garcia and Grainger 2005).

We have so far made no reference to anthropogenic climate change, very possibly the "elephant in the room". While subject to much recent public debate, there seems little doubt that climate change is a reality (Anderegg et al. 2010) and that it has the potential to cause profound changes in marine ecosystems. Global warming may lead to increased fishery yields in some parts of the world (Sherman et al. 2010) but a range of climate-related threats to fisheries has also been identified (Brander 2007). As the latter author comments, "*we have low confidence in predictions of future fisheries production because of uncertainty over future global aquatic net primary production and the transfer of this production through the food chain to human consumption*." Alongside global warming, ocean acidification threatens a range of marine organisms and associated fisheries, notably corals reefs (because deposition of calcium carbonate in coral skeletons is inhibited) and reef-associated fisheries (Hoegh-Guldberg et al. 2007).

Conclusion

Historically, scientific advice on marine fisheries in Europe has concerned the biological status of single fished stocks, and selection of appropriate catch or effort limits to ensure that they remain at healthy (sustainable) levels. More recently this has sometimes been expanded to include consideration of consequences for other target and non-target species and the wider ecosystem. Traditionally, fishery scientists then step back and let managers, regulators and politicians get on with the business of deciding if and how to implement recommendations, i.e. stepping back from the task of putting the scientific advice in its social, economic and political context, taking the view (and one long formalized in the system) that the role of scientists is to present facts and not opinions. However, evaluation of the social and economic impacts of proposed fishing regulations is very much within the remit of fishery science, even if not something that falls within the area of expertise of most biologists who participate in ICES Working Groups. Furthermore, it is a highly relevant scientific task to evaluate how implementation of recommendations might work, for example what regulatory, compliance and political issues might arise – and how these will, in turn, affect the fish stocks. Such considerations are currently finding their way into both EU policy and the way that ICES conducts its business. A key feature of these non-biological components of fishery assessment is that they require engagement with fishing industry stakeholders, for it is ultimately the uptake (compliance) by stakeholders which will secure the achievement of management goals. Less well recognized is the need to engage perhaps the most important stakeholders of all, the consumers. Informing and educating the public about the state of fish stocks could do much to create a backdrop conducive to achieving sustainability and provide greater incentives for all players – fishers, scientists, managers/regulators and politicians – to get it right.

At the time of writing, the EU is starting to address the concept of "MSY variants", whereby "traditional maximum sustainable yield" indicators of stock status are broadened into more general indicators of fishery status by incorporating additional biological, environmental, social and/or economic information. The consequences of using alternative MSY variants can be evaluated following an experimental approach with real fisheries (inherently slow and risky as well as being potentially unrepresentative) or through scenario modelling (and/or "pathway scenario" modelling). Successful multidisciplinary scenario modelling clearly needs a common currency and many strong assumptions – and the latter may lead to scepticism about its validity (e.g. Rose et al. 2010, Schmolke et al. 2010). However, strengths include the possibility to evaluate an almost infinite range of scenarios and options for communication, e.g. visualization of simulation outcomes.

Under this concept, the next step is to "develop an operational framework for the implementation of MSY variants in practical management settings". Arguably this approach, which separates the solution of governance issues from the optimization of fishery status, while realistic in that it implicitly recognizes the inertia of governance systems, is inherently limited. Ideally, the governance system would be treated as a variable alongside those that quantify environmental, social and economic sustainability. In this way inherently unworkable solutions may be avoided, and an achievable sub-optimal target for (say) ecological sustainability may be seen (rightly, in our opinion) as superior to an unachievable ecological panacea.

In the present review we have said little about fishery governance, while implying that governance issues may be amenable to scientific solutions, e.g. by extension of the management strategy evaluation (MSE) approach. It is worth remembering that MSE was developed under the auspices of the International Whaling Commission (IWC). To date, the Revised Management Procedure, designed to deliver sustainable whaling, has not been implemented. Reflecting strong differences of opinion between "pro-" and "anti-whaling" nations, there is no consensus within the IWC about how to proceed. In short, politics trumps science. Jentoft and Chuenpagdee (2009) argue that there are limitations to how rational and effective fisheries governance can be, and that some issues in fishery governance are inherently "wicked", i.e. they have no technical solution, are difficult to define, difficult to solve and likely to recur. In such cases, "governance must rely on the collective judgment of stakeholders involved, in a process that is experiential, interactive and deliberative" (Jentoft and Chuenpagdee 2009).

It is evident therefore that the fishery science of the near future should be inherently multidisciplinary, even if a substantial part of the new work that is needed could still, as currently, be undertaken at least partly by numerate biologists. Good simulation models are likely to be a key feature of multidisciplinary assessments, whether the core is a traditional ecosystem model (e.g. ECOPATH with ECOSIM) or something quite different. Ideas which behavioural and evolutionary ecologists imported from the human sciences, such as optimal foraging theory and game theory, can now be re-applied in something closer to their original contexts, alongside theory derived directly from the fields of economics and psychology (e.g. Clark and Munro 1975; Kahneman, and Tversky 1979), to predict and model the behaviour of fishers and fishing fleets under different management and regulatory regimes. Nevertheless, there is a need to involve economic, social and political scientists in this more broadly defined fishery assessment process. There are inevitable differences in methods and world views across disciplines (see Gould 2003, for a general historical perspective on this theme), and ensuring that they work together efficiently and harmoniously is itself a challenge.

Acknowledgements

The contents of this chapter draw on many discussions with colleagues, including those involved in INCOFISH (CEC INCO project, PL 003739, 2005-08, co-ordinated by Rainer Froese at IFM-GEOMAR), and on interviews with fishers in several European countries. GJP was supported by the ANIMATE project (MEXC-CT-2006-042337). CP was funded by a grant from the Portuguese Foundation for Science and Technology (SFRH/BD/17783/2004). MBS was supported by the Xunta de Galicia, Programa de Recursos Humanos (2010-104) and by the Spanish Ministry of Education, Programa Nacional de Movilidad de Recursos Humanos de Investigación (PR -2010- 0518).We thank Paul Fernandes (Marine Scotland Science) and Ratana Chuenpagdee (Memorial University of Newfoundland) for their helpful comments on the manuscript.

References

Alverson, D.L., Freeberg, M.H., Pope, J.G. and Murawski, S.A. (1994), A global assessment of fisheries bycatch and discards. 233 p. FAO Fisheries Technical Paper no. 339. FAO, Rome (http://www.fao.org/docrep/003/t4890e/T4890E00.HTM).

Amargós, F., Sansón, G., Jiménez Del Castillo, A., Fernández, A., Blanco, F. and Acosta De La Red, W. (2010), An experiment of fish spillover from a marine reserve in Cuba. Environmental Biology of Fishes, 87, pp. 363-372 (http://www.springerlink.com/content/54456lh752g37840/fulltext.pdf).

Anderegg, W.R.L., Prall, J.W., Harold, J. and Schneider, S.H. (2010), Expert credibility in climate change. Proceedings of the National Academy of Science of the United States of America (www.pnas.org/cgi/doi/10.1073/pnas.1003187107).

Anon (2010), The future of fisheries management in Scotland. Report of an Independent Panel. The Scottish Government, Edinburgh. 150 pp. (http://www.scotland. gov.uk/Resource/Doc/329048/0106408.pdf).

Arnason, R. (2000), Property rights as a means of economic organization. In Shotton, R. (ed.), Use of property rights in fisheries management. FAO Fisheries Technical Paper no. 404/1. Food and Agriculture Organization of the United Nations, Rome (http://www.fao.org/docrep/003/x7579E/x7579e03.htm).

– (2002), A review of international experience with ITQs: an annex to Future options for UK fish quota management. 64 p. CEMARE Report no. 58 (http://www.consult-poseidon.com/reports/CEMARE%20QM%20REPORT%20FINAL.pdf).

Beverton, R.J.H. and Holt, S.J. (1957), On the dynamics of exploited fish populations. 583 p. Ministry of Agriculture and Fisheries (U.K.) Fisheries Investigations Series II, vol. XIX.

Branch, T.A., Hilborn, R., Haynie, A.C., Fay, G., Flynn, L., Griffiths, J., Marshall, K.N., Randall, J.K., Scheuerell, J.M., Ward, E.J. and Young, M. (2006), Fleet dynamics and fishermen behaviour: lessons for fisheries managers. Canadian Journal of Fisheries and Aquatic Sciences, 63, pp. 1647-1668 (doi: 10. 1139/F06-072).

Brander, K.M. (2007), Global fish production and climate change. Proceedings of the National Academy of Science of the United States of America, 104 (50), pp. 19709-19714 (http://www.pnas.org/content/104/50/19709).

Burke, D.L. and Brander, G.L. (2000), Canadian Experience with Individual Transferable Quotas. In Shotton, R. (ed.), Use of property rights in fisheries management. FAO Fisheries Technical Paper no. 404/1. Food and Agriculture Organization of the United Nations, Rome (http://www.fao.org/docrep/003/x7579e/x7579e00.HTM).

Capasso, E., Jenkins, S.R., Frost, M. and Hinz, H. (2010), Investigation of benthic community change over a century-wide scale in the western English Channel. Journal of the Marine Biological Association of the United Kingdom, 90, pp. 1161-1172 (doi: 10.1017/S0025315409991020).

Catchpole, T.L., Frid, C.L.J. and Gray, T.S. (2005), Discards in North Sea fisheries: causes, consequences and solutions. Marine Policy, 29, pp. 421-430 (doi: 10.1016/j.marpol.2004.07.001).

Catchpole, T.L., Frid, C.L.J. and Gray, T.S. (2006), Importance of discards from the English *Nephrops norvegicus* fishery in the North Sea to marine scavengers. Marine Ecology Progress Series, 313, pp. 215-226 (http://www.int-res.com/articles/meps2006/313/m313p215.pdf).

Chapela, A., Gonzalez, A.F., Dawe, E.G., Rocha, F.J. and Guerra, A. (2006), Growth of common octopus (*Octopus vulgaris*) in cages suspended from rafts. Scientia Marina, 70, pp. 121-129 (http://www.icm.csic.es/scimar/index.php/secId/6/IdArt/3234/).

Chuenpagdee, R. and Bundy, A. (2006), What was hot at the fourth World Fisheries Congress? Fish and Fisheries, 7, pp. 147-150 (doi: 10.1111/j.1467-2979.2006.00211.x).

Clark, C.W. (2006), Fisheries bioeconomics: why is it so widely misunderstood?. Population Ecology, 48, pp. 95-98 (doi: 10.1007/s10144-006-0255-2).

Clark, C.W. and Munro, G.R. (1975), The economics of fishing and modern capital theory: A simplified approach. Journal of Environmental Economics and Management, 2, pp. 92-106 (doi: 10.1016/0095-0696(75)90002-9).

Clark, C., Munro, G. and Sumaila, U.R. (2005), Subsidies, buybacks, and sustainable fisheries. Journal of Environmental Economics and Management, 50, pp. 47-58 (doi: 10.1016/j.jeem.2004.11.002).

Cook, J.G. (1995), The International Whaling Commission's revised management procedure as an example of a new approach to fishery management. *Development in Marine Biology*, 4, pp. 647-657 (doi: 10.1016/S0163-6995(06)80064-8).

Costello, C., Gaines, S. and Lynham, J. (2008), Can catch shares prevent fisheries collapse? Science, 321, pp. 1678-1981 (doi: 10.1126/science.1159478).

Csirke, J. and Sharp, G.D. (1984), Reports of the Expert Consultation to Examine Changes in Abundance and Species Composition of Neritic Fish Resources. 102 pp. San Jose, Costa Rica, 18-29 April 1983. FAO Fisheries Reports, 291 (1) (http://www.fao.org/docrep/005/x6849e/x6849e00.HTM).

Cushing, D.H. (1975), Marine ecology and fisheries, 278 p. Cambridge University Press, Cambridge.

Dagne, T. (2009), Somalia: Conditions and Prospects for Lasting Peace. 26 p. Congressional Research Service, USA (http://www.fas.org/sgp/crs/row/RL33911.pdf).

ECOTRUST (2009), Briefing: A cautionary tale about ITQs in BC fisheries. 8 p. Ecotrust Canada. Intelligence on the Conservation Economy, issue 8 (http://www.ericennotamm.org/wp-content/uploads/2010/01/ITQ_Cautionary_Tale_FINAL2.pdf).

European Commission (2009), Green paper: Reform of the Common Fisheries Policy, COM (2009) 163 final. 28 p. Commission of the European Communities, Brussels (http://eur-lex.europa.eu/LexUriServ/LexUriServ.do?uri=COM:2009:0163:FIN:EN:PDF).

EUROSTAT (2009), Fishery statistics: Data 1995-2008. 63 p. EUROSTAT, European Commission, Luxemburg (http://epp.eurostat.ec.europa.eu/cache/ITY_OFFPUB/KS-DW-09-001/EN/KS-DW-09-001-EN.PDF).

FAO (2009), The state of world fisheries and aquaculture 2008. 176 p., FAO Fisheries and Aquaculture Department. Food and Agriculture Organization of the United Nations, Rome (ftp://ftp.fao.org/docrep/fao/011/i0250e/i0250e.pdf).

Frangoudes, K., Marugán, B. and Pascual-Fernández, J.J. (2008), From open access to co-governance and conservation: the case of women shellfish collectors in Galicia (Spain). Marine Policy, 32, pp. 223-232 (doi: 10.1016/j.marpol.2007.09.007).

Freire, J. and García-Allut, A. (2000), Socioeconomic and biological causes of management failures in European artisanal fisheries: the case of Galicia (NW Spain). Marine Policy, 24, pp. 375-384 (doi: 10.1016/S0308-597X(00)00013-0).

Froese, R. (2011), Fishery reform slips through the net. Nature 475, pp. 7.

Furness, R.W., Hudson, A.V. and Ensor, K. (1988), Interactions between scavenging seabirds and commercial fisheries around the British Isles. In Burger, J. (ed.), Seabirds and other marine vertebrates: competition, predation and other interactions, pp.240-268. Columbia University Press, New York, USA.

Garcia, S.M. and Grainger, R.J. (2005), Gloom and doom? The future of marine capture fisheries. Philosophical Transactions of the Royal Society B, 360, pp. 21-46 (doi: 10.1098/rstb.2004.1580).

Gardiner, P. and Kuperan-Viswanathan, K. (2004), Ecolabelling and fisheries management. WorldFish Centre Contribution no. 1714, Malaysia. 44 p. (http://www.worldfishcenter.org/resource_centre/WF_368.pdf).

Gelcich, S., Godoy, N., Prado, L. and Castilla, J.C. (2008), Add-on conservation benefits of marine territorial user rights fishery policies in Central Chile. Ecological Applications 18, pp. 273-281 (doi: 10.1890/06-1896.1).

Glover, J.B., Domino, M.E., Altman, K.C., Dillman, J.W., Castleberry, W.S., Eidson, J.P. and Mattocks, M. (2010), Mercury in South Carolina fishes, USA. Ecotoxicology 19(4), pp. 781-795 (doi: 10.1007/s10646-009-0455-6).

Greenstreet, S., Fraser, H. and Piet, G. (2009), Using MPAs to address regional-scale ecological objectives in the North Sea: modelling the effects of fishing effort displacement. ICES Journal of Marine Science, 66, pp. 90-100 (doi: 10.1093/icesjms/fsn214).

Halpern, B.S. (2003), The impact of marine reserves: Do reserves work and does reserve size matter? Ecological Applications, 13, pp. 117-137 (doi: 10.1890/1051-0761(2003)013[0117:TIOMRD]2.0.CO;2).

Halpern, B.S. and Warner, R.R. (2002), Marine reserves have rapid and lasting effects. Ecology Letters, 5, pp. 361-366 (doi: 10.1046/j.1461-0248.2002.00326.x).

Hardin, G. (1968), The Tragedy of the Commons. Science 162, pp. 1243-1248 (doi: 10.1126/science.162.3859.1243).

Hilborn, R. (2007), Reinterpreting the State of Fisheries and their Management. Ecosystems, 10, pp. 1362-1367 (doi: 10.1007/s10021-007-9100-5).

– (2010), An interview with Ray Hilborn (by William Anthony). ICES Insight 47, pp. 14-17 (http://www.ices.dk/products/Insight/INSIGHT%202010%20web.pdf).

Hilborn, R., Micheli, F. and De Leo, G.A. (2006), Integrating marine protected areas with catch regulation. Canadian Journal of Fisheries and Aquatic Sciences, 63, pp. 642-649 (doi: 10.1139/F05-243).

Hilborn, R., Punt, A.E. and Orensanz, J. (2004), Beyond band-aids in fisheries management: Fixing world fisheries. Bulletin of Marine Science, 74, pp. 493-507 (http://people.uncw.edu/scharff/courses/458/Hilborn%202004.pdf).

Hilborn, R. and Walters, C.J. (1992), Quantitative Fisheries Stock Assessment: Choice, Dynamics and Uncertainty. 570 p. Chapman and Hall, New York.

Hilborn, R., Walters, C.J. and Ludwig, D. (1995), Sustainable exploitation of renewable resources. Annual Review of Ecology and Systematics, 26, pp. 45-67 (doi: 10.1146/annurev.es.26.110195.000401).

Hoegh-Guldberg, O., Mumby, P.J., Hooten, A.J., Steneck, R.S., Greenfield, P., Gomez, E., Harvell, C.D., Sale, P.F., Eduwards, A.J., Caldeira, K., Knowlton, N., Eakin, C.M., Iglesias-Prieto, R., Muthiga, N., Bradbury, R.H., Dubi, A. and Hatziolos, M.E. (2007), Coral reefs under rapid climate change and ocean acidification. Science, 318, pp. 1737-1742 (doi: 10.1126/science.1152509).

ICES (2009a), Report of the Workshop on Sampling Methods for Recreational Fisheries (WKSMRF). 231 p., 14-17 April 2009, Nantes, France. ICES CM 2009\ACOM: 41. International Council for the Exploration of the Sea, Copenhagen (http://www.ices.dk/reports/ACOM/2009/WKSMRF/WKSMRF%202009.pdf)

– (2009b), ICES Science Plan 2009-2013. 21 p. International Council for the Exploration of the Sea, Copenhagen (http://www.ices.dk/assets/ssi/text/WhatsnewScience/ICES_Science_Plan__2009-2013.pdf).

– (2010), ICES Advice 2010. Books 1-9. International Council for the Exploration of the Sea, Copenhagen.

Jennings, S., Kaiser, M. and Reynolds, J.D. (2001), Marine fisheries ecology. 432 p. Blackwell Science, Oxford.

Jentoft, S. and Chuenpagdee, R. (2009), Fisheries and coastal governance as a wicked problem. Marine Policy, 33, pp. 553-560 (doi: 10.1016/j.marpol.2008.12.002).

Johansen, P., Muir, D., Asmund, G. and Riget, F. (2004), Human exposure to contaminants in the traditional Greenland diet. Science of the Total Environment, 331, pp. 189-206 (doi: 10.1016/j.scitotenv.2004.03.029).

Johnston, R., Wessells, C., Donath, H. and Asche, F. (2001), Measuring consumer preferences for ecolabeled seafood: An international comparison. Journal of Agricultural and Resource Economics, 26 (1), pp. 20-39 (http://ageconsearch.umn.edu/bitstream/31157/1/26010020.pdf).

Jones, J.B. (1992), Environmental-impact of trawling on the seabed – a review. New Zealand Journal of Marine and Freshwater Research, 26, pp. 59-67 (http://www.eurocbc.org/envimpact_trawlseabed_review.pdf).

Jones, P.J. (2006), Collective action problems posed by no-take zones. Marine Policy 30, pp. 143-156 (doi:10.1016/j.marpol.2004.10.003).

– (2007), Arguments for conventional fisheries management and against no-take marine protected areas: Only half of the story? Reviews in Fish Biology and Fisheries, 17, pp. 31-43 (doi: 10.1007/s11160-006-9016-8).

Kahneman, D. and Tversky, A. (1979) Prospect theory: An analysis of decision under risk. Econometrica, 47, pp. 263-292.

Kaiser, M.J. (2004), Marine protected areas: The importance of being earnest. Aquatic Conservation, Marine and Freshwater Ecosystems, 14, pp. 635-638 (doi: 10.1002/aqc.665).

– (2005), Are marine protected areas a red herring or fisheries panacea? Canadian Journal of Fisheries and Aquatic Sciences, 62, pp. 1194-1199 (doi: 10.1139/f05-056).

Kelleher, K. (2005), Discards in the world's marine fisheries. 131 p. An update. FAO Fisheries Technical Paper no. 470. FAO, Rome (http://www.fao.org/docrep/008/y5936e/y5936e00.htm).

King, M. (2007), Fisheries biology, assessment and management. 400 p. Blackwell Scientific Publications, Oxford.

Kurland, L.T., Fargo, S.N. and Siedler, H. (1960), Minamata disease. The outbreak of a neurologic disorder in Minamata, Japan, and its relationship to the ingestion of seafood contaminated by mercuric compounds. World Neurology 1, pp. 370-390.

Large, P.A., Graham, N.G., Hareide, N.-R., Misund, R., Rihan, D.J., Mulligan, M.C., Randall, P.J., Peach, D.J., McMullen, P.H. and Harlay, X. (2009), Lost and abandoned nets in deep-water gillnet fisheries in the Northeast Atlantic: retrieval exercises and outcomes. ICES Journal of Marine Science 66, pp. 323-333 (doi: 10.1093/icesjms/fsn220).

Matsuoka, T., Nakashima, T. and Nagasawa, N. (2005), A review of ghost fishing: scientific approaches to evaluation and solutions. Fisheries Science 71, pp. 691-702 (doi: 10.1111/j.1444-2906.2005.01019.x).

Mee, L.D., Jefferson, R.L., Laffoley, D. d'A. and Elliott, M. (2008), How good is good? Human values and Europe's proposed Marine Strategy Directive. Marine Pollution Bulletin 56, pp. 187-204 (doi: 10.1016/j.marpolbul.2007.09.038).

MSC (2009), Net benefits: The first ten years of MSC certified sustainable fisheries. Marine Stewardship Council, 70 p. (http://www.msc.org/documents/fisheries-factsheets/net-benefits-report/Net-Benefits-report.pdf).

Noble, T. (2003), Co-operating in fisheries management: trials and tribulations in Scotland. Marine Policy, 27(5), pp. 433-439 (doi: 10.1016/S0308-597X(03)00021-6).

Norse, E.A. and Waitling, L. (1998), Impacts of mobile fishing gear: The biodiversity perspective. In Fish habitat: essential fish habitat and rehabilitation, American Fisheries Society Symposium, 22, pp. 31-40 (www.vliz.be/imisdocs/publications/35278.pdf).

O'Connor, M. (2006), The 'Four Spheres' framework for sustainability. Ecological Complexity, 3, pp. 285-292 (doi: 10.1016/j.ecocom.2007.02.002).

Pauly, D. (1995), Anecdotes and shifting baselines syndrome of fisheries. Trends in Ecology and Evolution 10 (10), pp. 430 (doi: 10.1016/S0169-5347(00)89171-5).

– (2009a), Beyond duplicity and ignorance in global fisheries. Scientia Marina, 73, pp. 215-224 (doi: 10.3989/scimar.2009.73n2215).

– (2009b), Apocalypse Now: The end of fish. The New Republic, October 7, pp. 24-27 (http://www.tnr.com/article/environment-energy/aquacalypse-now).

Pauly, D., Christensen, V., Guénette, S., Pitcher, T., Sumaila, R., Walters, C., Watson, R. and Zeller, D. (2002), Towards sustainability in world fisheries. Nature, 418, pp. 689-695 (doi: 10.1038/nature01017).

Pew Environmental Group (2009), Overcapacity, what overcapacity? Pew Environmental Group, European Marine Programme, 4 p. (http://pewenvironmenteu.us/resources/Pew-Overcapacity-Briefing.pdf).

Pita, C. (2010), The human dimensions of marine protected areas: The Scottish fishing industry. PhD thesis in Social and Environmental Sustainability, University of Aberdeen. 260 p.

Pita, C., Dickey, H., Pierce, G.J., Mente, E. and Theodossiou, I. (2010a), Willingness for mobility amongst European fishermen. Journal of Rural Studies, 26, pp. 308-319 (doi: 10.1016/j.jrurstud.2010.02.004).

Pita, C., Pierce, G.J. and Theodossiou, I. (2010b), Stakeholders' participation in the fisheries management decision-making process: Fishers' perceptions of participation. Marine Policy, 34, pp. 1093-1102 (doi: 10.1016/j.marpol.2010.03.009).

Raaymakers, S. (2002), 1st International Ballast Water Treatment Standards Workshop, IMO London, 28-30 March 2001: Workshop Report. GloBallast

Monograph Series no. 4. IMO London (http://globallast.imo.org/monograph %204%20standards%20workshop.pdf).

Reijnders, P.J. (1986), Reproductive failure in common seals feeding on fish from polluted coastal waters. Nature, 324, pp. 456-457 (doi: 10.1038/ 324456a0).

Roberts, C.M., Hawkins, J.P. and Gell, F.R. (2005), The role of marine reserves in achieving sustainable fisheries. Philosophical Transactions: Biological Sciences 360, pp. 123-132 (doi: 10.1098/rstb.2004.1578).

Roheim, C. and Sutinen, J. (2006), Trade and marketplace measures to promote sustainable fishing practices, ICTSD Natural Resources, International Trade and Sustainable Development Series Issue Paper no. 3, International Centre for Trade and Sustainable Development and the High Seas Task Force, Geneva, Switzerland. 58 p. (http://ictsd.org/downloads/2008/06/roheim_sutinen _20061.pdf).

Rose, K., Allen, J., Artioli, Y., Barange, M., Blackford, J., Carlotti, F., Cropp, R., Daewel, U., Edwards, K., Flynn, K., Hill, S., HilleRisLambers, R., Huse, G., Mackinson, S., Megrey, B., Moll, A., Rivkin, R., Salihoglu, B., Schrum, C., Shannon, L., Shin, Y., Smith, S., Smith, C., Solidoro, C., John, M. and Zhou, M. (2010), End-to-end models for the analysis of marine ecosystems: challenges, issues, and next steps. Marine and Coastal Fisheries: Dynamics, Management, and Ecosystem Science, 2, pp. 115-130 (doi: 10.1577/C09-059.1).

Royal Society of Edinburgh (2004), Inquiry into the Future of the Scottish Fishing Industry. 107 p. The Royal Society of Edinburgh, Edinburgh (http://www. royalsoced.org.uk/enquiries/scottishfisheries/scottish_fishing_in dustry.pdf).

Rüdel, H., Fliedner, A., Kösters, J. and Schröter-Kermani, C. (2010), Twenty years of elemental analysis of marine biota within the German Environmental Specimen Bank: a thorough look at the data. Environmental Science and Pollution Research International, 17(5), pp. 1025-34 (doi: 10.1007/s11356-009-0280-8).

Sainsbury, K. (2010), Review of ecolabelling schemes for fish and fishery products from capture fisheries. 93 p. FAO Fisheries and Aquaculture Technical Paper no. 533. FAO, Rome. 2010 (http://www.fao.org/docrep/013/i1433e/ i1433e00.pdf).

Schmolke, A., Thorbek, P., DeAngelis, D.L. and Grimm, V. (2010), Ecological models supporting environmental decision making: a strategy for the future. Trends in Ecology and Evolution, 25, pp. 479-486 (doi: 10.1016/j.tree. 2010.05.001).

Seixas, S., Bustamante, P. and Pierce, G. (2005), Interannual patterns of variation in concentrations of trace elements in arms of *Octopus vulgaris*. Chemosphere, 59, pp. 1113-1124 (doi: 10.1016/j.chemosphere.2004.11.099).

Sherman, K., Belkin, I.M., Friedland, K.D., O'Reilly, J. and Hyde, K. (2009), Accelerated global warming and emergent trends in fisheries biomass yields. In Fisheries, Sustainability and Development: Fifty-two authors on coexistence and development of fisheries and aquaculture in developing and developed countries, pp. 69-86. Royal Swedish Academy of Agriculture and Forestry, Stockholm (http://www.ksla.se/sv/retrieve_file.asp?n=2099).

Sims, D.W. and Southward, A.J. (2006), Dwindling fish numbers already of concern in 1883. Nature, 439 (9), pp. 660 (doi: 10.1038/439660c).

Sørmo, E., Jenssen, B., Lie, E. and Skaare, J. (2009), Brominated flame retardants in aquatic organisms from the North Sea in comparison with biota from the high Arctic marine environment. Environmental Toxicology and Chemistry, 28 (10), pp. 2082-2090 (doi: 10.1897/08.452.1).

Starkey, D.J., Holm, P. and Barnard, M. (2007), Oceans past: History meets marine science. In Starkey, D.J., Holm, P. and Barnard, M. (eds.), Oceans past: management issues from the history of marine animal populations, pp. 1-5. Earthscan Research Editions, London (http://www.earthscan.co.uk/Portals/0/Files/Sample%20Chapters/9781844075270.pdf).

Stewart, G.B., Kaiser, M.J., Côté, I.M., Halpern, B.S., Lester, S.E., Bayliss, H.R. and Pullin, A.S. (2009), Temperate marine reserves: global ecological effects and guidelines for future networks. Conservation Letters, 2 (6), pp. 243-153 (doi: 10.1111/j.1755-263X.2009.00074.x).

Surís-Regueiro, J.C., Garza-Gil, M.D. and Varela-Lafuente, M.M. (2007), Prestige oil spill and its economic impact on the Galician fishing sector. Disasters, 31, pp. 201-215 (doi: 10.1111/j.1467-7717.2007.01004.x).

Symes, D. (2007), Fisheries management and institutional reform: a European perspective. ICES Journal of Marine Science, 64, pp. 1-7. (doi: 10.1093/icesjms/fsm007)

Symes, D. and Hoefnagel, E. (2010), Fisheries policy, research and the social sciences in Europe: Challenges for the 21st century. Marine Policy, 34 (2), pp. 268-275 (doi: 10.1016/j.marpol.2009.07.006).

Turner, R.E., Rabalais, N.N. and Justic, D. (2008), Gulf of Mexico hypoxia: alternate states and a legacy. Environmental Science and Technology, 42 (7), pp. 2323-2327 (doi: 10.1021/es071617k).

Ueno, D., Iwata, H., Tanabe, S., Ikeda, K., Koyama, J. and Yamada, H. (2002), Specific accumulation of persistent organochlorines in bluefin tuna collected from Japanese coastal waters. Marine Pollution Bulletin, 45 (1-12), pp. 254-261 (doi: 10.1016/S0025-326X(02)00109-1).

Valentin, A. and Spangenberg, J.H. (2000), A guide to community sustainability indicators. Environmental Impact Assessment Review, 20 (3), pp. 381-392 (doi: 10.1016/S0195-9255(00)00049-4).

Volpe, J.P. (2005), Dollars without Sense: The bait for big-money tuna ranching around the world. Bioscience, 55, pp. 301-302 (http://web.uvic.ca/~serg/papers/Volpe%202005%20Bioscience%20(55)%20301-302.pdf).

Wessells, C.R., Cochrane, K., Deere, C., Wallis, P. and Willmann, R. (2001), Product Certification and Ecolabelling for Fisheries Sustainability. FAO Fisheries Technical Paper. 422 p. FAO, Rome (ftp://ftp.fao.org/docrep/fao/005/y2789e/y2789e00.pdf).

Whittle, K.J., Anderson, D.A., Davis, H.K., Lavety, J., Mackie, P.R., McGill, A.S., Moffat, C.F., Shepherd, N.J. and McVicar, A.H. (1997), Tainting of caged salmon (*Salmo salar*) as a consequence of the Braer oil spill, pp. 665-679. In Luten, J.B., Borresen, T. and Oehlenschlager, J. (eds.), Seafood from producer to consumer, integrated approach to quality. Elsevier, Amsterdam.

Wilen, J.E. (2000), Renewable resource economists and policy: what difference have we made? Journal of Environmental Economics and Management, 39, pp. 306-327 (doi: 10.1006/jeem.1999.1110).

Wilen, J.E., Smith, M.D., Lockwood, D. and Botdford, L.W. (2002), Avoiding surprises: incorporating fisherman behaviour into management models. Bulletin of Marine Science, 70 (2), pp. 553-575 (http://www.ingentaconnect.com/con tent/umrsmas/bullmar/2002/00000070/00000002/art00012).

World Conservation Strategy (1980), Living Resource Conservation for Sustainable Development. 72 p.World Conservation Strategy, IUCN-UNEP-WWF.

Worm, B., Barbier, E., Beaumont, N., Duffy, J., Folke, C., Halpern, B., Jackson, J., Lotze, H., Micheli, F., Palumbi, S., Sala, E., Selkoe, K., Stachowicz, J. and Watson, R. (2006), Impacts of biodiversity loss on ocean ecosystem services. Science, 314, pp. 787-790 (doi: 10.1126/science.1132294).

Zeller, D., Cheung, W., Close, C. and Pauly, D. (2009), Trends in global marine fisheries – a critical view, pp. 87-107. In Fisheries, Sustainability and Development: Fifty-two authors on coexistence and development of fisheries and aquaculture in developing and developed countries. Royal Swedish Academy of Agriculture and Forestry, Stockholm (http://www.ksla.se/sv/retrieve_file.asp?n=2099).

Chapter 17

Perceived Importance of Fish Sustainability: The Impact of Consumers' Benefits and Risk Perception regarding Farmed and Wild Fish Consumption[1]

Ana Pinto de Moura, Luís Miguel Cunha[2], Margarida Bacelar, Rui Costa Lima

Abstract

Portugal is one of the major fish-consuming countries in the world, from where one may collect experiences that may be used as examples to improve fish consumption in other countries. The aim of this research is to explore the concept of sustainability as a new paradigm regarding fish consumption by comparing Portuguese consumers' views about the benefits and risks regarding wild and farmed fish consumption. A focus group was conducted with women from the Oporto metropolitan area previously identified as heavy consumers: > 5 fish meals/week. Narrative analysis was conducted based on Bardin content analysis. Findings indicated that focus group participants with a low education level, who were responsible for purchasing and preparing fish for consumption at home, were aware of farmed fish production systems. However, they do not perceive major differences between farmed and wild fish, revealing that both types of fish offer benefits and disadvantages. Nonetheless, some of these women are reluctant to substitute fish oil with vegetable oil on a farmed fish diet, considering this type of fish feed as unnatural. Although these findings cannot be generalized on a larger population, care should be takes as this is of major importance when promoting fish feeding, leading to a lower environmental impact.

1 The authors acknowledge financial support trough project PROAMBIENTE, financed within *Quadro de Referência Estratégico Nacional* – QREN *and Programa Operacional Regional do Norte* – ON2, supported by the European Fund for Regional Development – FEDER.

2 Corresponding author: lmcunha@fc.up.pt.

Introduction

In recent years, growing consumers' interest towards sustainable attributes and claims tends to increase the potential impact of sustainability considerations on consumer purchase decisions (Directorate General Environment 2008, 2009). The term "sustainable consumption" is defined in parallel to the Brundtland definition for sustainable development as: "the use of goods and services that respond to basic needs and bring a better quality of life, while minimising the use of natural resources, of toxic materials and emissions of waste and pollutants over the life-cycle, so as not to jeopardise the needs of future generations" (Norway Ministry of Environment 1994). In a more operational perspective, "sustainable consumption" is based on a decision-making process that takes the consumer's social responsibility and the needs of future generations into account in addition to individual needs and wants (Lindeman and Väänänen 2000; Meulenberg 2003; Moura and Cunha 2005). Although sustainability and ethics are of increasing consumer importance, little research has been conducted to reveal the association of consumers' perceptions towards these issues with seafood consumption behaviour (Verbeke et al. 2007b).

General consumption trends for the EU-15 countries reflect an increase in consumption of seafood products: the EU-15 average per capita consumption has risen from 23 kg/yr in 1989 to 26 kg/yr in 2010, and according to FAO projections, the EU-15 average per capita seafood products consumption will be 27 kg/yr in 2030 (Failler 2007). Fish demand is increasing as a result of the contribution of fish to healthy human diet, the increasing world population, higher living standards and the good overall image of fish among consumers (Olsen 2003; Trondsen et al. 2004; Verbeke et al. 2005; Verbeke and Vackier 2005). The potential heath benefits related to fish consumption are due to the presence of high-quality proteins, high concentrations of n-3 polyunsaturated fatty acids (PUFAs), eicosapentaenoic acid (EPA) and docosahexaenoic acid (DHA), and fish is a valuable source of a number of micronutrients, particularly retinol, vitamin D, vitamin E, iodine and selenium (Slimani et al. 2002; Nesheim and Yaktine 2007). On the other hand, seafood is also a potential source of human exposure to contaminants, responsible for immune suppression, inhibition of reproduction or carcinogenic effects. Although these contaminants are present in low levels in lakes, rivers, seas and oceans, the lipophilic environmental contaminants concentrate in fatty tissues of fish through bioaccumulation and biomagnification (Vrijens et al. 2002; Sidhu 2003; Cohen et al. 2005a,b; Hibbeln et al. 2007; Nesheim and Yaktine 2007). Additionally, a hypothesis emerged that methyl mercury may diminish the beneficial health effect of omega-3 PUFAs from fish (Sidhu 2003).

Moreover, the increase in demand that has led to a worldwide expansion of fishing fleets, together with higher fish capture efficiency has contributed to overfishing and to the risk of depletion of some natural fish stocks. The proportion of overexploited, depleted and recovering stocks has remained relatively stable in the last 10-15 years. It is estimated that in 2007 about 28% of stocks were overexploited, depleted or recovering from depletion; 58% of stocks were fully exploited and, therefore, producing catches that were at or close to their maximum sustainable limits, while only about 20% of stocks were moderately exploited or underexploited (FAO 2009). In response to these drivers, farmed fish, i.e. fish obtained through aquaculture production systems, has become a valuable alternative. Global aquaculture continues to be the fastest growing animal food-producing sector, from a production of less than 1 million tonnes per year in the early 1950s to 51.7 million tonnes reported in the 2006 production, which is equal to 47% of fish consumed globally in 2006 (FAO 2009). Nevertheless, the rapid rise of aquaculture has generated public concern (Naylor and Burke 2005). Experts consider public opposition as a very large threat to its future development (Hishamunda, and Poulain 2009). These concerns include, among others: the depletion of wild stock for producing fish feed; effluents discharge from farms, leading to degraded water quality; alteration or destruction of natural habitats and change in ecosystems; amplification and retransmission of diseases and parasites to the wild; and the escape of farmed fish which can have detrimental effects on wild fish populations through competition and interbreeding (Naylor and Burke 2005; FAO 2006).

As a result, consumers may feel confused twofold. On the one hand, consumers may balance between perceived benefits and risks from consuming fish in order to get the health benefits without exceeding tolerable intakes of chemical contaminants (Cohen et al. 2005a; Verbeke et al. 2005; Pieniak et al. 2008). On the other hand, consumers may balance between the potentially negative environmental impacts of farmed fish when compared with wild fish (Verbeke et al. 2007a,b).

The aim of this research was to explore Portuguese consumers' perceptions about the benefits and risks of fish consumption, while exploring differences on Portuguese consumers' views about wild and farmed fish.

It is noteworthy that Portuguese consumers present high levels of fish consumption. Portugal is the third major fish-consuming country in the world and the first in the 27 European Union member states (EU-27): in 2005, the Portuguese per capita consumption was 60 kg/yr versus 23.7 kg/yr for EU-27 (Failler 2007). Despite the traditional incorporation in the diet and the high levels of seafood consumption in Portugal, there is scarce information regarding consumer attitudes towards fish consumption in this country (Ernst and Young 2008).

Material and Methods

Women who were responsible for purchasing and preparing fish for consumption at home were selected to participate in a focus group. This was conducted to gain insights into their knowledge, attitudes and perceptions regarding benefits and risks associated with the inclusion of fish products as a regular part of their family's diet, as well as comparing their perceptions, preferences and behaviour concerning wild and farmed fish. This focus group included heavy seafood consumers, here defined as persons consuming seafood products more than five meals per week. The participants were recruited and rewarded by a professional food consumer research company (Sense Test, Portugal) using the above-mentioned selection criteria. Recruitment was based on the results of a simple questionnaire asking for details regarding fish purchasing and preparation experience, as well as for average seafood meals per week. The focus group promotes group interaction among participants as they query and explain themselves to each other, which offers valuable data through consensus and diversity and considerable insight into why certain opinions are held (Krueger 1994; Kitzinger 1995).

The focus group took place in a focus group room at a Portuguese food consumer research company (Sense Test, Portugal), during November 2010. The focus group was conducted by an experienced moderator: the first author of this paper. Prior to the focus group, participants completed a socio-demographic questionnaire to provide general information on personal characteristics. After an initial icebreaker introduction, participants were asked to provide a report on their two main meals (lunch and dinner) during the week prior to the day of the focus group session. The inclusion of exercises in the focus group session is suggested by the literature as a way to gather additional objective information or to promote discussion (Bloor et al. 2001). Subsequently, the focus group discussion began, considering themes taken from the literature (Verbeke et al. 2005; 2007a,b; Pieniak et al. 2008) together with concepts considered relevant to address the purpose of this exploratory study: i) benefits and barriers related to seafood consumption and seafood choice criteria; ii) general perceptions of farmed versus wild fish; iii) farmed versus wild fish risk perceptions. Participants were also evaluated regarding food neophobia and variety-seeking tendency, using the Portuguese version (Cadete et al. 2010) of both the Food Neophobia Scale (Pliner e Hobden 1992), and the VARSEEK scale (VanTrijp and Steenkamp 1992), respectively. This may be an important measure as people who are highly neophobic or with low variety-seeking tendency are less prone to adopt new products, namely new fish products.

The focus group session was video recorded for accuracy of transcription and analysis, following participants' permission, and the recordings were anonymously transcribed verbatim. Citations used in this study were translated into English by the research team from the discussion transcriptions. Those were analyzed based upon the core themes of the focus group guide, considering similarities and differences (Moura et al. 2010), based on the Bardin content analysis (Bardin 1977).

Results

A total of seven women aged between 32 and 48 (average: 38.7 ± 5.8 years), married or divorced, all mothers, with children aged 1-21 years old, from a middle-low socio-economic status, participated in this study. This target group is very relevant to investigate because it consists mainly of women of childbearing age and also, in most cases, representing households with young children/adolescents, who constitute a major audience for public health recommendations with respect to fish consumption (Verbeke et al. 2007b). One of the participants reported a single seafood meal during the previous week, justifying this irregular behaviour due to family problems, with the others ranging from 3 to 8 seafood meals taken during the week prior to the focus group session. Respondents were mainly neophilic (FNS ranging from 22 to 42, average 26.9 ± 7.0), presenting a large variety-seeking tendency (VARSEEK ranging from 35 to 48, average 43.3 ± 4.3) (see table 1).

Table 1: Socio-economic characteristic and personality traits of the focus group participants

Participant no.	**Age (years)**	**Total number of daughters and sons**	**Maximum level of education**	**Marital status**	**FNS* (ranging from 10 to 70 – extreme neophobic)**	**VARSEEK** (ranging from 8 to 56 – extreme variety seeking behaviour)**	**Reported number of fish meals during week before focus group**
1	32	2	Basic (9 yrs.) education	Married	26	35	4
2	32	2	Secondary (12 yrs.) education	Married	28	44	1
3	36	2	Secondary (12 yrs.) education	Divorced	22	47	6
4	40	2	Basic (9 yrs.) education	Married	23	45	3
5	42	2	Secondary (12 yrs.) education	Divorced	42	42	4
6	41	1	Secondary (11 yrs.) education	Married	24	48	7
7	48	2	Secondary (12 yrs.) education	Married	23	42	8

* *Food Neophobia Scale (Pliner and Hobden 1992), translated to Portuguese by Cadete, Cunha and Lima (2010).*

** *Variety seeking tendency scale (van Trijp and Steenkamp 1992), translated to Portuguese by Cadete, Cunha and Lima (2010), using a 7-point Likert scale.*

This relative homogeneity within the group ensures that participants felt comfortable speaking candidly about their fish-related consumption at home (Krueger and Casey 2000). A criterion for inclusion into the focus groups was that participants were mothers, considering that nowadays mothers continue to act as the family gatekeeper. Discussion evolved over 85 minutes. Focus group results are presented thematically.

Benefits and Barriers Related to Seafood Consumption and Seafood Choice Criteria

Two main reasons explain why participants prepare seafood meals for their families: generally, participants like the taste of fish/seafood products and these are perceived as a healthy food.

- "*First of all, because I enjoy it. My daughter used to love fish when she was little, but now she started to be fussy about it, but she still eats it! I try to vary a bit... Besides, one can diversify, exploring a bit. For instance, recently I discovered bullhead fillet ... and I started to cook it in many ways; it's easy and it cooks fast ...*". [P6]
- "*Fish is healthier. At home, everyone eats fish; it's healthier. Boiled fish and grilled fish is usually the option.*" [P7]
- "*I try to vary: sometimes meat, sometimes fish. The kids don't like it very much, but you know, we do an effort, we cook sardine, tuna [canned] but it's still fish; I try to vary. It will do you good, we need it, and so do they.*" [P5]

Participants separately mentioned fish as a rich source of iron, containing omega-3 fatty acids, with a protein content of higher quality than in meat. Participants noted that they tried to vary the family meals between meat and fish.

- "*I have always heard that fish contains a lot of iron; it's healthy, mainly for the kids.*" [P1]
- "*It's great for your health; it has proteins that don't exist in meat and it's important for us.*" [P3]
- "*You have got to eat fish; you have got to eat meat it's an option to vary.*" [P5]
- "*It's healthier, beginning with omega–3 fatty acids; it's a type of meal easier to digest. ... especially in the evening, fish is a lighter meal. It's essentially that, being healthy, red meat isn't.*" [P6]

On the contrary, the central barriers related to seafood consumption are associated to family preferences, with fish meals not being the preferred meal by children and husbands in some households.

- "*To be honest, I like fish but my husband doesn't, so, when I serve him fish he gets up, turns around, and takes a while to come back ...*" [P1]
- "*The kids. Between choosing a slice of meat or fish ... What they really like its turkey and chicken! They eat the rest, not that they like it. But they do eat it!*" [P6]

Additionally, consumers consider that fish meals take a long time to prepare and that it is inconvenient to preserve the fish meal prepared in advance for next day's meal. Moreover, participants refer to seafood market conditions as consumption barriers, such as availability of fresh fish as well as the price of fish.

- "*It's a little complicated to cook fish in the evening, for me, or my husband, to take it [to work] the next day.*" [P2]
- "*Sometimes it's the time it takes, because I have the kids playing sports. For example, Wednesday it's a day that I get home at 8 o'clock to cook, and I get tired of always making something easy and faster to prepare. I'll cook a soup with rice with something mixed in it ... It could be tuna ...*" [P4]
- "*Fish it's a little expensive. There are different types of fish. Some fish it's quite affordable, but there's fish that I really like which is somewhat expensive. Cooking a meal with fish, two pieces ... and off with 5 Euros ...*" [P3]
- "*Sometimes it's difficult to buy nearby, as there are only supermarkets and hypermarkets, and I can't get there very easily, and sometimes I notice that [available fish] it isn't that fresh.*" [P7]

When asked about the type of fish usually purchased for their families, participants said that their decision is influenced primary by appearance/freshness of the product and price, followed by fish availability at the store and family preferences.

- "*If I'm going to the supermarket, or to the hypermarket, I always check for promotions, usually a pricey fish that we won't buy, with the promotions it compensates. ... Sometimes I would like to eat fried plaice but I don't buy it because my boyfriend doesn't like it, so I buy horse mackerel, I already know he will eat it.*" [P3]
- "*It's the quality, mainly freshness. I go with something in my mind and I won't change opinion. Of course, the price also matters, but freshness it's above all!*" [P7]
- "*I go to the fishmonger and I check one by one, till I find the best. When we look at the fish ... generally, one can see when the sardine is fresh or not ... Sometimes it's because if feel like it.*" [P1]

Regarding place of purchase, there is no clear tendency among participants, as some participants prefer to purchase both fresh and frozen fish from organized retailing chains and others prefer to purchase from fishmongers or seafood specialty stores.

- "*Well, it's not always easy for me to buy, because nowadays you don't see fishmongers anymore. I buy more often at the supermarket, or when I go by the avenue, I'll go to the fishmonger. It depends on the fish, if I look and like it, I buy it fresh, otherwise I'll choose the frozen one. In the supermarket, it's always the frozen one. But I prefer when I go to the fishmonger at the avenue, if I like it [fresh fish] I buy it.*" [P1]
- "*I always have canned fish [at home], it's helpful, in a rush ... and it's in hypermarkets. Frozen fish also in hypermarkets. I get fresh fish depending on the places I pass by. ... There is a little market nearby my children's kindergarten, I stop by, acquire two or three varieties and I'll cook it that day and the day after, meanwhile I'll freeze the remaining for next time.*" [P3]

- "*I'll get fresh fish from the fishwife when she passes by our house, the rest I buy it in the hypermarket.*" [P5]

General Perceptions of Farmed versus Wild Fish

When asked about what comes to their mind first when they talk about aquaculture, "farmed production" is the image they most associate with aquaculture. One of the participants associated fish farming with laying hens in battery farming and with stressful environments. However, another participant noted that there are other aquaculture systems presently employed, namely the use of cages and tanks for marine cultivation in the open sea. Nevertheless, this participant claimed that one disadvantage associated with aquaculture, in her opinion: "*... are the genetic mutations that farmed fish are undergoing that are harmful to human health*" [P6], which one may consider as some kind of urban myth.

- "*It's like cages.*" [P3]
- "*I classify them as the chickens' aviaries, they are just kept there, though, not many weeks ago I saw a TV report, and there where some who said that this wrong idea exists, that fish raised in such conditions are not healthy, others say they are. Nowadays, technology is so developed that those fish are healthy.*" [P7]
- "*And probably they suffer from stress too, poor things, being closed in there.*" [P7]

Additionally, participants are aware of the existence of farmed fish in the market place, and noted that nowadays the majority of fish bought in retailing chains come from aquaculture production systems. In this context, when asked if they usually search for aquaculture fish when they buy fish products they stated that a major difficulty is to find wild fish in retailing chains. As a result, for participants there are no major differences between farmed and wild fish taste, mainly due to the lack of terms for comparison.

- "*The one that is for selling [farmed fish]; it's much more difficult to find wild fish.*" [P7]
- "*I have no preferences. I don't think I ever ate a wild sea bream. For instance, with sardine or horse mackerel, I would probably detect differences.*" [P6]
- "I never realized [flavour differences]" [P3]
- "*I cannot tell [regarding flavour differences]. My fishwife ... I never ask, she doesn't know.*" [P4]

Additionally, participants found no difference between quality of wild and farmed fish: with quality depending on freshness. By contrast, there is a general agreement that fish from farmed production is much cheaper than fish caught at sea.

- "*When the fish isn't fresh it's no good, whether it's from the sea or from aquaculture*" [P3]
- "*Everything depends on the way they manage to keep the fish fresh.*" [P4]

- "*The nutritional value of farmed fish is as good as any other fish. I heard a nutritionist saying on TV that at least it's as healthy as the other [wild fish], that we shouldn't have such a prejudice against farmed fish.*" [P7]
- "*And comparing prices... for instance, at the hypermarket, farmed sea bass and wild sea bass, really ... the differences in prices it's like double or more.*" [P3]

The most common farmed fish bought by participants at organized retailing chains included salmon and sea bass.

- "*Sea bass, salmon*" [P7]
- "*Salmon, sea bass, most of the time.*" [P3]
- "*Usually salmon.*" [P4]
- "*Sea bass*" [P1]

Considering the sustainability (environmental-friendly concept), there is no consensus in the group: some participants mentioned that farmed fish production is more sustainable whereas others considered wild fish capture as more sustainable. However, there is no significant discussion regarding this subject.

- "*Wild fish [is more environmentally friendly].*" [P3]
- "*It depends. Occasionally little sardines are captured, I don't know if that's good for the environment ... but they still come from the sea. That's not environmentally friendly.*" [P6]
- "*I think there's a higher control in aquaculture.*" [P2]

Curiously, participants when asked if they would buy farmed fish that would be exposed to changes in formulated diets based on fish oil and fish meal partially replaced by vegetable oils in order to better achieve sustainability, participants passionately noted that this new diet is unnatural and that these replacements may be translated into product changes.

- "*It's awkward. I don't know very well what those supplements of plant origin are ... it's a bit artificial.*" [P2]
- "*I believe that fish fed with fish, it's the one that will have the proper flavour. At that moment you're interfering with the product, isn't it?*" [P6]
- "*I speak for myself, I would relate the fish to an artificial thing ...*" [P1]
- "*I think it would make no difference. When they apply such changes, they've already studied and know ... replacing will have the same results than the other type of fishmeal ... I wouldn't stop buying because of that. It's like a vegetarian getting the same proteins, you find in meat, in others products of alternative sources ... I wouldn't restrain myself because of that. One must trust the producers; I know there are many ... you know. But I think you should have responsibility in your work ...*" [P3]

Farmed versus Wild Fish Risk Perceptions

With respect to the health and safety issues, participants face a perceived trade-off between the risk of contamination by pollutants (such as heavy metals and danger of oil being spilt into the sea), which are believed to be more present in wild fish versus the possible inadequate maintenance of water quality, which is believed to be more present in farmed fish.

- "*I don't know. While some are in the sea, I mean, you can't control oil spillages...*" *[P5]*
- "*And also, in the aquaculture pounds water isn't renewed constantly. It's better not to think about it.*" [P4]
- "*On the one hand aquaculture's controlled environment achieves the features they want, on the other hand, in the sea sometimes it may exist mercury and heavy metals, which won't be good for us.*" [P6]

Although participants in general perceived rather low risk of food poisoning from eating fish, the strategy to avoid safety incidents is to follow a varied diet, switching between meat and fish and between farmed fish and wild fish. However, one may suspect that participants mix safety with healthy concepts, considering that for a product to be healthy it must be safe.

- "*I think it's healthy.*" [P1]
- "*In time we developed some prejudice against that aquaculture pounds, like chickens in aviaries, many people have such a prejudice. They think they are raised too fast, that they are unhealthy fish. I believe that's a prejudgment.*" [P7]
- "*You have to find a balance. You can't eat too much fish or eat fish all the time ... So, cooking fish all the time will bring you no benefits. The same with meat. Diversify is the best option.*" [P3]
- "*It's easier to buy farmed fish ... it's safer. Exactly. Let's think water is not polluted, if it was they wouldn't develop so well, right.*" [P7]

When further questioned regarding advice they would dispense to stakeholders in order to minimize risks related to fish consumption, some of them claimed that the presence of a place of origin on fish information label would be recommendable. In this context, one participant has drawn a parallel with wine regions bearing a Protected Designation of Origin (PDO), suggesting a Coast Certification in order to help retailers buying fish with better quality. At the same time, participants emphasize the role of controlling bodies in order to control the quality of seafood. However, in this part of focus group discussion participants expressed serious doubts about the impact of their advice as long as others have more knowledge and influence.

- "*Information about fish origin it's the only thing. Or buying in places we think that have the highest quality.*" [P7]
- "*It's what I was saying ... like for the wine, you know, they have distinctive protected regions, certain places on the shore or so, could be regularly analyzed so one can say ... this one is from Peniche, and Peniche has a fishing zone that is ...*" [P6] "*Certified*" [P3]
- "*People responsible probably make a selection from were to collect the fish ...*" [P3]
- "*I think that a tight quality control must exist.*" [P5]
- "*But who are we to debate such things, with persons that have much more knowledge than us.*" [P4]

Discussion

The findings of this research emphasize that participants enjoy the taste of seafood products and they are strongly convinced that eating fish is healthy and nutritious, due to their content of proteins with high biological value, the presence of certain minerals and of omega-3 fatty acids, which is in accordance with scientific evidence about health benefits from fish consumption.

On the other hand, participants claim to be quite knowledgeable with respect to fish quality. This could be explained by the fact that knowledgeable and experienced consumers regarding fish have self-confidence in evaluating fish quality (Verbeke et al. 2007c; Brunsø et al. 2009), displayed a more favourable attitude, and higher fish consumption levels. In their choices, participants base their decisions on different attributes, using intrinsic attributes, like appearance and freshness, to assess fish quality. Price, fish availability (supply factors) and family preferences are also important criteria when shopping for fish or seafood products. Due to their knowledge regarding choice and preparation of fish and seafood, participants mentioned the use of different store types to buy fish as a function of their convenience and degree of processing (fresh, frozen, tinned), here the supermarket emerged as one outlet for buying both frozen and fresh fish. The growing role of supermarkets in the distribution of seafood products make seafood products more readily available to consumers and provide some sort of quality guarantee (Failler 2007). Despite fish meals not being the most preferred meal by family members, participants of this study mentioned that eating fish at home is a habit. In these families, preparation of a fish-based meal is a strong habit that has been formed. These results are in accordance with a previous Portuguese study that explored the factors that influence the consumption of fish among 8-14 year-old children, where habit emerges as a strong determinant of behaviour regarding fish consumption (Moura et al. 2011).

Moreover, participants were aware of the existence of farmed fish in the market place and noted that nowadays the majority of fish bought in retailing chains comes from aquaculture production systems. Despite participants' association of aquaculture with "farmed production", they match at the same level the taste, the healthiness and the safety of wild and farmed fish. These results are in clear contrast with others studies, from other countries, where consumers have little awareness about which fish species are sold as farmed or wild fish (Verbeke and Vackier 2005; Verbeke et al. 2005) and the wild fish attributes scored favourably compared to farmed fish (Arvanitoyannis et al. 2004; Verbeke et al. 2007a,b).

Furthermore, participants were convinced that both wild and farmed fish offer benefits and have disadvantages, revealing that consumers do not have strong specific preferences for wild or farmed fish. Consumers surrendered to market laws and are used to living with food scares and food crises (Cunha et al. 2010). In other words, participants do not feel scared when buying farmed fish and they relegate this information when cooking fish for the family. To avoid safety incidents, participants advocated the basic guideline of healthy eating: to eat a balanced and a varied diet (WHO/FAO 2003), varying, in this context, between meat and fish and between farmed fish and wild fish.

Considering sustainability, there is no consensus in the group: some participants mentioned that farmed fish production is more sustainable whereas others consider wild fish capture as more sustainable. However, there is no significant discussion within the subject, which could indicate that this issue is not at the top of the minds of participants or, alternatively, there is no real knowledge of the fact that aquaculture aims at providing an answer to the depletion of natural fish stocks. Curiously, participants are generally against the change of formulated diet based on vegetable oils in order to achieve better sustainability of farmed system production. This issue is in contrast with scientific evidence reporting that partial substitution of fish oil by vegetable oil in feeding mixtures avoids the use of unsuitable oil types, as well as chemicals and other non-biological elements, which may contaminate the catches. Moreover, it reduces the major problem that fish industry has to deal with regarding the shortage of by-catch to produce fish oil (Schacht et al. 2010). The main cause of this gap between participants' perception and scientific evidence seems to be the lack of knowledge concerning sustainable aquaculture, resulting in the use of their daily experience to judge seafood products and production systems.

Conclusions

The focus group discussion among women with a low education level who were responsible for purchasing and preparing fish for consumption at home indicates that they are aware of farmed fish production systems. However, they do not perceive major differences between farmed and wild fish concerning taste, safety and health-related attributes, revealing that both wild and farmed fish offer benefits and disadvantages. However, these women are reluctant to substitute fish oil with vegetable oil, considering this type of fish feed unnatural. These findings point to the need for more appropriate communication to consumers regarding sustainability in relation to wild and farmed fish (Verbeke et al. 2007b).

The authors stress that this is an exploratory study and findings cannot be generalized over a larger population. Additional studies, covering different socio-economic status, are being thought through and organized.

References

Arvanitoyannis, I.S., Krystallis, A., Panagiotiaki, P. and Theodorou, A.J. (2004), A marketing survey on Greek consumers' attitudes towards fish. Aquaculture International, vol. 12, pp. 259-279.

Bardin L. (1977), L'Analyse de Contenu, Presses Universitaires de France (PUF), Paris.

Bloor, M., Frankland, J., Thomas, M. and Robson, K. (2001), Focus Group in Social Research. Sage, London.

Bouzan, C., Choen, J.T., Connor, W.E., Kris-Etherton, P.M., Gray, G.M., Köning, A., Lawrence, R.S., Savitz, D.A. and Teutsch, S.M. (2005), A quantitative analysis of fish consumption and stroke risk. American Journal of Preventive Medicine, vol. 29, no. 4, pp. 347-352.

Brunsø, K., Verbeke, W., Olsen, S.O. and Jeppesen, L.F. (2009), Motives, barriers and quality evaluation in fish consumption situations: Exploring and comparing heavy and light users in Spain and Belgium. British Food Journal, vol. 111, no. 7, pp. 699-716.

Cadete, S., Cunha, L.M. and Lima, R.C. (2010), Translation into Portuguese and exploratory application of the food neophobia scale and of the variety seeking tendency scale. Sensiber 2010 – 6th Ibero American Sensory Analysis Symposium, 19-21 August 2010. S. Paulo, Brazil.

Cohen, J.T., Bellinger, D., Connor, W.E., Kris-Etherton, P.M., Lawrence, R.S., Savitz, D.A., Shaywitz, B.A., Teutsch, S.M. and Gray, G.M. (2005a), A quantitative risk-benefit analysis of changes in population fish consumption. American Journal of Preventive Medicine, vol. 29, no. 4, pp. 325-334.

Cohen, J.T., Bellinger, D.C., Connor, W.E., and Shaywitz, B.A. (2005b), A quantitative analysis of prenatal intake of n-3 polyunsaturated fatty acids and cognitive development. Journal of Preventive Medicine, vol. 29, no.4, pp. 366-374.

Cunha, L.M., Moura, A.P., Lopes, Z., Santos, M.C. and Silva, I. (2010), Public perceptions of food-related hazards: an application to Portuguese consumers. British Food Journal, vol. 112, no. 5, pp. 522-543.

Directorate General Environment (2008), Attitudes of European Citizens towards the Environment, Special Eurobarometer 295, European Commission, Directorate General Environment (available at: http://ec.europa.eu/public_opinion/archives/ebs/ebs_295_en.pdf – accessed 5 November 2010).

– (2009), Europeans' Attitudes towards the Issue of Sustainable Consumption and Production, The Analytical Report, Eurobarometer 256, European Commission, Directorate General Environment (available at: http://ec.europa.eu/public_opinion/flash/fl_256_en.pdf – accessed 5 November 2010).

Ernst & Young (2008), Enquête d'Image sur la Perception des Produits de la Pêche et de l'Aquaculture, Rapport final, DG MARE, European Commission, Fisheries (available at: http://ec.europa.eu/fisheries/documentation/studies/image_survey/survey_sector_image_summary_fr.pdf – accessed 5 November 2010).

Failler, P. (2007), Future Prospects for Fish and Fishery Products, 4. Fish Consumption in the European Union in 2015 and 2030, Part 1. European Overview, FAO Fisheries Circular no. 972/4, Food and Agricultural Organisation (FAO), Rome (available at: ftp://ftp.fao.org/docrep/fao/010/ah947e/ah947e.pdf – accessed 5 November 2010).

Food and Agriculture Agency of the United Nations (FAO) (2006), State of World Aquaculture 2006, Fisheries Technical Paper no. 500. FAO, Rome, Italy (available at: ftp://ftp.fao.org/docrep/fao/009/a0699e/a0699e.pdf – accessed 5 November 2010).

– (2009), The State of World Fisheries and Aquaculture 2008. FAO, Rome, Italy (available at: ftp://ftp.fao.org/docrep/fao/011/i0250e/i0250e.pdf – accessed 5 November 2010).

Hibbeln, J.R., Davis, J.M., Steer, C., Emmett, P., Rogers, I., Williams, C. and Golding, J. (2007), Maternal seafood consumption in pregnancy and neurodevelopment outcomes in childhood (ALSPAC study): an observational cohort study. Lancet, vol. 369, no. 9561, pp. 578-585.

Hishamunda, N. and Poulain, F. (2009), Prospective Analysis of Aquaculture Development: The Delphi Method, Fisheries Technical Paper no. 521. FAO, Rome, Italy (available at: http://www.fao.org/docrep/012/i0975e/i0975e.pdf – accessed 5 November 2010).

Kitzinger, J. (1995), Qualitative Research. Introducing Focus Groups, BMJ, vol. 311, pp. 299-302.

Köning, A., Bouzan, C., Cohen, J.T., Connor, W.E., Kris-Etherton, P.M., Gray, G.M., Lawrence, R.S., Savitz, D.A. and Teutsch, S.M. (2005), A quantitative analysis of fish consumption and coronary heart disease mortality. American Journal of Preventive Medicine, vol. 29, no. 4, pp. 335-346.

Krueger, R. (1994). Focus Groups: A Practical Guide for Applied Research, 2nd ed. Sage Publications. Thousand Oaks, CA.

Krueger, R. and Casey, M.A. (2000), Focus Groups: A Practical Guide for Applied Research, 3rd ed., Sage Publications. Thousand Oaks, CA.

Lindeman, M. and Väänänen, M. (2000), Measurement of ethical food choice motives. Appetite vol. 34, no. 1, pp. 55-59.

Meulenberg, M. (2003), Consumer and citizen, meaning for the market of agricultural products and food products. TSL, vol. 18, pp. 43-56 (in Dutch). In Verbeke, W., Vanhonacker, F., Sioen, I., Van Camp, J. and De Henauw, S. (2007b), Perceived importance of sustainability and ethics related to fish: a consumer behaviour perspective. AMBIO: A Journal of the Human Environment, vol. 37, no. 7, pp. 580-585.

Moura, A.P. and Cunha, L.M. (2005), Why consumers eat what they do: an approach to improve nutrition education and promote healthy eating. In Consumer Citizenship: Promoting New Responses. Taking Responsibility, Doyle, D. (ed.), Forfatterne, vol. 1, pp. 144-156.

Moura, A.P., Cunha, L.M., Azeiteiro, U.M. and Lima, R.C. (2011), Acceptance of fish and fish products by Portuguese young consumers: an exploratory study based on mothers' evaluation. Seventh International Conference on Culinary Arts and Sciences, ICCAS 2011. Bournemouth University, UK. In press.

Moura, A.P., Cunha, L.M., Azeiteiro, U.M., Aires, L., Graça, P. and de Almeida, M.D.V. (2010), Food consumer science post-graduate courses: comparison of face to face versus online delivery systems. British Food Journal, vol. 112, no. 5, pp. 544-556.

Naylor, R. and Burke, M. (2005), Aquaculture and ocean resources: raising tigers of the sea. Annual Review of Environment and Resources, vol. 30, pp. 185-218.

Norwegian Ministry for the Environment (1994), Symposium on Sustainable Consumption, Oslo, Norway.

Nesheim, M.C. and Yaktine, A.L. (eds.) (2007), Seafood Choices: Balancing Benefits and Risks, Institute of Medicine of the National Academies. The National Academies Press, Washington, D.C. (available at: http://www.google.com/books?id=f79PXLsAA1MC&printsec=frontcover&hl=pt-PT#v=onepage&q&f=false – accessed 5 November 2010).

Olsen, S.O. (2003), Understanding the relationship between age and seafood consumption: the mediating role of attitude, health involvement and convenience. Food Quality and Preference, vol. 14, pp. 119-209.

Pieniak, Z., Verbeke, W., Scholderer, J., Brunsø, K. and Olsen, S.O. (2008), Impact of consumers' health beliefs, involvement and risk perception on fish consumption: a study in five European countries. British Food Journal, vol. 110, no. 9, pp. 898-915.

Pliner, P. and Hobden, K. (1992), Development of a scale to measure the trait of food neophobia in humans. Appetite, vol. 19, pp. 105-120.

Schacht, K., Leal Filho, W., Koppe, W., Struksnaes, G. and Busch-Stockfisch, M. (2010), Sustainability as a new paradigm regarding food consumption. British Food Journal, vol. 112, no. 5, pp. 476-488.

Sidhu, K.S. (2003), Health benefits and potential risks related to consumption of fish or fish oil. Regulatory Toxicology and Pharmacology, vol. 38, pp. 336-344.

Slimani, N., Fahey, M., Welch, A.A., Wirfält, E., Stripp, C., Bergström, E., Linseisen, J., Schulze, M.B., Bamia, C., Chloptsios, Y., Veglia, F., Panico, S., Bueno-de-Mesquita, H.B., Ocké, M.C., Brustad, M.E., Lund, E., González, C.A.,

Barcos, A., Berglund, G., Winkvist, A., Mulligan, A., Appleby, P., Overvad, K., Tjønneland, A., Clavel-Chapelon, F., Kesse, E., Ferrari, P., Van Staveren, W.A. and Riboli, E. (2002), Diversity of dietary patterns observed in the European Prospective investigation into Cancer and Nutrition (EPIC) project. Public Health Nutrition, vol. 5, no. 6B, pp. 1311-1328.

Trondsen, T., Braaten, T., Lund, E. and Eggen, A.E. (2004), Consumption of seafood the influence of overweight and health benefits. Food Quality and Preference, vol. 15, pp. 361-374.

van Trijp, H.C.M. and Steenkamp, J.-B.E.M. (1992), Consumers' variety seeking tendency with respect to foods: measurement and managerial implications. European Review of Agricultural Economics, vol. 19, no. 2, pp. 181-195

Verbeke, W. and Vackier, I. (2005), Individual determinants of fish consumption: application of the theory of planned behaviour. Appetite, vol. 44, pp. 67-82.

Verbeke, W., Sioen, I., Brunsø, K., De Henauw, S. and Van Camp, J. (2007a), Consumer perception versus scientific evidence of farmed and wild fish: exploratory insights from Belgium. Aquaculture International, vol. 15, pp. 121-136.

Verbeke, W., Sioen, I., Pieniak, Z., Van Camp, J. and De Henauw, S. (2005), Consumer perception versus scientific evidence about health benefits and safety risks from fish consumption. Public Health Nutrition, vol. 8, no. 4, pp. 422-429.

Verbeke, W., Vanhonacker, F., Sioen, I., Van Camp, J. and De Henauw, S. (2007b), Perceived importance of sustainability and ethics related to fish: a consumer behaviour perspective. AMBIO: A Journal of the Human Environment, vol. 37, no. 7, pp. 580-585.

Verbeke, W., Vermier, I. and Brunsø, K. (2007c), Consumer evaluation of fish quality as basis for fish market segmentation. Food Quality and Preference, vol. 18, pp. 651-661.

Vermeir, I. and Verbeke, W. (2008), Sustainable food consumption among young adults in Belgium: Theory of planned behaviour and the role of confidence and values. Ecological Economics, vol. 64, pp. 542-553.

Vrijens, B., De Henauw, S., Dewettinck, K., Talloen, W., Goeyens, L., De Backer, G. and Willems, J.L. (2002), Probabilistic intake assessment and body burden estimation of dioxin-like substances in background conditions and during a short food contamination episode. Food Additives and Contaminants, vol. 19, no. 7, pp. 687-700.

World Health Organization/Food and Agriculture Agency of the United Nations (WHO/FAO) (2003), Diet, Nutrition and the Prevention of Chronic Disease, report of a joint WHO/FAO expert consultation, WHO Technical Report Series No 916, World Health Organization, Geneva (available at: http://www.who.int/hpr/NPH/docs/who_fao_expert_report.pdf – accessed 5 November 2010).

Chapter 18

Collaborative Governance for the Preservation and Valorization of the Ecosystem Services and Biodiversity in Serra da Aboboreira

–

Sustainable Planning Instruments and Biodiversity Conservation

Alcide Gonçalves, Ulisses Miranda Azeiteiro, João Honrado

Abstract

The deliberative democracy implies new public participation forms that enhance greater citizen's involvement in the complex processes of decision-making.

On the other hand, this new way of governance will demand that the public administration in the territory and the democratic institutions to adopt new management styles allowing a wider shared power, which makes the State and citizens partners.

Another issue is several environmental strategies and policies that reinforce the citizen's need of being more collaborative and participant in actions towards the biodiversity conservation and the maintenance of the ecosystems services whose aim is to achieve sustainable development and to ensure a common future.

The present paper has the mission to attempt a proposal of participatory management for Serra da Aboboreira (Baixo-Tâmega) through a *Plano de Parceria* (Partnership Plan) as a tool for good governance.

Introduction

Governance in Portugal is now taking its first steps and is passing through the right formats for its implementation. Although Portuguese legislation has been updated as far as participation and public involvement exist as an active way necessary towards sustainable development and to the well-functioning of democratic institutions, its realization seems to be very far from what is expected in the perspectives and practices of politicians, technical staff and citizens. Furthermore, they do not seem to be prepared for its full adoption.

After reading all the specific literature on the subject, the result is that planning, project and the socio-territorial issues are the only ones that have been getting the most experience within the governance framework. That is to say that in the platform of governance, the territorial administrative bodies have been asking specialized teams for interactive participation processes, for the elaboration of adjusted participation processes to complex questions, so that the results recognize the citizen need to be closer to power levels.

As far as territorial management is concerned, governance lies in a web composed of entities and tutelary institutions that sometimes are themselves difficult to coordinate in due time, this being one of the main obstacles to governance. There is a myriad of organizations, rules and legislation that strangle the ecosystem services, biodiversity and environment and prevent the effectiveness of collaborative management.

In terms of nature conservation, in which ICNB [Institute for Biodiversity and Nature Conservation] is the national authority, its management is being carried out by programmes and projects under the umbrella of communitarian directives whose aim is primarily the recovery of endangered habitats and/or endangered species. That is the case of Programa Lince [Lynx Programme] and the Projects *Peneireiro-das-torres* [*Lesser Kestrel* Project], *Signatus,* Life *Priolo* [Life *Bullfinch*], Life *Estepárias* [Life *Steppe Birds*] amongst others.

Most of the work undertaken is due to the role played by environmental NGOs and other environmental associations that are developing actions and initiatives to promote and protect nature and the environment, mobilizing institutions and civil society in general, affirming themselves and their important role in public-decision contexts.

The fact is that they have been the driving force of active participation processes and have contributed to citizenship holding important tasks both in the ways of citizens' involvement and in environmental awareness and education by increasing critical consciousness by the general public.

'Collaborative Governance' will thus have to emerge in this highly uncertain and complex context as the unique reality in each geographical zone and this way accomplishing the mission to reach sustainable development. Following this, and in quite a pioneering way, this paper will issue a proposal of collaborative management – through a *Partnership Plan,* integrative for *Serra da Aboboreira* and for the preservation of its environmental, natural and landscape values.

As far as programmes, projects and other initiatives are concerned, partnerships have been carried out based on a prospect of financial and logistic investment and also by means of technical, scientific and research supports. Focused on territorial sustainable management, grounded on active participation by EVERYONE, no matter their level of interest, the Partnership Plan breaks with the classic way of territorial management, fully centred in the State's power and in a *top-down* vision of governance by turning it into a *bottom-up* practice.

The main purpose is to implement and promote a new philosophy/mentality of what may lead to sustainable development of the territory, and not only basing the management in this new way of power practice (*bottom-up)*, being fully aware that the path lies in the articulation of both ways, *top-down* and *bottom-up*.

> "The issue is to create functional structures born out of a common project. This kind of organization allows the gathering of a wide range of people (representing institutions or even themselves personally) and institutions and there is no gap between the decision makers (politicians), the operational staff (the technicians) and the receivers (the citizens)" (Guerra 2007).

Conscious of several problematic issues placed by the Partnership Plan, namely due to the model that supports it not being defined as an instrument of territorial management in the Portuguese legal framework, it is inevitable that a profound thinking of the governance model would take place in all its strands – from the form, up to the tools, across all contents.

Governance, Active Citizenship and Active Public Participation

Governance and the Questions on Active Participation

Nowadays, we are global witnesses of a turning point in the political, economic and social ways of governing and in democracy itself as a consequence of *environmental* concerns, globalization effects and the emergence of informational societies, themselves leveraged by technological development. Citizens are increasingly called to cooperate and to become involved in the major issues, namely environmental, in order to minimize their effects/impacts and to achieve the defined objectives and goals.

Efforts and commitment by governments and governmental bodies (part of a representative democracy concept) seem to be insufficient to carry out these goals and objectives outlined by several global policies. This leads us to question the participative forms of populations by democratic system tools [the vote].

On the other hand, the challenges derived from various economical and social dynamics and the complex interaction coming from them (result of the globalization) provoke changes in the frame of significances of the rights and of individual and collective responsibilities, getting the global/local scales closer in terms of performance and impacts, "*Right from the beginning, because that interaction between local and global makes us agents (and not simply passive spectators) both of the proximities and of the asymmetries between global North and South, what became an essential element of the citizen nature within our time*" (MNE; ME 2009).

The conventional concept of *citizenship* is thus expanding its "*inclusion/exclusion distinct frontiers matching with the belonging bonds to a national community*" (MNE; ME 2009), to a responsible bond belonging to a wider community of an international and planetary scope. This new concept of *citizenship* is now solely *inclusive* and demanding an efficient and effective level of action/participation from each and every citizen.

This conceptual change brings a transformation to the social fabric, giving it a major responsibility which is already a result of the transformation of the social power into political power. Thus, communities' opinions and needs have a higher strength and direct influence on decision-making processes.

Following on from this, *active citizenship* implies that citizens go beyond the vote (common way of political participation in the western world) and involve themselves in political, economic, social and cultural strands of the society of which they are a part and which gives way to a partnership, to comprehensive communication, to effectiveness of decision-making and to political legitimacy, whenever needed, whose only purpose is collective well-being.

On the other hand, the new democratic method paradigm, named participative democracy, implies another governing vision and format. This means that new ways of shared power in decision-making are required and encouraged through various processes together with a reversal of direction (way) in what concerns power practice, from *top-down* to *bottom-up*.

Governance based on such a *bottom-up* way of power practice allows the citizen a major approach to the political power, actions/projects that are much closer and more adapted to local contexts and, last but not least, a more accurate response to the needs of the local populations with a better outcome.

However, in order to acquire civic skills citizens need to be given opportunities to express their opinions, to learn from one another, to commit in discussion and deliberation, to read, hear and question the experts, political candidates and opinions of people they trust and, finally, learn from ways based on freedom of speech.

> "Citizenship is a set of rights and obligations that engages all the members of a political community no matter the sex, age, ethnic, educational level, religion or other criteria that can restrain a responsible sense of belonging and that may exclude people from freely practicing political rights and civic obligations" (Guerra 2006: 97).

Despite that, volunteerism from each and every one of us will not be enough as a way to show the qualification as citizens. It will be fundamental and necessary to adjust the participation, in all its dimensions, to the processes and also to articulate it with the democratic system installed or to be achieved. The degree of participation must be coherent with the desires and goals meant for society and the democratic institutions to reach as their answer (or contextualized response) to the ongoing decision-making political process goals.

Participation *per se* is equally not sufficient for the process to become democratic. The *form* – who and how, and the *content* – in which we are participating, what interests, which conflicts and consensus, and how complex they become in the role play, are extremely relevant in this articulation between the process of actions/institutions (Guerra 2007; Vasconcelos 2003).

Before starting the participative process, which is meant to be a constructive one, it would be wise and productive if a joint reflection is made, and that the direction to follow is defined in detail as far as what active participation concerns.

Active Public Participation Needs Empowerment

According to Dinet (1997 cit. Ninacs 2002: 17), *it is not enough to be an actor of its development it must also be a real author.* But being author of the development, it requires from the individuals, organizations and communities (1) knowledge, (2) a critical decision-making consciousness about the nature of the individual and collective problems, and (3) the need of a collective action for overcoming problems.

Following this, the empowerment process respects a set of mechanisms through which people, organisms and communities gain maturity and enables themselves for the decision making process (Menezes 2007).

As training methodology it will allow the individual to access the resources and to develop their personal capacities towards active participation, construction of their own lives and that of the community regarding economical, social and political dimensions. Thus the individual and/or the group become more autonomous and less dependent on external forces.

Zimmerman (1995) explains through individual, organizational and community analysis levels how empowerment throughout the process develops personal and group capacities and skills that will, afterwards, support the different participation forms. That process is illustrated in Table 1.

What concerns the individual level, the result is the development of the personal control perception, participation with the others in order to achieve goals and critical consciousness of the factors responsible for the real difficulties.

In organizational terms, processes and structures stress the participation of the members to improve organization efficiency. And last but not least, at the community level, the empowerment process permits the development of joint actions with the purpose of improving conditions of community life. Among these joint actions are the constitution and organization of partnerships (Menezes 2007).

The development of any of the levels will be fundamental to the capacity-building process. Only with informed citizens can an active participation and co-responsibility in changing processes be promoted.

Table 1: A Comparison of Empowering Processes and Empowered Outcomes Across Levels of Analysis, Zimmerman (1995)

Levels of Analysis	Process	Outcomes
Individual	Helping others gain control over their lives. Receiving help from others to gain control Mutual help	Sense of control Critical awareness Participatory behaviors
Organizational	Provoding opportunities for members to develop and practice skills. Erecting participatory decision-making structures Sharing responsibilities and leadership.	Effective resource management Lingages with other organizations Influence in policy decisions or creation of alternative service
Community	Providing equal access to resources. Allowing expression of diverse opinion. Building participatory structures in community institutions.	Organizations working together to exert control over policy decisions Collective efforts to maintain or improve quality of live Residents' participatory skills

Ecosystem Services and Biodiversity Conservation as a Leverage for Sustainable Management and Development

Ecosystem services have been defined as the benefits that humans take from ecosystems and that jointly contribute to human well-being (Costanza et al. 1997; Daily 1997). Considered as an ensemble of direct and indirect benefits, ecosystem services are now a key topic in discussions on environmental policy, land management and sustainable development (Daily et al. 2009). Results from assessments on several scales, from global to local, stressing the fact that biodiversity and ecosystem services are declining at dramatic rates, have also contributed to this novel focus on ecosystem functioning and on the impacts of biodiversity loss on the long-term provision of ecosystem services (MEA 2003; Pereira et al. 2009).

In terms of their diversity, ecosystem services have been categorized into four main groups (MEA 2003; Pereira et al. 2009). Provisioning services are those related to the production of goods, usually with market value, such as food, water, fibre, wood and timber. Regulating services are those related to prevention or mitigation of catastrophic or otherwise harmful events and their effects, including climate regulation, flood or drought prevention, wildfire prevention, pest or disease control, etc. Cultural services are the benefits related to recreation, tourism, culture, science and education activities valued by society.

The current and long-term provision of all these services is dependent on a set of ecological processes, such as photosynthesis, water and nutrient cycle regulation, soil formation, and habitat for biodiversity, which are considered a fourth category of services provided by ecosystems (Supporting services; MEA 2003; Pereira et al. 2009). While provision, regulating and cultural services are related to more immediate uses and benefits from ecosystems, supporting services deal with the fundamental processes behind services, and so they are most relevant for discussions and decisions on sustainability of ecosystem functioning and of ecosystem services.

Biodiversity is a key element of ecosystems, determining the structure and functioning of ecosystems in a wide range of environmental conditions. Plants, animals, fungi, lichens, algae and bacteria are key players in most ecological processes. Biodiversity is therefore crucial to feed and maintain the provision of ecosystem services. This has been a major breakthrough in discussions between scientists, politicians and other stakeholders concerning biodiversity and nature conservation, since it has shifted the focus from pure ethical reasons for preserving nature to more utilitarian, anthropocentric, and therefore societally more relevant, arguments in favour or biodiversity and ecosystem integrity.

Ecosystem services thus provide an overarching conceptual framework, encompassing many areas and topics of societal interest and relevant for policy and land management. In this regard, ecosystem services have been used to formalize alternative options on land management for political decision and strategic planning, as well as to analyze trade-offs among such options (Daily et al. 2009). As a socio-ecological theoretical framework, ecosystem services also provide a powerful communication tool to address sustainability, nature conservation and governance with policy makers, stakeholders and the general public.

Management among Governance – New Challenges for the Territory Management Authorities

We are witnessing the emergence of a new model of public administration, called new public or estate service partner of the citizen, as the outcome role played by the government (in its various levels of practice) as one of the actors in this scenario of governance.

This new public service model emerges from the traditional one (contraposition of laws and administrative procedures specifically to values and people, separation of politicians and public administrators, affirming themselves by the rationality of the "administrative men") and evolves from it by integrating subsequent positive aspects. On the dichotomy between good public management and good democracy, this model chooses the last one although not neglecting good management (adapt. Bilhim 2004).

> "This new vision of public administration known as 'new public service' presents a set of lessons that challenge traditional practices:
>
> - Serve citizens and not clients
> - Search for public interest
> - Valuing citizenship and not the entrepreneurial capacity
> - Think strategically and act democratically
> - Accountability is not easy
> - Serve more than to order
> - Valuing people not only productivity
>
> Consequently, local governance must focus on this new context marked by key concepts such as decentralization, deregulation, delegation, competitiveness, market, reinvention, life and service quality (measured by the degree of satisfaction of the citizens/clients and collaborators), digital platforms (information management), management centred on competence and on people. There are no prêt-a-porter models. There are sources of inspiration that each local elector and each leader of local Administration will have to cook its own way" (Denhardt s/d).

One of the purposes of governance is the sharing of power and the approach of citizens to the political act. In reference to the Arnstein model (1969) (fig. 1) that gives us a better understanding of the involvement levels of the citizens concerning public participation or their political practice, we find that it is precisely from partnership (level 6 on the upward ladder) that citizens can reach a major degree of power and where power through negotiation is distributed. The share of accountability on decision-making is also a fact.

Above this level, only power delegation and the citizens' control can legitimize major degrees of power, where higher responsibility or more specific tasks can be taken and also, as a consequence, a major governance level can be seen.

The public-private partnership is nowadays presented as the new paradigm of public management. However "to be efficient it requires simultaneously a long term strategic vision and management and technical skills from the local municipalities" (CEE 2006).

The architecture of the participative model in territorial management is complex, and puts new challenges to the supervisory institutions and organizations. They face a range of theoretical and practical issues that, at the current time, can disrupt its full application and operationalization. It is necessary to find ways to adjust to each territory and also to adjust skills and acquire others which will take some time!

Also methodologies and territory planning, in their task of elaborating the various territorial management tools (Plans), will themselves have to be redefined. They can no longer remain at the technocratic and political spheres of centralized government culture and limit active involvement of civil society in decision-making.

Fig. 1: Ladder of participation in citizen involvement in planning processes in United States, Arnstein (1969)

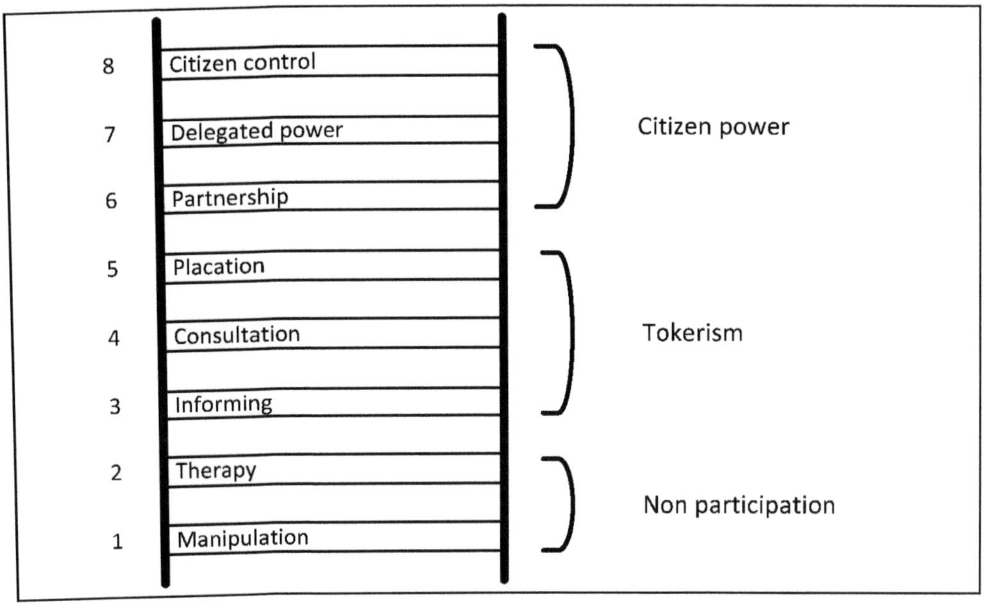

Literature states that the "Paper Plans" generation, despite its technical validity and the pertinence of the followed strategies, as main instruments of territorial management has encountered lots of obstacles to its implementation (the divorce between these entities and all the ones that use, occupy and transform the territory in question) and to the satisfaction of the needs in terms of sustainable development.

This is why effective participation of civil society is confessedly one of the most important elements to the implementation of sustainable development from the territorial perspective and is irrefutably the direction to be followed!

We still stand on experimental ground when it comes to adjusting the model to local specificities in order to assure that it will serve the specific needs of each region or municipality.

There is however a certainty regarding what the characteristics must be of the governance model concerning its form and purposes, and that it must be separated from the electoral political processes (and why not tell it!).

The territorial management models of governance, aiming at the achievement of local sustainable development, cannot be complacent with the possible changes on the party governance of the institutions. The model's goals must themselves be strategic and transversal and aim at the long term and, on the other hand, meet the "new public service".

Serra da Aboboreira – Sustainable Planning Instruments and Biodiversity Conservation

Serra da Aboboreira – Ecosystem Services and Biodiversity

"Serra da Aboboreira" is a small mountain located at the westernmost end of the Alvão-Marão mountain range in Northern Portugal (fig. 2). It is dominated by granitic bedrock and acidic soils, reaching a maximum elevation of around 900m. The climate is temperate submediterranean, with two dry summer months and a strong oceanic influence. Nonetheless, snowfall is common at the peak of winter, and in summer maximum temperatures may reach over 35°C.

Fig. 2: Geographic location of the "Serra da Aboboreira" at the westernmost end of the Alvão-Marão mountain range in Northern Portugal (left), and dominant land cover types (right)

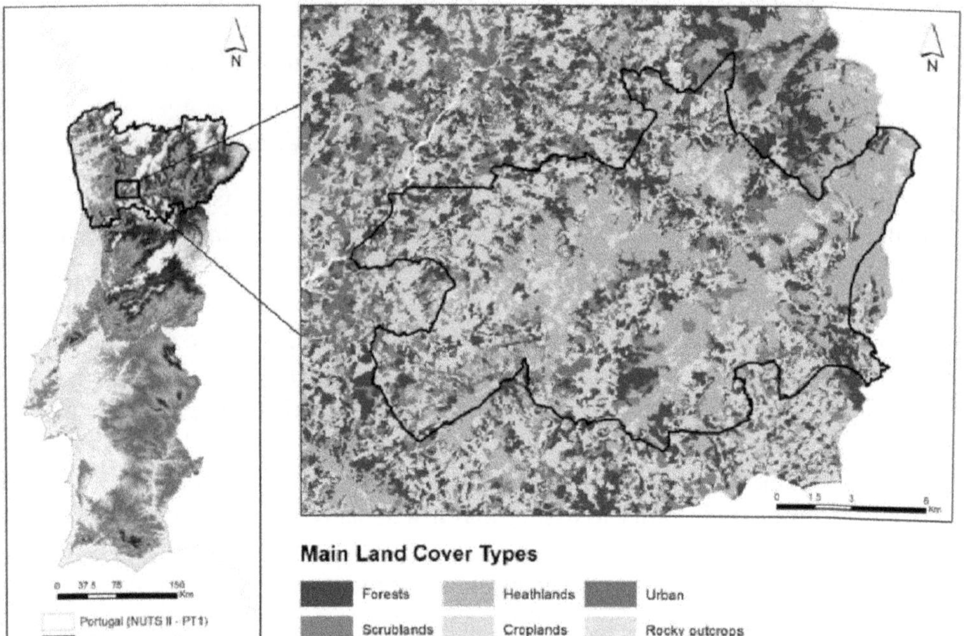

Biogeographically, the Aboboreira is located at the southwest end of the Eurosiberian (Atlantic) region, close to the transition towards the Mediterranean region of Europe. Climax forest vegetation is dominated by Pedunculate-oak (*Quercus robur*), joined by Pyrenean-oak (*Quercus pyrenaica*) in mountain areas and by Cork-oak (*Quercus suber*) in lowlands. These oak woodlands are the most widespread and conservation-relevant native ecosystem type in the area, but heathlands and *Cytisus* scrublands are also common in summit areas where wildfires and soil erosion are more common.

The most significant ecological feature of the Aboboreira is the occurrence of highly diverse, small-scale landscapes, in which crop fields, meadows, scrublands and woodlands are arranged in close mosaics (fig. 5.1 and 5.2). These small-scale landscapes are known for their high nature value, related to high levels of biodiversity and to the provision of valuable ecosystem services (Santos 2009; Honrado et al. 2010). Relevant ecosystem services provided by these landscapes include provisioning services (mostly from agriculture), supporting and regulating services (from soil formation and protection to natural hazard mitigation), and also cultural services (related to cultural value, recreation, tourism, science and education).

Fig. 3: An example of the high nature value, small-scale landscapes typical of the region

The most important activities and land uses that shaped these landscapes throughout centuries were agriculture (both annual and perennial crops) and cattle raising (usually extensive grazing). However, such landscapes are changing rapidly due to severe demographic declines and ageing of the rural population. There is thus a difficult and urgent challenge of making these socio-ecological systems economically and socially viable again, without affecting their intrinsic natural value. The "ecosystem service" paradigm may prove to be a valuable asset and tool to address the future of the "Serra da Aboboreira" within a sustainability framework. Nonetheless, adequate mechanisms are still missing to deliver operational and socially viable opportunities for human welfare based on managing for ecosystem services in these multi-functional landscapes.

Serra da Aboboreira Partnership Plan – a Sustainable Development Management Proposal

Due to Serra da Aboboreira's geographical and economical features, the investment in the territorial development should be based on promoting a strategy of economical and patrimonial goods that fight against the more structural issues such as the abandonment of villages and of their traditional agricultural systems, and being complementary to the conservation and preservation strategies of those ecosystem services and of the Serra da Aboboreira biodiversity.

This is thought to be achieved through the creation of a Partnership Plan that would allow the accomplishment of this mission by the settled commitment between all interested parties, from the Central and Local Administration, several entities, companies, institutions, citizens, tourists, to various others whose will and citizenship feeling would convey in promoting local development.

In order that the Partnership Plan for Serra da Aboboreira (PPSA) could respond to all these challenges, it must be structured around 3 pillars (fig. 4):

- 'Collaborative Governance', which includes aspects of participation, collaboration and decision, ensuring the efficient and active involvement of the stakeholders or key actors;
- 'Active Citizenship' that aims to establish awareness, consciousness and training of citizens and organizations, i.e. its capacity-building and empowerment;
- 'Sustainable Management' that promotes the preservation and valorization of the ecosystems services and biodiversity of the ridged areas.

Fig. 4: Partnership Plan for Serra da Aboboreira (PPSA) – collaborative governance model towards sustainable development

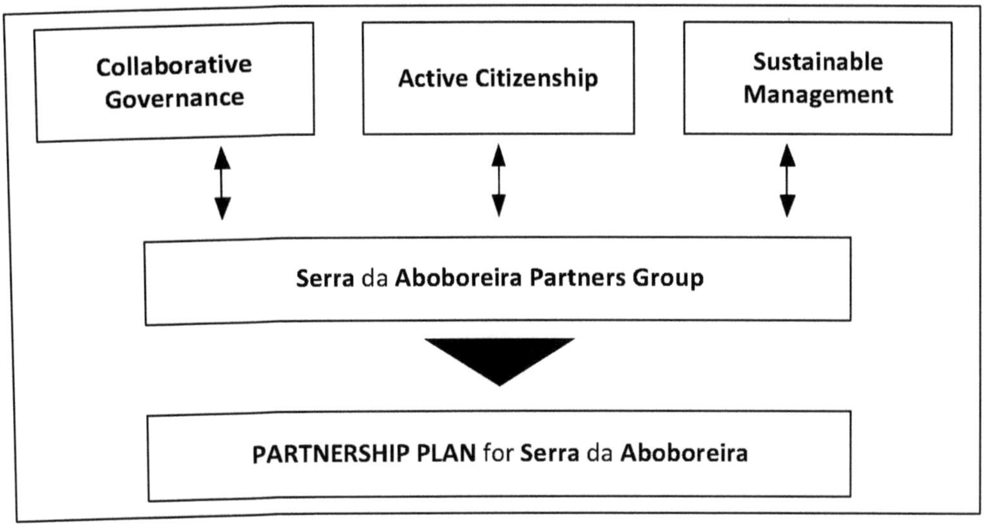

The Partnership Plan thus intends to put together the territorial management (aiming at local sustainable development that will respond to the Serra da Aboboreira challenges) with a strong sense of co-responsibility from several actors on choosing the path towards future development, transforming the territory into a shared space of rights and duties where active citizenship and deliberative democracy encounter a common denominator. It also has the purpose of creating a governance model that articulates the *bottom-up* and *top-down* styles, where all citizens side-by-side with the territory supervisory administrative authorities become partners and are involved in an active way in several issues facing territorial planning – power redistribution.

Afterwards, the proposed Partnership Plan has, as a main goal, the creation of a management model that: (1) aims at joint management between public and private sectors and mobilizes all economical sectors based on an integrated vision of the territory, one of its actions being the territory's promotion (transversal action); (2) integrates the real needs of the various stakeholders; (3) makes the power closer to the citizens through *bottom-up* decision-making processes, (4) achieves an effective and efficient territorial management; (5) fosters local identity, and; (6) promotes local sustainable development.

In terms of procedures and formalities, an elaboration of this Partnership Plan includes all work from the launch of the project with the formation of the Serra da Aboboreira Partners Group (S-APG) until the phase of the elaboration of the project/plan and its implementation, maintenance and conservation.

In consequence, all entities (governmental, local authorities, associations of all nature, promoting entities of leisure and tourism, etc.) with special emphasis on the population, will be called to actively participate in all these works and stages of the process, paying their contribution to what will be definite and finally the consensual gained result.

The municipalities of Amarante, Baião and Marco de Canaveses, or in its representation the *Associação de Municípios do Baixo-Tâmega* (AMBT) [Municipality Association of the Low-Tâmega – MALT], will be another member in this management model – PPSA, among a wider group of partners that will constitute the managing entity for Serra da Aboboreira, having in mind the governance model. However, it must be underlined that as partners the functions and responsibilities of the local authorities are not eliminated or invalidated, they are enhanced. The PPSA is launched with a presentation of the project by the AMBT whose mission is the preservation and valorization of the ecosystems services and biodiversity of Serra da Aboboreira.

The fact that the AMBT is the one to launch the process will reinforce the unicity of the project. On the other hand, assuming that the project results come from a political will and its mayor's initiative, it is the first step to a commitment inside the partnership to be proposed. Nevertheless, the political aspect must itself be very well clarified among residents, namely in terms of its objectives and true intention, what is planned to be achieved with the Project of Valorization of the Ridge Mountain. All rhetorical questions will have to be clearly answered. We stress the mandatory principle of transparency throughout the whole process!

The entire construction process of the partnership management model – Phases 1 and 2 – is based on interactive active participative methodologies such as workshops, minutely designed by the facilitator team conducting the process. After launching the project, the AMBT would participate in all the ongoing works in the same way and at the same level as the other partners.

The plan's implementation will be a *continuum* during all the settled period (20 to 25 years) although every 4 years a follow-up must be done regarding the achievement of the specific and general goals/objectives, the approved budget, actions taken and the results. The outcome of this analysis will end up in new documents containing revised and updated goals to be achieved in the following period (next 4 years), as well as the definition of the projects and actions to be taken.

The fact that the model plan for sustainable management is based on those pillars of collaborative governance and active citizenship only allows us, in theoretical terms, to draw its architecture just up to the phase of the plan elaboration and its respective commitment statement – Phase 2. Understandably, regarding questions related to functions and competences, in future allocated to the S-APG, and also because of its tie to the process, it is not the model's role to define objectives, strategies and/or action plans or activities to be developed by the project.

The one and only objective previously settled is that 'The Partnership Plan' will become the desired sustainable management tool, and that will accomplish the sustainable development purposes.

Partnership Plan – how to Implement

The construction of the Serra da Aboboreira governance model, that will support/frame the Partnership Plan, demands a set of phases (fig. 5) to be taken, beginning with the project/idea launching – collaborative management of the Serra (Ridge Mountain), up to the formation of the Partners' Group which will be in charge of the strategy to be implemented for the Serra da Aboboreira sustainable development.

The Plan's contents (future vision of the Serra da Aboboreira, definition of goals and/or actions proposals), as already mentioned, will be the focus of specific work between the partners that will articulate the problems, visions, interests and intentions together. There is, though, certainty about the arduous task and obstacles to overcome faced by the Partners' Group due to the complex territory in question.

The elaboration of the Partnership Plan has three fundamental phases to be followed:

PRE-PHASE – corresponds to the launch of the project/idea and has the purpose of preservation and conservation of the ecosystems services and biodiversity that characterize Serra da Aboboreira.

This will be the starting point to diffuse, motivate and begin to involve the future partners in the project and also provide the means to arouse interest and extras of the various actors.

The presentation format will have to be carefully planned: (1) the main goal to be achieved; (2) the characteristics and contexts of the various target publics and; (3) the instruments and adequate means to make the communication process more effective and fruitful.

Emphasis must be given to the importance of the adopted interactive formats that will promote the empowerment and capacity-building of the participants (to form "full body" citizens). It is acknowledged in literature that decision-making processes held in complex and uncertain contexts must involve the stakeholders in the process right from the very beginning. It is also a fact that these processes are collective apprenticeship ones and so the participative process must start here.

PHASE 1 – The listening and negotiation processes among the stakeholders (direct and indirect ones) will be held at this stage by gathering of information.

As a result of literature on this matter, the use of semi-structured interviews is very common at this stage. Nevertheless, the resources to be adopted will have to be defined by the expert team that will be in charge of the whole process.

Fig. 5: Output of each phase of the interactive participative process – from the process/project start-up to the partnership-building

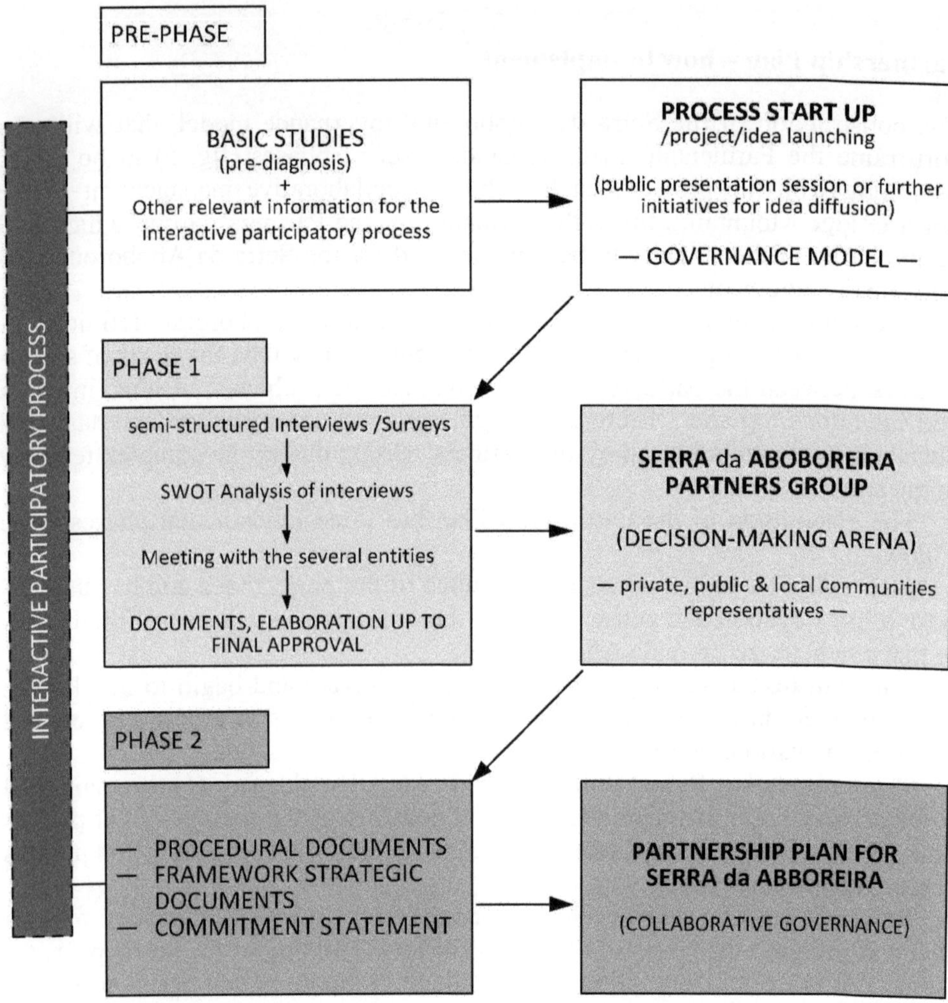

The analysis of the actors for the partners' group formation is crucial and therefore it is fundamental to proceed to their identification, to acknowledge the property ownership situation (owner, renter, others), what functions do they have (producers/

consumers, politicians/decision-makers/technicians), main problems experienced, their positioning towards the project, what are the interests, among other parameters.

A first draft is thus elaborated where all the data is processed and systematized in order to be presented to the tutelary entities. Their feedback will be given to the other actors for discussion.

The same procedure will be followed and repeated up to the moment the partners become totally and completely clarified about what is expected for the future Partnership Plan. As a result, this phase could be extended even more within the process but, even so, giving up time will be important, considering its importance and relevance.

Only then will a final document be made that will contain the partners' proposal, the one that will be in the process. This document, though, will have to be ratified by the tutelary entity(ies).

PHASE 2 – Once the definition of the partners (public and private representatives and local communities' ones) and the starting point have been made, [meaning, the built scenario made through the analysis of the listening and conversation process between the partners (previous phase)], it is now the moment to work on the plan's architecture.

It is from inside this decision-making arena (S-APG) that the joint vision for Serra da Aboboreira will come out (e.g. until 2030 and beyond), as well as the short, mid and long-terms goals to be achieved, the outline structural axis, and where all the action plan (pillars, measures and actions) to be implemented will be thought out.

Important, too, is the assignment of functions and roles to each partner representative in the field, who will be in charge of the implementation and follow-up of the defined management plan, so that they are aware of the actions they will be expected to make.

The monitoring issues (e. g. reports) and the plan revision as well as the establishment of their periodicity will have to de indicated in the procedural documents elaborated in the sphere of the S-APG. The procedural documents' answer to the questions *What*, *How*, *When* and *Why*.

The documents to be produced by the partners' group include the framework strategic documents that refer to the politics, plans, maps and orientation principles applicable to Serra da Aboboreira, which will set the management plan in motion. These documents will be the output of the interaction with current sectoral politics (accomplishment of the integration principle), legal stipulations, among other strategic documents regarding territorial planning. But, above all, they will be the result of the discussion and reflection of the decision-making arena together with the regionally posed challenges on the way to the territorial sustainable development.

Other equally important aspects to be debated and outlined in the process are:

- definition of the key principles (e.g. openness, inclusivity, integrity and accountability principles, among others) that will characterize the governance model;
- the establishment of clear interaction rules for the meetings with the key actors;
- the forms the decision-making must have within the partner's group (by deliberation, voting or other);
- public or local communities' participation (accessibility and transparency principle) in the elaboration of the proposals;
- among several other rules with relevance to the future partnership context, and that will be compiled in one proper single document – governance codes, the Commitment Statement.

The marketing and communication strategies will be part of the project's current management and will have to be in accordance with ongoing achievements. The sharing of a common goal is reflected in all efforts undertaken towards success.

The collaborative management model – Partnership Plan – must guarantee the continuity in time, so that it can be an effective and efficient instrument, as well as serving the purpose of sustainable development. It must not, in any way, comply with the electoral and political agenda. As such, the model must be recognized as an instrument of higher level, of transversality to various politics and strategies of territorial planning and of the preservation of the natural and landscape heritage of Serra da Aboboreira.

To be clearer about the several previously described procedures, it must be noted that the model construction process itself must be separate from the project/idea. This means that the leader team will assume the role of organization and orientation of the whole interactive participatory process, allowing the empowerment, capacity-building and legitimizing of key actors' decisions. The Partners' Group will take decisions regarding the projects (fig. 6).

Fig. 6: Technical Team process leading – orientation of the interactive participatory process

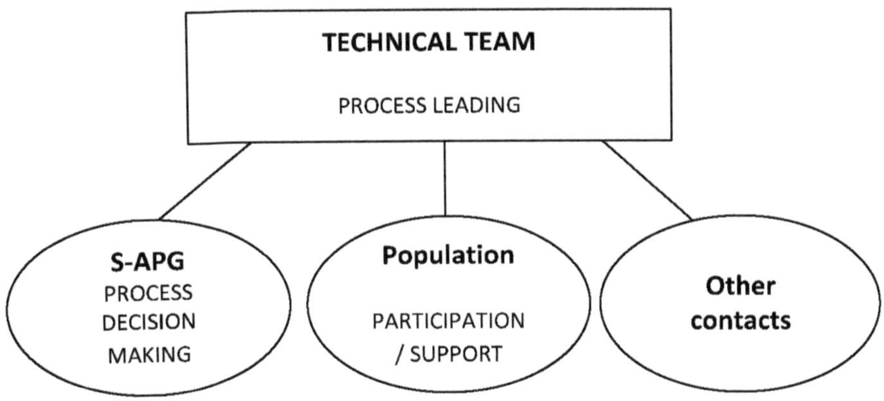

Among the various tasks, the technical team will be responsible for making the interviews with the various actors and to make the correspondent SWOT analysis, the data processing throughout the process, the organization of workshops on specific and/or broader scope themes, or other interactive formats chosen accordingly to the expected goals, to meet with the partners and simultaneous work with local population (participation and support), also developing other contacts with key entities from the local community and/or public administration, every time needed.

Legal Framework

If we search in our current legal framework, it will be quite impossible to find an appropriate proposed territorial management model – Partnership Plan! This new partnership concept will require innovative legislation consistent with these new governance model precepts.

Although there is already some evolution in legislation, on the way towards the change of management and sustainable development paradigm the legal rules remain based on a partnership vision of the project's viability.

Nevertheless, we must mention that there are some regulations among current legislation that can serve these purposes and consequently complement a new legal and judicial framework of that same paradigm.

The model itself is complex, and it will require the gathering of legislation concerning territorial management instruments – regional municipal, conservation of nature and natural resources, of the principles of governance, citizenship and of sustainable development.

The foundations that define environmental policy within Portuguese law are found in the Lei nº 11/87, de 7 Abril – Lei de Bases do Ambiente (Law nº11/87, from 7 April – Law on the Environment), and it is under this umbrella that the various instruments of environment policy and territorial planning are defined [article 27, Chapter IV]. However, this legal document does not include the management by partnership as an instrument of territorial management.

What becomes very clear from article 40º of the Chapter VII of this Law, and matches the goals of sustainable development, is the citizens' together with public and private institutions' duty of collaboration on the creation of an ecologically balanced environment, as well as the improvement of life standard of living (point 1); the promotion of participation by private entities in initiatives of declared interest in the pursuit of the purposes stipulated in the current law, namely the national and local environmental protection associations, of the natural and built heritage (point 3); attributing the municipalities with a special role on these promotional actions.

More recently, and further to the matter of nature conservation, stems the D.L. nº142/2008, de 24 Julho [Act of Law nº142/2008, from 24 July] which stipulates the legal framework for nature conservation and biodiversity (MAOTDR 2008) and within which the proposed model – SAPP – is likely to be strongly based and framed.

However, this possibility can only be viable in case the model concerned becomes part of the Sistema Nacional de Áreas Protegidas (SNAP) [National System of Classified Areas] [NSCA]. This means that it will demand the categorization and framing of Serra da Aboboreira in one of the typologies of protected areas so that it is included under the protection of the correspondent economical and financial regime (article 35º/2, Chapter V).

Regarding this issue, the proposal for the future judicial and legal framework for Serra da Aboboreira leaves open the possible classification status, e.g. Protected Landscape, once it is considered that this decision, on the one hand, will be the entire responsibility of the Partners' Group (GPS-A in Portuguese) and, on the other hand, must comply with the vision for the Serra da Aboboreira and with the proposal/approval of what will be defined by the partners themselves.

It is thought that the process for the status attribution should emerge from the collective will of the S-APG (*bottom-up*) and that it will reflect the clear intensions and properly defined goals regarding the path to be followed to reach the Serra da Aboboreira sustainability.

At the level of territorial management itself, the current regulatory framework is made through the Decreto-Lei nº380/1999, 22 Setembro (Act of Law nº380/1999, 22 September) [establishment of the applicable regime for the territorial management instruments] and it provides, as a management instrument, a hierarchy of plans whose methodologies of elaboration are deeply referred to in the technical and political arenas.

Regarding public participation (article 6º – Right to Participate)even with the passivity of the provided formats, the law is clear and its particularities must be underlined in the context of governance. Following on from this, the reference to participation directs to three aspects, namely:

- citizenship practice – "citizens and associations have the right to participate in the elaboration, changing, revision, execution and evaluation of the territorial management instruments", in national, regional and municipal contexts;
- duty of the territorial administrative institutions – "the responsible public entities have the duty to balance the presented proposals" (consultation and public discussion);
- be open and inclusive – "All citizens ...."

Still, in the scope of the territorial management system, the DL nº316/2007, de 19 Setembro (Act of Law nº316/2007, from 19 September) introduces an update to the DL nº380/99, de 22 Setembro (Act of Law nº380/99, from 22 September), related to the information access (transparency and publication) aiming at public participation. Thus, the various competent institutions/public bodies regarding the elaboration of the plans must spread the information and provide the documents via the Internet (MAOTDR 2007).

The issue of public participation gains even more predominance and expression with the Agenda 21, considered to be the proper environmental management instrument for the promotion of local sustainability.

It basically defines itself as being the local expression and "an instrument to the sustainability management of a place, starting from the diagnosis of the current/ reference situation and establishing goals to be achieved in terms of environmental protection, socio-economical development and social cohesion, developed by the local actors together with the citizens and civil society" (APA, www.apambiente.pt), being an accountable municipalities' task.

In fact, it is the instrument that aims in practice to involve and mobilize the citizen, legitimizing his political act and social action. Nevertheless, we are too far away from the paradigm of the defined achievements of the proposed management model – an effective governance of the shared powers (*top-down* and *bottom-up* in articulation).

> "*While the concept of participation directs to a linear scheme of power practice, the partnership one opens doors to practices based on the principle of interaction between populations, technicians, decision-makers and administrators to whom the references are not so much the power (its conquest, share and conservation) but the action itself, meaning that the*

effective resolution of the problems are consensually identified because they are worked within the communication mechanisms between the partners" (Guerra 2007: 101).

As such, we consider this instrument (Local Agenda 21) as the beginning of the main course towards participative democracy, opening an even wider space for the exercise of the citizen double responsibility – right and duty to participate enshrined on the Portuguese Republic Constitution.

There is still hope that with the revision of the Estratégia Nacional de Conservação da Natureza e da Biodiversidade (MARN 2001) [National Strategy for the Nature and Biodiversity Conservation] that ran until 2010, the possibility of management by partnership will become integrated, and that partnership is taken as being on the active participation level, the 6th level of the Arnstein ladder, defining its contours in detail and unequivocally.

References

APA (Agência Portuguesa do Ambiente). Instrumentos de Gestão Ambiental. Ministério do Ambiente e do Ordenamento do território (www.apambiente.pt – acedido em Outubro 2010).

Arnstein, S.R. (1969), A Ladder of Citizen Participation, JAIP, vol. 35, no. 4, July 1969, pp. 216-224.

Bilhim, J. (2004), A governação nas autarquias locais, SPI – Sociedade Portuguesa de Inovação, Porto.

CCE (COMISSÃO DAS COMUNIDADES EUROPEIAS) (2006), A Política de Coesão e as cidades: Contribuição das cidades e das aglomerações para o crescimento e o emprego nas regiões, COMUNICAÇÃO DA COMISSÃO AO CONSELHO E AO PARLAMENTO EUROPEU, COM (2006) 385 final, {SEC(2006) 928} Bruxelas.

Costanza, R., d'Arge, R., Groot, R. d., Farber, S., Grasso, M., Hannon, B., Limburg, K., Naeem, S., O'Neill, R.V., Paruelo, J., Raskin, R.G., Sutton, P. and Belt, M.v.d. (1997), The value of the world's ecosystem services and natural capital. Nature 387, 253-260.

Daily, G.C. (1997), Nature's Services: Societal Dependence on Natural Ecosystems. Island Press, Washington, DC.

Daily, G.C., Polasky, S., Goldstein, J., Kareiva, P.M., Mooney, H.A., Pejchar, L., Ricketts, T.H., Salzman, J. and Shallenberger, R. (2009), Ecosystem services in decision making: time to deliver. Front Ecol Environ 7, 21-28.

Guerra, Isabel C. (2006), Participação e Acção Colectiva – Interesses, Conflitos e Consensos, Princípia Editora, Lda, Estoril, 1ª edição.

– (2007), Fundamentos e Processos de Uma Sociologia de Acção – O Planeamento em Ciências Sociais, Princípia Editora, Lda, Estoril, 2ª edição revista e aumentada.

Honrado, J. (coord.) (2009), O património natural e cultural como factor de desenvolvimento e competitividade territoriais no Baixo Tâmega. Relatório final (1ª fase), Faculdade de Ciências da Universidade do Porto.

Honrado, J.P. (coord.) (2010), Património natural das Serras da Aboboreira, do Castelo e do Marão: Perspectivas de conservação e valorização. Final report of project "O património natural como factor de desenvolvimento e competitividade territoriais no Baixo Tâmega". Faculdade de Ciências da Universidade do Porto.

MAOTDR (2007), Regime Jurídico da Avaliação Ambiental de Planos e Programas (aplicação no âmbito do sistema de gestão territorial)

– (2008), Regime Jurídico da Conservação da Natureza e da Biodiversidade.

MARN (2001), Estratégia Nacional de Conservação da Natureza e da Biodiversidade (ENCNB) (2001-2010).

MEA (Millennium Ecosystem Assessment) (2003), Ecosystems and Human Well-Being. A Framework for Assessment. Island Press.

Menezes, I. (2007), Intervenção Comunitária – Uma perspectiva psicológica, Legis Editora, 1ª Edição.

MNE; ME (2009), Estratégia Nacional de Educação para o Desenvolvimento (ENED) (2010-2015).

Ninacs, W. (2002), Le Pouvoir dans la participation au développement Local. In Tremblay, M., Tremblay, P.A., Tremblay, S. (2002), Développement Local, Économie Social et Démocratie. Presses de L'Université di Québec, pp. 15-39.

Pereira, H.M., Domingos, T., Vicente, L. and Proença, V. (eds.) (2009), Ecossistemas e Bem-Estar Humano. Avaliação para Portugal do Millennium Ecosystem Assessment. Escolar Editora. 734 pp.

Santos, C.M.C. (2009), Integrating biodiversity monitoring and landscape monitoring in high nature value farmlands. Master thesis, Faculdade de Ciências da Universidade do Porto.

Vasconcelos, Lia T. (2003), O Exercício da Cidadania pela Participação Popular. In Associação para a Gestão do Forum Social Português 2003 (org.). Um Outro Mundo É Possível … Mas Não Será Fácil, pp. 21-27. Edições Colibri, Lisboa 2006 (ISBN13:978-972-772-663-9).

Zimmerman, Marc A. (1995), Empowerment Theory for Rehabilitation Research: Conceptual and Methodological Issues, University of Michigan

About the Authors

Ana Sofia Almeida, is presently a Secondary School teacher in Portugal. She has concluded her Master thesis in 2006, as a researcher in CESAM (Centre for Environmental and Marine Studies, University of Aveiro). Her lines of interest are connected to Environmental Education and Ecology. As a teacher, she has always dedicated her attention to Educational subjects, concerning a connection between school curricula and investigative research.

CESAM – Centre for Environmental and Marine Studies
Department of Biology
University of Aveiro
3810-193 Aveiro
Portugal
Contact e-mail: anasofiaalmeida@msn.com

Maria Filomena Madeira Ferreira Amador, holds a PhD on Philosophy and Science Education from the Universidad Complutense (Madrid – Spain) (1994), and a BSc in Geology from the Universidade de Lisboa (1978). She is currently Assistant Professor with Aggregation in the Department of Science and Technology at Universidade Aberta and a researcher at the Centro de Geologia from the University of Porto, Portugal. Her main research and teaching areas include environmental education, science education, history and philosophy of science, and e-learning. She is now the vice-coordinator of of the Doctoral program in Social Sustainability and Development.

Departamento de Ciências e Tecnologia
Universidade Aberta
Rua Escola Politécnica nº 141-147
1269-001 Lisbon
Portugal
Contact e-mail: famador@univ-ab.pt

Ulisses Miranda Azeiteiro, was educated at the Aveiro University, Portugal, from which he received his Master Science Degree in 1994, Coimbra University, Portugal, where he received his Doctorate (PhD) in Ecology in 1999 and Universidade Aberta, Portugal, where he received his Aggregation in Biology in 2006. He holds an Assistant Professor teaching position at the Universidade Aberta (Assistant Professor with Aggregation and Tenure),

Portugal, in the Department of Exact Sciences and Technology, where he teaches *Environmental Education*, *Biology*, *Biodiversity* and *Research Methodologies*. He is Senior Researcher at the *Centre for Functional Ecology* of Coimbra University where he is the Coordinator of the Group *Ecology and Society* and Vice-Coordinator of the Group *Ecology of Marine and Coastal Ecosystems*.

Universidade Aberta
Rua do Ameal 752
4200-055 Porto
Portugal
Contact e-mail: ulisses@univ-ab.pt

Margarida Bacelar, got her degree on Marine Biology, in 2008, at the University of the Azores. Afterwards, she started working in research projects on aquaculture, specifically on animal nutrition. She is currently undertaking her MSc in Consumer Science and Nutrition, at the University of Porto.

REQUIMTE
University of Porto
Porto
Portugal

DGAOT
Faculty of Sciences
University of Porto
Porto
Portugal
Contact e-mail: margarida.bacelar@gmail.com

Fernando Caetano has a *BSc* in Technological Chemistry from the University of Lisbon, Portugal and made his Chemical Engineering *MSc* and *PhD* at the Higher Technical Institute from the Technical University of Lisbon (IST-UTL), Portugal, where he keeps part of his research activities, at Centro de Química Estrutural, in thermophysical properties of liquids. Currently is an Assistant Professor at Universidade Aberta, the public distance learning university in Portugal, where he teaches chemistry and technological related topics in Environmental Sciences courses from the 1st to the 3rd cycles (PhD doctoral program) and his main interests are teaching of environmental and experimental sciences in e-learning.

Departamento de Ciências e Tecnologia
Universidade Aberta
Rua Escola Politécnica nº 141-147
1269-001 Lisbon
Portugal
Contact e-mail: fcaetano@univ-ab.pt

Ana Capelo, PhD is undertaking a post doctoral programme in science education in the CESAM/University of Aveiro, Portugal. Her research interests include education for sustainable development and science education. She also has interest in biotechnology and biology of stress.

Department of Biology & CESAM
Laboratory de Biotecnology & Cytomics
University of Aveiro
Campus Universitário de Santiago
3810-193 Aveiro
Portugal
Phone: +351 234370766
Contact e-mail: anacapelo@ua.pt

Sandra Sofia Ferreira da Silva Caeiro, holds a *PhD* on Environmental Engineering from the New University of Lisbon (2004), a *MSc* in Sciences of Coastal Zones from the University of Aveiro (1996) and a *BSc* in Environmental Engineering from the New University of Lisbon (1992). She is currently Assistant Professor in the Department of Science and Technology at Universidade Aberta and a researcher at Institute of Marine and Environmental Research and at the Education Development Research Unit, Portugal. Her main research and teaching areas include environmental risk assessment, environmental management tools, environmental citizenship and participation, sustainable consumption and e-learning in science education. She is now the coordinator of a B.Sc. programme on Environmental Sciences.

Instituto do Mar e do Ambiente CMA/IMAR
Coimbra
Portugal

Universidade Aberta
Portugal
Contact e-mail: scaeiro@univ-ab.pt

Sónia Costa, his presently a PhD student at University of Coimbra and a secondary school teacher in Portugal. Her main area of research includes the effects of contaminants in estuarine species. Parallel, she is interested in Sustainable Development and Environmental Education.

CFE – Centre for Functional Ecology
Department of Life Sciences
University of Coimbra
P.O.box 3046
3001-401 Coimbra
Portugal
Contact e-mail: soniacosta@ci.uc.pt

Luís Miguel Cunha, is a Food Engineer with a PhD on Food Science and Technology from ESB-UCP. He is an Associate Professor (w/tenure) at the Faculty of Sciences, University of Porto, were he is the Director of the MSc course degree in Consumer Science and Nutrition. He is a researcher at the Food Quality and Safety research group from the Associated Laboratory REQUIMTE. His research focuses on minimally processed produce, on consumer attitudes towards food safety and quality and on consumer oriented sensory evaluation of foods.

REQUIMTE
University of Porto
Porto
Portugal
DGAOT
Faculty of Sciences
University of Porto
Rua do Campo Alegre s/n
4169-007 Porto
Portugal
Contact e-mail: lmcunha@fc.up.pt

Daniela de Figueiredo, was graduated in Biology at the University of Aveiro, Portugal, in 1999. In 2000, she completed a post-graduation in Applied Ecology at Faculty of Sciences of University of Porto, Portugal, and obtained a master degree in Molecular Microbiology at University of Aveiro, Portugal, in 2003. In 2010, she concluded the PhD in Biology (on the subject of cyanobacterial occurrence in Portuguese water bodies) at the University of Aveiro, Portugal. At the present time, she is a post-doctoral fellow at the University of Aveiro and she is still developing research on the subject of cyanobacterial blooms. Other research interests include aquatic microbiology and environmental education/scientific communication. The list of publications in ISI journals extends up to sixteen papers as well as four book chapters and thirty eight communications in conferences.

CESAM – Centre for Environmental and Marine Studies
Department of Biology
University of Aveiro
3810-193 Aveiro
Portugal
Contact e-mail: dfigueiredo@ua.pt

Daniel Gil-Pérez, Professor of Science Education at the University of València, Spain. His researches and publications deal with different science teaching/learning problems such as teachers' conceptions about the nature of science, laboratory work, paper and pencil problem-solving or assessment and evaluation. All these researches are linked and aim to contribute to the construction of a coherent body of knowledge in the field of Science Education. Nowadays his main line of research is on Education for Sustainability. He has directed 19 PhD theses in science education and he is directing at the present 2 other theses. He is the author of more than 30 books and has published in international journals such as International Journal of Science Education, Science Education, Studies in Science Education, Bulletin de l'Union des Physiciens, Aster, La Fisica nella Scuola, Enseñanza de las Ciencias, Science & Education, Revista de Enseñanza de la Física, etc.

Contact e-Mail: luis@ua.pt
Web: www.uv.es/gil
e-mail: luís@ua.pt&#8221

Alcide Gonçalves, graduate in Landscape Architecture at Universidade de Évora. Environment Citizenship and Participation Master Degree, Universidade Aberta. Teacher at the Department of Plant Sicences and Environment at Escola Superior Agrária do Instituto Politécnico de Viana do Castelo. Team member at the CFE – Centre for Functional Ecology, Department of Life Sciences, Universidade de Coimbra. On progress PhD on Social Sustainability and Development at Universidade Aberta.

Escola Superior Agrária – Instituto Politécnico de Viana do Castelo
Departamento de Ciências da Planta e do Ambiente
Refóios do Lima
4990-706 | Ponte de Lima
phone: 258 909 740 | 967 288 263 | 967 288 264
Contact e-mail: alcide@esa.ipvc.pt

Ana Marta Gonçalves, earned her degree in Biology at the University of Coimbra, Portugal, in 2004. In 2006 she holds a MSc in Ecology, University of Coimbra. Presently is a doctoral candidate at the University of Coimbra (Centre for Functional Ecology, Department of Life Sciences). Her current research interests are focused on spatio-temporal dynamics and trophic ecology of estuarine and marine (small-sized) zooplankton and the effects of climate variability on the structure of these assemblages. In parallel, she has collaborated in some Environmental Education projects which have yielded several contributions in educational book chapters. She has also developed several works and collaborations in projects within Ecotoxicology and Biology of Stress.

CFE – Centre for Functional Ecology
Department of Life Sciences
University of Coimbra
P.O. box 3046

3001-401 Coimbra
Portugal
Contact e-mail: ammendes@student.zoo.uc.pt

Fernando Gonçalves is presently Associate Professor with Habilitation at the Department of Biology, University of Aveiro, Portugal. Since he was invited to join an academic career (in 1995), he has established his own research team within the Ecotoxicology and Biology of Stress research field and produced more than 115 scientific papers in SCI journals, 4 books and 21 chapters. His research interests lie in the frontier between Ecology and Toxicology, including the ecological aspects of man-induced change. As a teacher, he has always devoted attention to Educational issues, particularly those related to Environmental Education.

Departamento de Biologia
CESAM – Centro de Estudos do Ambiente e do Mar
Universidade de Aveiro
Campus Universitário de Santiago
3810-193 Aveiro
Portugal
http://www.cesam.ua.pt/fernandogoncalves
Contact e-mail: fjmg@ua.pt

João Honrado got his PhD in Biology at the University of Porto in 2003, with a thesis on vegetation ecology. He is currently Auxiliary Professor of ecology and biogeography at the University of Porto, and leads the Biodiversity and Conservation Ecology group at CIBIO-UP, a research centre devoted to the study of all aspects of biodiversity and genetic resources. His research focuses on describing and modelling the spatial patterns of biodiversity at multiple scales, and on forecasting the dynamics of such patterns resulting from environmental changes and management options. In support of land management and governance, he is also interested in developing cost-efficient monitoring schemes to tackle the impact of changes in social-ecological systems on biodiversity, ecosystems and their services.

Faculdade de Ciências da Universidade do Porto
Edifício FC4 (Departamento de Biologia)
Rua do Campo Alegre, s/n
4169-007 Porto
Portugal
Contact phone: +351 220 402 815
Contact email: jhonrado@fc.up.pt

Walter Leal Filho, (BSc, PhD, DPhil, DSc, DL, DEd., DLitt) is a Senior Professor at London Metropolitan University, UK and heads the Research and Transfer Centre "Applications of Life Sciences" of the Hamburg University of Applied Sciences, where he is in charge of various sustainability-related projects across the world

Research and Transfer Centre „Applications of Life Sciences"
Faculty of Life Sciences
Hamburg University of Applied Sciences
Lohbruegger Kirchstraße 65, Sector S4 / Room 0.38
21033 Hamburg, Germany
phone: +49-40-42875-6313, fax: +49-40-42875-6079
contact e-mail: walter.leal@haw-hamburg.de

Marina Prieto Afonso Lencastre, was graduated as a psychologist at the Faculty of Psychology and Education of the University of Louvain-la-Neuve (Belgium) and holds a PhD in psychology at the same University, where she worked on theoretical questions concerning evolutionary epistemology and human behaviour. She is Senior Professor at the department of Education of the Faculty of Psychology and Educational Sciences of the University of Porto and lectured animal and human ethology, environmental ethics and education, bio-psycho-social perspectives on human behaviour and communication. She is currently doing research on evolutionary psychology, evolutionary psycho-pathology and psychodynamics in relation with psychotherapy and is Invited Professor in Psychology at the Catholic University in Portugal (Porto) and Professor at the University of Fernando Pessoa.

Faculdade de Psicologia e Ciências da Educação
Universidade do Porto
Rua do Dr. Manuel Pereira da Silva
4200-392 Porto
Portugal
Contact e-Mail: mar_afonso@hotmail.com

Rui Costa Lima, got his degree on Food Engineering, in 1996, at UCP-ESB. Since then he has been working on sensory analysis, being the one of the co-founders and manager of Sense Test, Lda. the first Portuguese company dedicated to sensory analysis, founded in 2000. His main interests are consumer evaluation of food quality and attitudes towards food consumption.

Sense Test, Lda.
Vila Nova de Gaia
Portugal
Contact e-mail: rcl@sensetest.pt

Betina da Silva Lopes, holds a Master degree in Communication and Education in Science from the University of Aveiro, Portugal. She is currently studying for a PhD in Didactics in association with the Research Centre for Didactics and Technology in Teacher Education (CIDTFF). The PhD project aims to investigate the relationship between Preferential Teaching Approaches and Questioning practices of Biology University teachers.

Department of Education
University of Aveiro
Portugal
Contact e-mail: blopes@ua.pt

Fernanda Luísa Figueiredo Lopes is a teacher in a Basic and Secondary School. She has concluded her Master thesis in 2004 as a researcher in IMAR, Faculty of Science and Technology, University of Coimbra. Her lines of interest are connected to experimental and investigative work in Science teaching and to Environmental Education and Ecology. As a teacher, she has always dedicated her attention to Educational subjects, concerning a connection between school curricula and investigative research.

Marine and Environmental Research Centre (IMAR-CMA) & Department of Life Sciences
University of Coimbra
Apartado 3046
3001-401 Coimbra
Portugal
Contact e-mail: fjmg@ua.pt

Evangelos Manolas was born in Naxos, Greece, in 1961. He has received a Bachelor of Arts in Sociology from the University of Essex (1983), a Master of Arts in International Relations from the University of Kent at Canterbury (1985) and a Philosophy Doctorate from the University of Aberdeen (1989). He is Assistant Professor of Sociology and Environmental-Forest Education in the Department of Forestry and Management of the Environment and Natural Resources at the Democritus University of Thrace.

Department of Forestry and Management of the Environment and Natural Resources
Democritus University of Thrace
193 Pantazidou Street
68200 Orestiada
Greece
Contact e-mail: emanolas@fmenr.duth.gr

Ana Paula Teixeira Martinho holds a *PhD* on Environmental Engineering from the New University of Lisbon (2003), a *MSc* in Sanitary Engineering from the New University of Lisbon (1999) and a *BSc* in Environmental

Engineering from the New University of Lisbon (1993). She is currently Invited Assistant Professor in the Department of Science and Technology at Universidade Aberta and a researcher at the Education Development Research Unit, Portugal. Her main research and teaching areas include environmental impact assessment, integrated waste management, environmental ethics and environmental citizenship and participation, and e-learning in science education. She is now the vice-coordinator of *MSc* in Environmental Citizenship and Participation.

Department of Sciences and Technology
Universidade Aberta
R. Escola Politécnica, 147
1269-001 Lisboa
Portugal
Contact e-mail: aptm@univ-ab.pt

Isabel P. Martins, Portuguese, born in 1948. Full Professor of the University of Aveiro. PhD in Education (Didactics of Sciences); Bachelor in Chemistry; Scientific Coordinator of the Research Centre for Didactics and Technology in Teacher Education; Vice-Rector for Post-Graduation Studies and Scientific Affairs, University of Aveiro, 2004-2010; Supervisor of PhD Thesis and Master Dissertations; Author of the National Chemistry Curriculum for Secondary Schools (Levels 10-11-12); Invited Speaker in National and International Conferences about Science Teaching and Learning and Teacher Education; Author and Co-author of many articles and chapters of books; Chair of National and International Conferences; Principal Researcher of Scientific Projects; Referee of Scientific Journals; Member of National and International Evaluation Panels.

Contact e-mail: imartins@ua.pt

Luís Marques, graduation in Geology at the University of Coimbra. PhD in earth science education at the University of Keele. Retired. Main research areas – teaching and learning in outdoor learning environments, practical work, curriculum materials development in a Science-Technology-Society context. Supervision of several PhD thesis and Pos-doc projects. Coordinator of several national research projects and participation in many others, national and international. Publications in national and international journals.

Contact e-Mail: luis@ua.pt

Sofia Marques is presently a teacher in a Basic and Secondary School, and in the High School of Education from Viseu Polytechnic Institute (Portugal). She has concluded her Master thesis in Communication and Education in

Science, as a researcher in CESAM (Centre for Environmental and Marine Studies, University of Aveiro). Her lines of interest are connected to Environmental Education and Ecology. As a teacher, she has always dedicated her attention to Educational subjects, concerning a connection between school curricula and investigative research.

Aurora Coelho Moreira, holds a Master degree in Communication and Education in Science from the University of Aveiro, Portugal. She is currently studying for a PhD in Didactics in association with the Research Centre for Didactics and Technology in Teacher Education (CIDTFF). The PhD project aims to develop teaching, learning and assessment strategies for promoting students questioning and learning in Higher Education.

Department of Education
University of Aveiro
Portugal
Contact e-mail: aurora.moreira@ua.pt

Paula Bacelar-Nicolau, *PhD* in Microbial Ecology at University of North Wales, Bangor, UK, is Assistant Professor at Universidade Aberta, Portugal. Her main research interests include e-learning in biological and environmental sciences, problem-based-learning, and microbial ecology in marine/coastal systems. She integrated the Coordinating team of the *MSc* in Environmental Citizenship and Participation at Universidade Aberta between 2005 and 2010.

Department of Sciences and Technology
Universidade Aberta
R. Escola Politécnica 147
1269-001 Lisboa
Portugal
Contact e-mail: pnicolau@univ-ab.pt

Carla Padrel de Oliveira, Vice Rector of Universidade Aberta, has a *BSc* in Chemistry from the University of Lisbon, Portugal (1987) and a *PhD* on Chemical Engineering from Imperial College of Science, Technology and Medicine, London (1991). Currently is an Associate Professor in the Department of Science and Technology at Universidade Aberta and a researcher at the Centro de Química Estrutural from Universidade Técnica de Lisboa, Portugal. Her main research interests include sustainability sciences and e-learning and environmental science teaching. She is now the Coordinator of the Doctoral program in Social Sustainability and Development.

Departamento de Ciências e Tecnologia
Universidade Aberta
Rua Escola Politécnica nº 141-147
1269-001 Lisbon
Portugal
Contact e-mail: cpadrel@univ-ab.pt

Miguel Ângelo Pardal, holds a PhD in Biology, University of Coimbra, 1999. The main research topic adresses estuarine and marine ecology, with participation in several EU and national projects. More than 140 papers were already published focusing the impact of eutrophication and nutrient dynamics in estuarine systems; macrobenthos, fish and plankton communities; the effect of extreme climate events in estuarine populations; the effect of contaminants (e.g. heavy metals and PCBs) on the marine trophic webs. Presently is an Associated Professor at the University of Coimbra (Portugal) and integrates the Centre for Functional Ecology.

CFE – Centre for Functional Ecology
Department of Life Sciences
University of Coimbra
P.O.box 3046
3001-401 Coimbra
Portugal
www.http://cfe.uc.pt
Contact e-mail: mpardal@ci.uc.pt

Maria Arminda Pedrosa, PhD is Assistant Professor in the Department of Chemistry, Faculty of Sciences and Technology, University of Coimbra, Portugal. Her research interests include education for sustainable development, science/chemistry education and science teacher education.

R&D Unity 70/94 Molecular Physical Chemistry,
Department of Chemistry, Faculty of Sciences and Technology,
University of Coimbra (FCTUC),
Rua Larga à Universidade,
3004-335 Coimbra,
Portugal
Phone: +351 239 854454
Contact e-mail: apedrosa@ci.uc.pt

Maria Helena Pedrosa de Jesus, holds a PhD in Science Education from the University of East Anglia, U.K. She is currently an Associate Professor, with Agregação, at the Department of Education, University of Aveiro, Portugal. Her main research interests lies in the teaching and learning of science, both in higher and secondary education, looking at how questions promote active learning, critical thinking and the understanding of students' learning process. Questions are also used for mentor training to promote critical reflection and professional development. The present focus of her research is on questioning and Approaches to develop the Scholarship of Teaching and Learning in Higher Education.

Department of Education
University of Aveiro
Portugal
Contact e-mail: hpedrosa@ua.pt

Joana Luísa Pereira is presently a Post-Doctoral researcher in CESAM (Centre for Environmental and Marine Studies) and the Department of Biology, at the University of Aveiro, after concluding her PhD thesis on *Daphnia* responses to environmental stress. Her current research interests relate to the ecotoxicological assessment of contaminants, the effects of environmental global change on the aquatic ecosystem structure and function, and the genetically-driven responses of zooplankton populations to environmental stress. Her graduation in Biology was primarily devoted to education; since then, she collaborated in several projects dedicated to Environmental Education and Sciences (Biology) Teaching, being a co-author of several contributions in books on this subject.

CESAM & Departamento de Biologia da Universidade de Aveiro
3810-193 Aveiro
Portugal
Contact e-mail: jpereira@ua.pt

Ruth Pereira is assistant researcher at CESAM, Centro de Estudos do Ambiente e do Mar, Department of Biology, University of Aveiro, Portugal. Within its main expertise field, the risk assessment of contaminants and of contaminated areas, she is author of more than fifty papers in scientific international journals. Further she is coordinator and collaborator in several national and international projects, and supervisor of several post-doc, PhD and MSc students. She was also teacher in the High School of Education from Instituto Piaget (Viseu, Portugal) where she has gained interest in experimental science teaching, being editor/author/co-author of two books about practical science activities and environmental education.

Departamento de Biologia
CESAM – Centro de Estudos do Ambiente e do Mar
Universidade de Aveiro
Campus Universitário de Santiago
3810-193 Aveiro
Portugal
Contact e-mail: ruthp@ua.pt

Graham Pierce is a Professor of Zoology at the University of Aberdeen, based at the University's Oceanlab. During 2007-10 he was Marie Curie professor at Instituto Español de Oceanografia in Vigo, Spain. He has worked in marine biology and ecology for 25 years and is author or co-author of around 190 peer-reviewed papers. Since the early 1990s he has been involved in European collaborative research, coordination and training projects in fisheries and aquaculture, addressing questions related to sustainable use of marine living resources, including human and environmental aspects of sustainability. He is currently President of the Cephalopod International Advisory Council and a former chair of the International Council for the Exploration of the Sea's Cephalopod Working Group (ICES WGCEPH).

Oceanlab
University of Aberdeen
Main Street
Newburgh
Aberdeenshire AB41 6AA
UK
Contact e-mail: g.j.pierce@abdn.ac.uk

Ana Pinto de Moura, is a Food Engineer from the Portuguese Catholic University (ESB-UCP), with a PhD from the INPL-Nacy, France. She is an Assistant Professor at Universidade Aberta (Open University of Portugal), where she coordinates the distance learning (online) MSc course degree in Food Consumption Sciences. She is a researcher at Food Quality and Safety research group from the Associated Laboratory REQUIMTE. Her main research topics relate to consumer behaviour towards food. She has specific interest in sustainable food consumption, food risk perception, attitudes towards food and nutrition, and perception of nutrition information.

Open University of Portugal (Universidade Aberta)
Rua do Ameal 752
4200-055 Porto
Portugal
Contact e-mails: apmoura@univ-ab.pt and anapintodemoura@gmail.com

Manuela Pinto, graduated in Educational Sciences, is a collaborator of the Portuguese Open University (Universidade Aberta). The Authors of this Edition sincerely acknowledge the careful and competent help provided by Ms. Manuela Pinto in the Peer Review process, Communication with authors and text editing.

Dr. Cristina Pita, PhD in Social and Environmental Sustainability, is an interdisciplinary researcher working on issues of Environmental Sustainability in the Marine Environment at the University of Aberdeen. She has been involved in several EU projects focusing on the ecosystem approach to fisheries management. Her research activity focus on the human dimensions of fisheries and marine protected areas (MPAs), small-scale fisheries management, fisheries and aquaculture socio-economics, sustainable coastal activities, co-management and community-based resource management, stakeholders participation in the decision-making process and governance.

School of Biological Sciences
University of Aberdeen
UK

João Praia, Associate professor of the University of Porto, with PhD from University of Aveiro. Retired. Participation in several science education research projects with the financial support of FCT. Main research interests – epistemology an history of science, rationale for new science teaching perspectives, education environmental issues such as, sustainability of the Planet and its implication in terms of values and attitudes. This last dimension is being developed in articulation with the Spanish research group from University of Valènçia. Publications in national and international journals, co-author of two books and project assessor of FCT during several years.

Contact e-mail: jfpraia@sapo.pt

Rui Ribeiro is Associate Professor with "Agregação" at the University of Coimbra, Portugal; Scientific Coordinator of the Ecotoxicology and Environmental Risk Assessment unit of the IMAR (Institute of Marine Research), Portugal. His main areas of interest are Stress Ecology, Environmental Physiology and Ecotoxicology. His research group has been pursuing several approaches aiming at (i) reducing uncertainties in Ecological Risk Assessments of contaminated sites, particularly in the tropics; (ii) evaluating microevolution due to pollution in both soil and aquatic populations and studying heritability of resistance to metals; and (iii) developing both laboratory and in situ ecologically relevant short-term sublethal toxicity assays representing different functions in freshwater and estuarine ecosystems.

Marine and Environmental Research Centre (IMAR-CMA) & Department of Life Sciences
University of Coimbra
Apartado 3046
3001-401 Coimbra
Portugal

Patrícia Sá, Portuguese, born in 1977; Researcher of the Research Centre for Didactics and Technology in Teacher Education, University of Aveiro; PhD in Education (Didactics of Sciences); Supervisor of PhD Thesis and Master Dissertations; Speaker in National and International Conferences about Science Teaching and Learning and Teacher Education; Author and Co-author of articles and chapters of books; Researcher of several Scientific Projects.

Contact e-mail: f2390@ua.pt

Dr. M. Begoña Santos PhD in Zoology (University of Aberdeen, UK, 1998). Research Scientist at IEO since 2006. She works in the fields of ecology, abundance estimation and feeding ecology of marine mammals and also on fisheries of small pelagic fish in the Northeast Atlantic. She has participated in more than 20 international projects including studies on ecology of pelagic fish and their predators and also projects on fish stock assessment and fisheries management. She has published more than 50 articles in peer-reviewed journals. She is a member (alternate) of the ICES Scientific Committee and of several ICES Expert Groups.

Centro Oceanográfico de Vigo
Instituto Español de Oceanografía
P.O.box 1552
36200 Vigo
España

Maria Conceição Santos, PhD and DSc is Associate Professor with "Habilitation" in the Department of Biology, University of Aveiro, and senior researcher of CESAM, Portugal. She's the head of the Laboratory of Biotechnology and Cytomics at the University of Aveiro. Her research interests and expertise cover developmental biology, (eco)toxicology and biotechnology. She also has interests in biology/biotechnology education.

Department of Biology & CESAM
Laboratory de Biotecnology & Cytomics
Universidade de Aveiro
Campus Universitário de Santiago
3810-193 Aveiro
Portugal
Phone: +351 234370766
e-mail: csantos@ua.pt

Sónia Isabel Fernandes Borges Pena Seixas, is actually Assistant Professor at Universidade Aberta (Portugal). Is graduated (5 years) in Biology by Lisbon University, Master (2 years) in Marine and Coastal Resources by Algarve

University and PhD. in Biology by Lisbon University. Worked several years for a private enterprise of aquaculture and in several environment impact assessments. Teaching at university since 1995. Participated in several studies on the marine environment.

Departamento de Ciências e Tecnologia
Universidade Aberta
Rua Escola Politécnica 147
Lisboa
Portugal
Contact e-mail: sonia@univ-ab.pt

António Veríssimo is Assistent Professor, Life Sciences Department, University of Coimbra Present research interests: Epidemiology, ecology and population genetics of *Legionella*. Molecular evolution of virulence traits in *Legionella pneumophila*. Environmental microbiology. Genetic and functional diversity of microbial communities: a) in non-saline alkaline environments; b) in ground-water; c) in composted municipal sludge; d) in heavy metal contaminated environments. Science education.

CESAM & Departamento de Biologia da Universidade de Aveiro
3810-193 Aveiro
Portugal

Amparo Vilches, High school science teacher and also responsible for science teachers' training courses at the University of Valencia. Her research interests and publications are centred in Science-Technology-Environment-Society relationships, and more particularly in Education for Sustainability. She has directed 5 PhD theses in science education and she is directing 5 other theses. She is the author and co-editor of several books and has published in international journals such as Science Education, Science & Education, Science & Technological Education, Educación Química, Enseñanza de las Ciencias, Revista de Enseñanza de la Física, etc.

Web: www.uv.es/vilches
Contact e-mail: Amparo.Vilches@uv.es

Thematic Index

Umweltbildung, Umweltkommunikation und Nachhaltigkeit
Environmental Education, Communication and Sustainability

Herausgegeben von / Edited by Walter Leal Filho

Band 1 Walter Leal Filho (Hrsg.): Umweltschutz und Nachhaltigkeit an Hochschulen. Konzepte-Umsetzung. 1998.

Band 2 Walter Leal Filho / Farrukh Tahir (eds.): Distance Education and Environmental Education. 1998.

Band 3 Walter Leal Filho (ed.): Environmental Engineering: International Perspectives. 1998.

Band 4 Walter Leal Filho / Mauri Ahlberg (eds.): Environmental Education for Sustainability: Good Environment, Good Life. 1998.

Band 5 Walter Leal Filho (ed.): Sustainability and University Life. 1999. 2nd, revised ed. 2000.

Band 6 Wout van den Bor / Peter Holen / Arjen Wals / Walter Leal Filho (eds.): Integration Concepts of Sustainability into Education for Agriculture and Rural Development. 2000.

Band 7 Manfred Oepen / Winfried Hamacher (Eds.): Communicating the Environment. Environmental Communication for Sustainable Development. 2000.

Band 8 Walter Leal Filho (ed.): Communicating Sustainability. 2000.

Band 9 Walter Leal Filho (ed.): Environmental Careers, Environmental Employment and Environmental Training. International Approaches and Contexts. 2001.

Band 10 Rolf Jucker: Our Common Illiteracy. Education as if the Earth and People Mattered. 2002.

Band 11 Walter Leal Filho (ed.): Teaching Sustainability at Universities. Towards curriculum greening. 2002.

Band 12 Walter Leal Filho (ed.): International Experiences on Sustainability. 2002.

Band 13 Lars Wohlers (Hrsg.): Methoden informeller Umweltbildung. 2003.

Band 14 Ulisses Azeiteiro / Fernando Conçalves / Walter Leal Filho / Fernando Morgado / Mário Pereira (eds.): World Trends in Environmental Education. 2004.

Band 15 Walter Leal Filho / Arnolds Ubelis (eds.): Integrative approaches towards sustainability in the Baltic Sea Region. 2004.

Band 16 Walter Leal Filho / Michael Littledyke (eds.): International Perspectives in Environmental Education. 2004.

Band 17 Daniela Krumland: Beitrag der Medien zum politischen Erfolg. Forstwirtschaft und Naturschutz im Politikfeld Wald. 2004.

Band 18 Walter Leal Filho / Bernd Delakowitz (Hrsg.): Umweltmanagement an Hochschulen: Nachhaltigkeitsperspektiven. 2005.

Band 19 Walter Leal Filho / Volker Lüderitz / Gunther Geller (Hrsg.): Perspektiven der Ingenieurökologie in Forschung, Lehre und Praxis. 2006.

Band 20 Walter Leal Filho (ed.): Handbook of Sustainability Research. 2005.

Band 21 Walter Leal Filho / Dieter Greif / Bernd Delakowitz (eds.): Sustainable Chemistry and Biotechnology – A Contribution to Rivers Management. 2006.

Band 22 Walter Leal Filho / David Carpenter (eds.): Sustainability in the Australasian University Context. 2006.

Band 23 Walter Leal Filho / Arnolds Ubelis / Dina Berzina (eds.): Sustainable Development in the Baltic and Beyond. 2006.

Band 24 Walter Leal Filho (ed.): Innovation, Education and Communication for Sustainable Development. 2006.

Band 25 Walter Leal Filho / Mario Salomone (eds.): Innovative Approaches to Education for Sustainable Development. 2006.

Band 26 Walter Leal Filho / Franziska Mannke / Philipp Schmidt-Thomé (eds.): Information, Communication and Education on Climate Change – European Perspectives. 2007.

Band 27 Ulisses Miranda Azeiteiro / Fernando Gonçalves / Ruth Pereira / Mário Jorge Pereira / Walter Leal Filho / Fernando Morgado (eds.): Science and Environmental Education. Towards the Integration of Science Education, Experimental Science Activities and Environmental Education. 2008.

Band 28 Walter Leal Filho / Nils Brandt / Dörte Krahn / Ronald Wennersten (eds.): Conflict Resolution in Coastal Zone Management. 2008.

Band 29 Walter Leal Filho / Franziska Mannke (eds.): Interdisciplinary Aspects of Climate Change. 2009.

Band 30 Mahshid Sotoudeh: Technical Education for Sustainability. An Analysis of Needs in the 21st Century. 2009.

Band 31 Walter Leal Filho (ed.): Sustainability at Universities – Opportunities, Challenges and Trends. 2009.

Band 32 Walter Leal Filho (ed.): World Trends in Education for Sustainable Development. 2011.

Band 33 Fernando Gonçalves / Ruth Pereira / Walter Leal Filho / Ulisses Miranda Azeiteiro (eds.): Contributions to the UN Decade of Education for Sustainable Development. 2012.

www.peterlang.de

www.ingramcontent.com/pod-product-compliance
Ingram Content Group UK Ltd.
Pitfield, Milton Keynes, MK11 3LW, UK
UKHW050954290926
15470UKWH00006B/1813

9 783631 613474